DISCRETE MATHEMATICS AND ITS APPLICATIONS

Series Editor KENNETH H. ROSEN

ALGORITHMIC COMBINATORICS ON
PARTIAL WORDS

FRANCINE BLANCHET-SADRI

CRC Press
Taylor & Francis Group
Boca Raton London New York

CRC Press is an imprint of the
Taylor & Francis Group, an **informa** business

A CHAPMAN & HALL BOOK

DISCRETE MATHEMATICS AND ITS APPLICATIONS

Series Editor KENNETH H. ROSEN

ALGORITHMIC COMBINATORICS ON
PARTIAL WORDS

CRC Press
Taylor & Francis Group
6000 Broken Sound Parkway NW, Suite 300
Boca Raton, FL 33487-2742

First issued in paperback 2019

ISBN-13: 978-1-4200-6092-8 (hbk)
ISBN-13: 978-0-367-38825-6 (pbk)

Library of Congress Cataloging-in-Publication Data

Blanchet-Sadri, Francine.
  Algorithmic combinatorics on partial words / Francine Blanchet-Sadri.
    p. cm. -- (Discrete mathematics and its applications)
  Includes bibliographical references and index.
  ISBN 978-1-4200-6092-8 (hardback : alk. paper)
    1. Computer algorithms. 2. Computer science--Mathematics. 3. Combinatorial analysis. I. Title. II. Series.

QA76.9.A43B53 2007
005.1--dc22   2007031486

Visit the Taylor & Francis Web site at
http://www.taylorandfrancis.com

and the CRC Press Web site at
http://www.crcpress.com

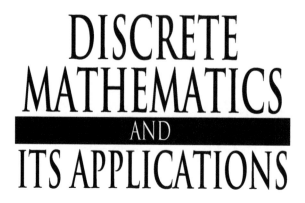

DISCRETE MATHEMATICS
AND
ITS APPLICATIONS

Series Editor
Kenneth H. Rosen, Ph.D.

Continued Titles

Dedication

To my children: Ahmad, Hamid and Mariamme

Contents

List of Tables

List of Figures

Preface

Over the last few years the discrete mathematics and theoretical computer science communities have witnessed an explosive growth in the area of algorithmic combinatorics on words. *Words*, or strings of symbols over a finite alphabet, are natural objects in several research areas including automata and formal language theory, coding theory, and theory of algorithms. Molecular biology has stimulated considerable interest in the study of *partial words* which are strings that may contain a number of "do not know" symbols or "holes." The motivation behind the notion of a partial word is the comparison of genes. Alignment of two such strings can be viewed as a construction of two partial words that are said to be *compatible* in a sense that will be discussed in Chapter 1. While a word can be described by a total function, a partial word can be described by a partial function. More precisely, a partial word of length n over a finite alphabet A is a partial function from $\{0, \ldots, n-1\}$ into A. Elements of $\{0, \ldots, n-1\}$ without an image are called holes (a word is just a partial word without holes). Research in combinatorics on partial words is underway [10, 15, 16, 17, 18, 19, 20, 21, 22, 23, 24, 25, 26, 27, 28, 29, 30, 31, 32, 33, 34, 35, 36, 37, 38, 39, 40, 41, 42, 43, 105, 111, 124, 130, 131] and promises a rich theory as well as substantial impact especially in molecular biology, nano-technology, data communication, and DNA computing [104]. Partial words are currently being considered, in particular, for finding good encodings for DNA computations. Courses, covering different sets of topics, are already being taught at some universities. The time seems right for a book that develops, in a clear manner, some of the central ideas and results of this area, as well as sets the tone of research for the next several years. This book on algorithmic combinatorics on partial words addresses precisely this need.

An effort has been made to ensure that this book is able to serve as a textbook for a diversity of courses. It is intended as an upper-level undergraduate or introductory graduate text in algorithms and combinatorics. It contains a mathematical treatment of combinatorics on partial words designed around algorithms and can be used for teaching and research. The chapters not only cover topics in which definitive techniques have emerged for solving problems related to partial words but also cover topics in which progress is desired and expected over the next several years. The principal audience we have in mind for this book are undergraduate or beginning graduate students from the mathematical and computing sciences. This book will be of interest to students, researchers, and practitioners in discrete mathematics and theoretical computer science who want to learn about this new and exciting class

of partial words where many problems still lay unexplored. It will also be
of interest to students, researchers, and practitioners in bioinformatics, computational molecular biology, DNA computing, and Mathematical Linguistics
seeking to understand this subject. We do assume that the reader has taken
some first course in discrete mathematics.

BOOK OVERVIEW

The book stresses major topics underlying the combinatorics of this emerging class of partial words. The contents of the book are summarized as follows:

- Part I concerns basics. In Chapter 1, we fix the terminology. In particular, we discuss compatibility of partial words. The compatibility
  relation considers two strings over the same alphabet that are equal except for a number of insertions and/or deletions of symbols. It is well
  known that some of the most basic combinatorial properties of words,
  like the conjugacy ($xz = zy$) and the commutativity ($xy = yx$), can be
  expressed as solutions of word equations. In Chapter 2, we investigate
  these equations in the context of partial words. When we speak about
  such equations, we replace the notion of equality ($=$) with compatibility
  ($\uparrow$). There, we solve $xz \uparrow zy$ and $xy \uparrow yx$.

- Part II which consists of Chapters 3, 4 and 5 focuses on three important concepts of *periodicity* on partial words: one is that of *period*, an
  other is that of *weak period*, and the last one is that of *local period*
  which characterizes a local periodic structure at each position of the
  word. These chapters discuss fundamental results concerning periodicity of words and extend them in the framework of partial words. These
  include: First, the well known and basic result of Fine and Wilf [77]
  which intuitively determines how far two periodic events have to match
  in order to guarantee a common period; Second, the well known and
  fundamental critical factorization theorem [49] which intuitively states
  that the minimal period (or global period) of a word of length at least
  two is always locally detectable in at least one position of the word resulting in a corresponding *critical factorization*; Third, the well known
  and unexpected result of Guibas and Odlyzko [82] which states that the
  set of all periods of a word is independent of the alphabet size.

- Part III covers primitivity. *Primitive words*, or strings that cannot be
  written as a power of another string, play an important role in numerous research areas including formal language theory, coding theory, and
  combinatorics on words. Testing whether or not a word is primitive can
  be done in linear time in the length of the word. Indeed, a word is primitive if and only if it is not an inside factor of its square. In Chapter 6,

we describe in particular a linear time algorithm to test primitivity on partial words. The algorithm is based on the combinatorial result that under some condition, a partial word is primitive if and only if it is not compatible with an inside factor of its square. The concept of *speciality*, related to commutativity on partial words, is foundational in the design of the algorithm. There, we also investigate the number of primitive partial words of a fixed length over an alphabet of a fixed size. The zero-hole case is well known and relates to the Möbius function. There exists a particularly interesting class of primitive words, the *unbordered* ones. An *unbordered* word is a string over a finite alphabet such that none of its proper prefixes is one of its suffixes. In Chapter 7, we extend results on unbordered words to unbordered partial words.

- Part IV relates to coding. Codes play an important role in the study of the combinatorics on words. In Chapter 8, we introduce *pcodes* that play a role in the study of combinatorics on partial words. Pcodes are defined in terms of the compatibility relation. We revisit the theory of codes of words starting from pcodes of partial words. We present some important properties of pcodes, describe various ways of defining and analyzing pcodes, and give several equivalent definitions of pcodes and the monoids they generate. It turns out that many pcodes can be obtained as antichains with respect to certain partial orderings. We investigate in particular the Defect Theorem for partial words. We also discuss two-element pcodes, complete pcodes, maximal pcodes, and the class of circular pcodes. In Chapter 9, using two different techniques, we show that the pcode property is decidable.

- Part V covers further topics.

  Chapter 10 continues the study of *equations* on partial words, study that was started in Chapter 2. As mentioned before, an important problem is to decide whether or not a given equation on words has a solution. For instance, the equation $x^m y^n = z^p$ has only periodic solutions in a free monoid, that is, if $x^m y^n = z^p$ holds with integers $m, n, p \geq 2$, then there exists a word w such that x, y, z are powers of w. This result, which received a lot of attention, was first proved by Lyndon and Schützenberger [109] for free groups. In Chapter 10 we solve, among other equations, $x^m y^n \uparrow z^p$ for integers $m \geq 2, n \geq 2, p \geq 4$.

  Chapter 11 introduces the notions of binary and ternary *correlations*, which are binary and ternary vectors indicating the periods and weak periods of partial words. Extending the result of Guibas and Odlyzko of Chapter 5, we characterize precisely which of these vectors represent the period and weak period sets of partial words and prove that all valid correlations may be taken over the binary alphabet. We show that the sets of all such vectors of a given length form distributive lattices under suitably defined relations. We also show that there is a well defined minimal set of generators for any binary correlation of length n and demonstrate that these generating sets are the primitive subsets of $\{1, 2, ..., n-1\}$. Lastly, we investigate the number of partial word correlations of length n.

The notion of an *unavoidable* set of words appears frequently in the fields of mathematics and theoretical computer science, in particular with its connection to the study of combinatorics on words. The theory of unavoidable sets has seen extensive study over the past twenty years. In Chapter 12, we extend the definition of unavoidable sets of words to unavoidable sets of partial words. We demonstrate the utility of the notion of unavoidability on partial words by making use of it to identify several new classes of unavoidable sets of full words. Along the way we begin work on classifying the unavoidable sets of partial words of small cardinality. We pose a conjecture, and show that affirmative proof of this conjecture gives a sufficient condition for classifying all the unavoidable sets of partial words of size two. Finally, we give a result which makes the conjecture easy to verify for a significant number of cases.

KEY FEATURES

Key features of the book include:

- The style of presentation emphasizes the understanding of ideas. Clarity is achieved by a very careful exposition, based on our experience in teaching undergraduate and graduate students. Worked examples and diagrams abound to illustrate these ideas. In the case of concept definitions, we have used the convention that terms used throughout the book are in **boldface** when they are first introduced in definitions. Other terms appear in *italics* in their definition.

- Many of the algorithms are presented first through English sentences and then in pseudo code format. In some cases the pseudo code provides a level of detail that should help readers interested in implementation.

- There are links to many World Wide Web server interfaces that have been established for automated use of programs related to this book. The power of these internet resources will be demonstrated by applying them throughout the book to understand the material and to solve some of the exercises.

- Bibliographic notes appear at the end of each chapter.

- Exercises also appear at the end of each chapter. Practice through solving them is essential to learning the subject. In this book, the exercises are organized into three main categories: exercises, challenging exercises and programming exercises. The exercises review definitions and concepts, while the challenging exercises require more ingenuity. This wealth of exercises provides a good mix of algorithm tracing, algorithm design, mathematical proof, and program implementation. Some of the exercises are drills, while others make important points about the material covered in the text or introduce concepts not covered there at all. Several exercises are designed to prepare the reader for material covered later in the book.

- At the end of the book, solutions or hints are provided to selected exercises to help readers achieve their goals. They are marked by the symbols $\boxed{\text{S}}$ and $\boxed{\text{H}}$ respectively. Some solutions can be found in the literature (the reference that solves the exercise is usually cited in the bibliographic notes).

Sections of the book can be assigned for self study, some sections can be assigned in conjunction with projects, and other sections can be skipped without danger of making later sections of the book incomprehensible to the reader. The bibliographic notes also provide tips for further reading. The following drawing depicts the interdependency of chapters.

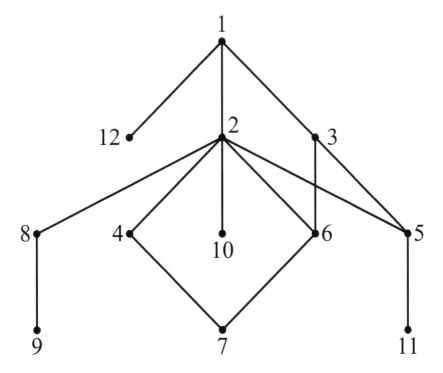

WEBSITES

I believe that without collaboration with Ajay Chriscoe on the paper entitled *Local periods and binary partial words: an algorithm* published in *Theoretical Computer Science* I would not be writing this preface, so I am thankful to have had that opportunity. Ajay spent countless hours helping me design the version of the algorithm that appears in Chapter 5. It is Ajay who decided to establish a World Wide Web server interface at

http://www.uncg.edu/mat/AlgBin

for automated use of the program. Many other websites have been established by Ajay and other students for research related to this book:

http://www.uncg.edu/mat/bintwo
http://www.uncg.edu/mat/border
http://www.uncg.edu/mat/cft
http://www.uncg.edu/mat/research/cft2
http://www.uncg.edu/mat/research/correlations
http://www.uncg.edu/mat/research/equations
http://www.uncg.edu/mat/research/finewilf
http://www.uncg.edu/mat/research/finewilf2
http://www.uncg.edu/mat/research/finewilf3
http://www.uncg.edu/mat/pcode
http://www.uncg.edu/mat/primitive
http://www.uncg.edu/mat/research/primitive2
http://www.uncg.edu/mat/research/unavoidablesets
http://www.uncg.edu/cmp/research/bordercorrelation
http://www.uncg.edu/cmp/research/correlations2
http://www.uncg.edu/cmp/research/finewilf4
http://www.uncg.edu/cmp/research/freeness
http://www.uncg.edu/cmp/research/tilingperiodicity
http://www.uncg.edu/cmp/research/unavoidablesets2

The bintwo website was designed by Brian Shirey; the border by Margaret Moorefield; the cft by Ajay Chriscoe; cft2 by Nathan Wetzler; correlations by Joshua Gafni and Kevin Wilson; equations by Dakota Blair, Craig Gjeltema, Rebeca Lewis and Margaret Moorefield; finewilf by Kevin Corcoran; finewilf2 by Taktin Oey and Tim Rankin; finewilf3 by Deepak Bal and Gautam Sisodia; pcode by Margaret Moorefield; primitive by Arundhati Anavekar and Margaret Moorefield; primitive2 by Brent Rudd; finewilf4 by Travis Mandel and Gautam Sisodia; unavoidablesets by Justin Palumbo; bordercorrelation by Emily Clader and Olivia Simpson; correlations2 by Justin Fowler and Gary Gramajo; `freeness` by Robert Mercaş and Geoffrey Scott; `tilingperiodicity` by Lisa Bromberg and Karl Zipple; and `unavoidablesets2` by Tracy Weyand and Andy Kalcic. Other websites related to material in this book are emerging even as I write this.

An accompanying website has been designed by Brian Shirey at

`http://www.uncg.edu/mat/research/partialwords`

that contains information on partial words. In addition, a website at

`http://www.uncg.edu/mat/reu`

has been designed by Margaret Moorefield for my NSF supported project *Algorithmic Combinatorics on Words*.

I also received invaluable expert technical support from Richard Cheek and Brian Shirey. It is Brian who helped with the drawing of most of the figures in the book. In addition, Shashi Kumar's help with LATEX typesetting is much appreciated.

COMMENTS/SUGGESTIONS

If you find any errors, or have any suggestions for improvement, I will be glad to hear from you. Please send any comments to me at

`blanchet@uncg.edu`,

or at

Francine Blanchet-Sadri,
Department of Computer Science,
University of North Carolina,
P.O. Box 26170,
Greensboro, NC 27402-6170,
USA

I thank in advance all readers interested in helping me make this a better book.

ACKNOWLEDGEMENTS

I have been fortunate to have been awarded several grants from the National Science Foundation. The material in this book is based upon work supported by NSF under Grant No. CCR–9700228 *RUI: Decipherability of*

Codes and Applications, Grant No. CCF–0207673 *RUI: Computing Patterns in Strings*, and Grant No. DMS–0452020 *REU Site: Algorithmic Combinatorics on Words*.

The present book started its life thanks to: Jean-Eric Pin from LIAFA: Laboratoire d'Informatique Algorithmique: Fondements et Applications of University Paris 7, Paris, France who provided a lot of encouragement and a stimulating work environment during my stays in 2000, 2004 and 2005, and Pál Dömösi from the University of Debrecen, Debrecen, Hungary who was very kind in sending me the book of Shyr *Free Monoids and Languages*, some sections of my book being heavily based on, and in giving me a draft of his book *Primitive Words and Context-Free Languages* during my stay in Debrecen in 2005. Many thanks also to Yulia Gamzova in giving me research papers during her visit to UNCG in 2005. I wish to thank Carlos Martin Vide for the opportunity of lecturing on partial words during the 1st term of the 5th International PhD School in Formal Languages and Applications in Tarragona, Spain (May 12–13, 2006). The following people were very kind in sending me research papers: Peter Leupold, Gerhard Lischke, Robert Mercaş, and Arseny Shur.

Research Assignments in 2000 and 2005 from the University of North Carolina at Greensboro are gratefully acknowledged. It is UNCG that provided research assistantships to Robert Hegstrom, Phuongchi Thi Le, Donald Luhmann, and Brian Shirey to work with me on various research projects. I also wish to thank many colleagues, in particular Paul Duvall, Nancy Green and Shan Suthaharan who have given me assistance with my first REU summer on Algorithmic Combinatorics on Words and with several theses related to this book. I am grateful to my students who helped me proofread an early draft and helped me in several ways: Charles Batts, Jonathan Britton, Archana Dattatreya, Malcolm Gethers, Craig Gjeltema, Zhiyong Guo, Robert Mercaş, John Mitaka, Robert Misior, Ji-Young Oh, Aparna Pisharody, Stephanie Rednour, Charles Renn, and Adrian Rudd. Special thanks are due to Charles Renn who helped me with Chapters 1, 2 and 3. In addition, he made many figures and provided several helpful comments and suggestions. I had many helpful discussions with Stephanie and Robert, who also provided crucial assistance in LATEX typesetting matters. Archana and Aparna also made many contributions, some of which have been incorporated into the exercises.

My research assistants Mihai Cucuringu, Crystal Davis, Robert Mercaş, Margaret Moorefield, and Gautam Sisodia provided invaluable ideas that have expanded many parts of this book. I also wish to thank other research assistants who helped me in different research projects related to combinatorics on words: Steve Adkins, Pheadra Agius, Arundhati Anavekar, Jerry Barnes, Cheng-Tao Chen, Ajay Chriscoe, Stacy Duncan, Dale Gaddis, Firdouse Gama, Robert Hegstrom, Tracey Howell, Bobbie Jobe, Phuongchi Thi Le, Donald Luhmann, Liem Mai, Crystal Morgan, Baouyen Phan, Brian Shirey, Steven Wicker, Lili Zhang, and Xin-Hong Zhang.

I thank my co-authors: Arundhati Anavekar, Deepak Bal, Dakota Blair,

Jonathan Britton, Lisa Bromberg, Naomi Brownstein, Cheng-Tao Chen, Ajay Chriscoe, Emily Clader, Kevin Corcoran, Mihai Cucuringu, Crystal Davis, Joel Dodge, Stacy Duncan, Justin Fowler, Joshua Gafni, Gary Gramajo, Robert Hegstrom, Andy Kalcic, Rebeca Lewis, Donald Luhmann, Travis Mandel, Robert Mercaş, Margaret Moorefield, Jenell Nyberg, Taktin Oey, Justin Palumbo, Tim Rankin, Geoffrey Scott, Brian Shirey, Olivia Simpson, Gautam Sisodia, Nathan Wetzler, Tracy Weyand, Kevin Wilson, Jeffery Zhang, and Karl Zipple. Many of them provided data that appear throughout the book: Ajay Chriscoe, Mihai Cucuringu, Joshua Gafni, Margaret Moorefield, Nathan Wetzler, and Kevin Wilson.

Finally, I wish to thank my husband Fereidoon and my children Ahmad, Hamid and Mariamme, for providing the support without which this book could not have been written.

Francine Blanchet-Sadri   September 27, 2007

Part I

BASICS

Chapter 1

Preliminaries on Partial Words

In this chapter, we give a short review of some basic notions on partial words that will be used throughout the book.

1.1 Alphabets, letters, and words

Let A be a nonempty finite set of symbols, which we call an *alphabet*. An element $a \in A$ is called a *letter*. A *word* over the alphabet A is a finite sequence of elements of A.

Example 1.1
The following sets are alphabets:

$$A = \{a, b, c, n\}$$

$$B = \{0, 1\}$$

The sequence of letters *banana* is a word over the alphabet A, as well as the word *cbancb*. Over the alphabet B, the sequences 0, 1, and 01010111110 are words. □

For any word u, $\alpha(u)$ is defined as the set of distinct letters in u. We allow for the possibility that a word consists of no letters. It is called the *empty word* and is denoted by ε.

Example 1.2
Consider the words $u = banana$, $v = aaccaaa$, and the empty word ε. Then,

$$\alpha(u) = \{a, b, n\}$$

$$\alpha(v) = \{a, c\}$$

$$\alpha(\varepsilon) = \emptyset$$

□

The set of all words over A is denoted by A^* and is equipped with the associative operation defined by the concatenation of two sequences. We use multiplicative notation for concatenation. For example, if $u = aaa$ and $v = bbb$ are words over an alphabet A, they are members of A^*, and the word $uv = aaabbb$ is also a member of A^*. The empty word is the neutral element for concatenation, as any word u concatenated with the empty word is simply itself again ($u\varepsilon = \varepsilon u = u$).

The set of nonempty words over A is denoted by A^+. Thus we have $A^+ = A^* \setminus \{\varepsilon\}$.

Notice that for any two words u and v in either A^* or A^+, their product uv is also in the same set. The only difference between these sets is that the empty word ε is an element of A^* and not A^+. We note that the set A^+ is equipped with the structure of a semigroup. It is called the *free semigroup* over A. The set A^*, with its inclusion of the empty word, is equipped with the structure of a monoid. It is called the *free monoid* over A.[1]

For a word u, we can write the i-power of u, where

$$u^i = \underbrace{uuu \ldots u}_{i \text{ times}}$$

We can also define u^i recursively with the following definition:

$$u^i = \begin{cases} \varepsilon & \text{if } i = 0 \\ uu^{i-1} & \text{if } i \geq 1 \end{cases}$$

Example 1.3
Let a and b be letters in an alphabet A. Then,

$$a^6 = aaaaaa$$

$$(aba)^3 = (aba)(aba)(aba) = abaabaaba$$

At this point, we define a word u to be *primitive* if there exists no word v such that $u = v^i$ with $i \geq 2$.

Example 1.4
The word $u = abaaba$ is not primitive, as shown here:

$$u = abaaba = (aba)^2 = v^2 \text{ where } v = aba$$

---

[1]A *semigroup* is a nonempty set together with a binary associative operation. A *monoid* is a semigroup with identity.

The word $aaaaa = a^5$ is also clearly not primitive, whereas the word $aaaab$ is primitive.

We also note here that the empty word ε is not primitive, for

$$\varepsilon = \varepsilon^i \text{ for all } i$$

☐

1.2   Partial functions and partial words

To students of mathematical sciences, the concept of a function is a familiar one. We refine that concept with the following definition.

DEFINITION 1.1   *Let f be a function on a set X. If f is not necessarily defined for all $x \in X$, then f is a **partial function**. The **domain** of f, $D(f)$, is defined as*

$$D(f) = \{x \in X \mid f(x) \text{ is defined}\}$$

*A partial function where $D(f) = X$ is a **total function**.*

The "usual" idea of a function is captured in the definition of total function above, because we typically state a function only on a set of input values for which the function is defined. With the notion of a partial function, we allow for the possibility that for certain values the function may **not** be defined.

Example 1.5
In Figure 1.1, we have a graphical representation of a partial function f on the set $\{0, 1, 2, 3, 4\}$ to the set $\{a, b, c\}$. Note that $D(f) = \{0, 1, 3\}$. ☐

In the context of our discussion about words, total functions allow us to refer to specific letter positions within a given word in the following manner. A word of length n over A can be defined by a total function $u : \{0, \ldots, n - 1\} \to A$ and is usually represented as

$$u = a_0 a_1 \ldots a_{n-1} \text{ with } a_i \in A$$

Example 1.6
Let $u : \{0, 1, 2, 3\} \to \{a, b, c\}$ be the total function defined below:

$$u(0) = a$$
$$u(1) = c$$
$$u(2) = a$$
$$u(3) = b$$

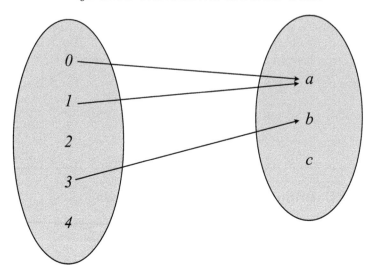

FIGURE 1.1: A picture of a partial function.

The word described by this function is therefore $u = acab$. Also, note that the letter indices of a word begin at zero.   ☐

Partial functions allow us to extend the above definition to words that are "incomplete," that is, words that have missing letters. For example, suppose that u is a word of length 5 over an alphabet A, but that the letters in the second and fourth positions are unknown. Using a partial function, we can define a function $u : \{0, 1, 2, 3, 4\} \rightarrow A$ and then acknowledge that $u(2)$ and $u(4)$ are undefined. We make the following definition.

DEFINITION 1.2   *A* **partial word** *(or,* **pword***) of length n over A is a partial function $u : \{0, \ldots, n-1\} \rightarrow A$. For $0 \le i < n$, if $u(i)$ is defined, we say that i belongs to the* **domain** *of u (denoted by $i \in D(u)$). Otherwise we say that i belongs to the* **set of holes** *of u (denoted by $i \in H(u)$).*

Just as every total function is a partial function, every total word is itself a partial word with an empty set of holes. For clarity, we sometimes refer to words as *full words*. For any partial word u over A, $|u|$ denotes its length. Clearly, $|\varepsilon| = 0$.

Example 1.7
Let the function $u : \{0, 1, 2, 3, 4\} \rightarrow A$ be a partial function where $u(2)$ and $u(4)$ are undefined. Therefore,

$$D(u) = \{0, 1, 3\} \text{ and } H(u) = \{2, 4\}$$

It follows that u is a partial word with holes in the second and fourth positions and $|u| = 5$. Example 1.5 and Figure 1.1 are examples of such a partial function. ▯

Example 1.8
Let the word u be given by Example 1.6. Then $|u| = 4$,

$$D(u) = \{0, 1, 2, 3\} \text{ and } H(u) = \emptyset$$

and u is clearly a full word. ▯

We denote by $W_0(A)$ the set A^*, and for $i \geq 1$, by $W_i(A)$ the set of partial words over A with at most i holes. This leads to the nested sequence of sets,

$$W_0(A) \subset W_1(A) \subset W_2(A) \subset \cdots \subset W_i(A) \subset \cdots$$

We put $W(A) = \bigcup_{i \geq 0} W_i(A)$, the set of all partial words over A with an arbitrary number of holes.

Now that we have defined the notion of a partial word, we are in need of a method to represent partial words. In particular, we need a way to represent the positions of the holes of a partial word. In order to do this, we introduce a new symbol, $\diamond$, and make the following definition.

DEFINITION 1.3 *If u is a partial word of length n over A, then the* **companion** *of u, denoted by $u_\diamond$, is the total function $u_\diamond : \{0, \ldots, n-1\} \to A \cup \{\diamond\}$ defined by*

$$u_\diamond(i) = \begin{cases} u(i) & \text{if } i \in D(u) \\ \diamond & \text{otherwise} \end{cases}$$

We extend our definition of $\alpha(u)$ for any partial word u over an alphabet A in the following way:

$$\alpha(u) = \{a \in A \mid u(i) = a \text{ for some } i \in D(u)\}$$

It is important to remember that the symbol $\diamond$ is not a letter of the alphabet A. Rather, it is viewed as a "do not know" symbol, and its inclusion allows us to now define a partial word in terms of the total function $u_\diamond$ given in the definition.

Example 1.9
The word $u_\diamond = abb\diamond b\diamond cb$ is the companion of the partial word u of length 8 where $D(u) = \{0, 1, 2, 4, 6, 7\}$ and $H(u) = \{3, 5\}$. Note that

$$u_\diamond(1) = u(1) = b \text{ because } 1 \in D(u) \text{ and}$$

$$u_\diamond(3) = \diamond \text{ while } u(3) \text{ is undefined}$$

$$\alpha(u) = \{a, b, c\}$$

☐

The bijectivity of the map $u \mapsto u_\diamond$ allows us to define for partial words concepts such as concatenation and powers in a trivial way. More specifically, for partial words u, v, the concatenation of u and v is defined by $(uv)_\diamond = u_\diamond v_\diamond$, and the i-power of u is defined by $(u^i)_\diamond = (u_\diamond)^i$.

Example 1.10
Let u and v be partial words, with their companions $u_\diamond = a\diamond$ and $v_\diamond = b\diamond c$. The partial word uv is formed in terms of the companions in the expected way:

$$(uv)_\diamond = u_\diamond v_\diamond = a\diamond b\diamond c$$

Similarly, powers are formed in terms of the companions as well,

$$(u^3)_\diamond = (u_\diamond)^3 = (a\diamond)^3 = a\diamond a\diamond a\diamond$$

☐

With the operation now defined for partial words, the set $W(A)$ becomes a monoid under the concatenation of partial words (ε serves as identity). For convenience, we often drop the word "companion" from our discussion, and we consider a partial word over A as a word over the enlarged alphabet $A \cup \{\diamond\}$, where the additional symbol $\diamond$ plays a special role. Thus, we say for instance "the partial word $\diamond ab\diamond b$" instead of "the partial word with companion $\diamond ab\diamond b$."

1.3   Periodicity

Periodicity is an important concept related to partial words, and we introduce two formulations of periodicity in this section.

DEFINITION 1.4   *A* (**strong**) **period** *of a partial word u over A is a positive integer p such that $u(i) = u(j)$ whenever $i, j \in D(u)$ and $i \equiv j \bmod p$.* [2] *In such a case, we call u p-periodic.*

---

[2] Throughout the book, $i \bmod p$ denotes the remainder when dividing i by p using ordinary integer division. We also write $i \equiv j \bmod p$ to mean that i and j have the same remainder when divided by p; in other words, that p divides $i - j$ (for instance, $12 \equiv 7 \bmod 5$ but $12 \not\equiv 7 \bmod 5$ ($2 = 7 \bmod 5$)).

Notice that nothing in the definition precludes a partial word from having more than one period. The set of all periods of u will be denoted by $\mathcal{P}(u)$. However, we will often want to refer to the *minimal* period of a partial word. We represent this minimal period by $p(u)$.

Example 1.11
Consider these examples of partial words and their periods:

$$u = ababab \text{ is 6-periodic, 4-periodic, and 2-periodic, and } p(u) = 2$$
$$v = a\diamond\diamond a\diamond b \text{ is 6-periodic, 4-periodic, and 3-periodic, and } p(v) = 3$$
$$w = bb\diamond b \text{ is 4-, 3-, 2-, and 1-periodic, and } p(w) = 1$$

As seen above, any partial word u is trivially $|u|$-periodic, showing $\mathcal{P}(u)$ is never empty. ⬚

Frequently, it is much easier to determine if a partial word u is p-periodic by writing, in order, the letters of u into p columns. If every letter in each column is the same, ignoring holes, then u is p-periodic.

Example 1.12
We use the partial words of the previous example and disregard the trivial period. We see that u is indeed 4-periodic and 2-periodic by writing

$$\begin{array}{c} a\ b\ a\ b \\ a\ b \end{array} \quad \text{and} \quad \begin{array}{c} a\ b \\ a\ b \\ a\ b \end{array}$$

Similarly, we verify that v is 4-periodic and 3-periodic:

$$\begin{array}{c} a\ \diamond\ \diamond\ a \\ \diamond\ b \end{array} \quad \text{and} \quad \begin{array}{c} a\ \diamond\ \diamond \\ a\ \diamond\ b \end{array}$$

⬚

In partial words, the presence of holes gives us an opportunity to define another type of periodic behavior.

DEFINITION 1.5 *A **weak period** of u is a positive integer p such that $u(i) = u(i + p)$ whenever $i, i + p \in D(u)$. In such a case, we call u **weakly p-periodic**. We denote the set of all weak periods of u by $\mathcal{P}'(u)$ and the minimal weak period of u by $p'(u)$.*

As before, it is much easier to identify if a partial word u is weakly p-periodic by writing u into p columns. However, now we only require that letters in a column be the same if there is no hole between them in that column. Letters in columns with holes need to be the same if they are consecutive.

Example 1.13

Let $u = abb\diamond bbcbb$. We write

$$
\begin{array}{ccc}
a & b & b \\
\diamond & b & b \\
c & b & b
\end{array}
$$

The partial word u is weakly 3-periodic but is not 3-periodic (this is because a occurs in position 0 while c occurs in position 6).   ⧠

It is clear that if a partial word u is p-periodic, then u is weakly p-periodic, and hence $\mathcal{P}(u) \subset \mathcal{P}'(u)$ for any partial word u. The converse of this statement holds only for full words, however, and thus we see that for full words there is no distinction between periods and weak periods.

Example 1.14

In Example 1.11, we determined that for $v = a\diamond\diamond a\diamond b$,

$$\mathcal{P}(v) = \{3, 4, 6\}$$

This partial word v is also weakly 1-periodic, and therefore,

$$\mathcal{P}'(v) = \{1, 3, 4, 6\} \text{ and}$$

$$p(v) = 3 \text{ and } p'(v) = 1$$

   ⧠

Another difference between full words and partial words that is worth noting is the fact that even if the length of a partial word u is a multiple of a weak period of u, then u is not necessarily a power of a shorter partial word.

Example 1.15

For the full word, $v = ababab$, v is clearly 2-periodic and $v = (ab)^3$. However, recall the weakly 3-periodic word u from Example 1.13, $u = abb\diamond bbcbb$. The partial word u is not the power of a shorter partial word.   ⧠

1.4   Factorizations of partial words

Given two subsets X, Y of $W(A)$, we define

$$XY = \{uv \mid u \in X \text{ and } v \in Y\}$$

We sometimes write $X \sqsubset Y$ if $X \subset Y$ but $X \neq Y$. For a subset X of $W(A)$, we use the notation $\|X\|$ for the cardinality of X.

Example 1.16

Let $X = \{\varepsilon, a, ac\}$ and $Y = \{b, bb\}$. Then XY is the following set,

$$\{b, bb, ab, abb, acb, acbb\}$$

Note that this "set product" is not commutative, as YX equals

$$\{b, bb, ba, bba, bac, bbac\}$$

☐

Given a subset X of $W(A)$, we can apply the previous idea and form the set product of a set with itself.

Example 1.17

Let $X = \{a, b\}$. We can then construct the following sequence of sets:

$$X = X^1 = \{a, b\}$$
$$XX = X^2 = \{aa, ab, ba, bb\}$$
$$XXX = X^3 = \{aaa, aab, aba, abb, baa, bab, bba, bbb\}$$
$$\vdots$$

For completion, we define $X^0 = \{\varepsilon\}$.

☐

In general, for a subset X of $W(A)$ and integer $i \geq 0$, we denote by X^i the set

$$\{u_1 u_2 \ldots u_i \mid u_1, \ldots, u_i \in X\}$$

We denote by X^* the submonoid of $W(A)$ generated by X, or $X^* = \bigcup_{i \geq 0} X^i$ and by X^+ the subsemigroup of $W(A)$ generated by X, or $X^+ = \bigcup_{i > 0} X^i$.

DEFINITION 1.6 *A* **factorization** *of a partial word u is any sequence $u_1, u_2, \ldots, u_i$ of partial words such that $u = u_1 u_2 \ldots u_i$. We write this factorization as $(u_1, u_2, \ldots, u_i)$. A partial word u is a* **factor** *of a partial word v if there exist partial words x, y (possibly equal to ε) such that $v = xuy$. The factor u is* **proper** *if $u \neq \varepsilon$ and $u \neq v$. The partial word u is a* **prefix** *(respectively,* **suffix***) of v if $x = \varepsilon$ (respectively, $y = \varepsilon$).[3] We occasionally use*

---

[3]Notation: If the partial word x is a prefix of y, we sometimes write $x \preceq_p y$ or simply $x \preceq y$. We can write $x \prec y$ when $x \neq y$.

the notation $u[i..j)$ to represent the factor of the partial word u starting at position i and ending at position $j - 1$. Likewise, $u[0..i)$ is the prefix of the partial word u of length i, and $u[j..|u|)$ is the suffix of u of length $|u| - j$.

Factors of a partial word u are sometimes called *substrings* of u. It is immediately seen that there may be numerous factorizations for a given partial word.

Example 1.18

Let $v = abc\diamond ab$. The following are two factorizations of v:

$$(ab, c\diamond, a, b)$$
$$(a, bc\diamond, ab)$$

In addition, we call the factorizations $(\varepsilon, abc\diamond ab)$ and $(abc\diamond ab, \varepsilon)$ trivial. The prefixes of v are ε, a, ab, abc, $abc\diamond$, $abc\diamond a$, and $abc\diamond ab$. Likewise, the suffixes of v are ε, b, ab, $\diamond ab$, $c\diamond ab$, $bc\diamond ab$, and $abc\diamond ab$. □

For partial words u and v, the unique *maximal common prefix* of u and v is denoted by $\mathrm{pre}(u, v)$.

Example 1.19

Let $u = a\diamond bcb$ and $v = a\diamond bbab$. The common prefixes of u and v are $\varepsilon, a, a\diamond, a\diamond b$, the latter being $\mathrm{pre}(u, v)$. □

By definition, each partial word u in X^* admits at least one factorization $u_1, u_2, \ldots, u_i$ whose elements are all in X. Such a factorization is called an *X-factorization*.

For a subset X of $W(A)$, we denote by $F(X)$ the set of all factors of elements in X. More specifically,

$$F(X) = \{u \mid u \in W(A) \text{ and there exist } x, y \in W(A) \text{ such that } xuy \in X\}$$

We denote by $P(X)$ the set of all prefixes of elements in X and by $S(X)$ the set of suffixes of elements in X:

$$P(X) = \{u \mid u \in W(A) \text{ and there exists } x \in W(A) \text{ such that } ux \in X\}$$
$$S(X) = \{u \mid u \in W(A) \text{ and there exists } x \in W(A) \text{ such that } xu \in X\}$$

If X is the singleton $\{u\}$, then $P(X)$ (respectively, $S(X)$) will be abbreviated by $P(u)$ (respectively, $S(u)$).

1.5   Recursion and induction on partial words

We begin this section with the concept of the *reversal* of a partial word, and use this concept to illustrate recursion and induction with partial words.

DEFINITION 1.7   *If $u \in A^*$, then the **reversal** of the word $u = a_0 a_1 \ldots a_{n-1}$ is $rev(u) = a_{n-1} \ldots a_1 a_0$ where $a_i \in A$ for all i. The **reversal** of a partial word u is $rev(u)$ where $(rev(u))_\diamond = rev(u_\diamond)$. The **reversal** of a set $X \subset W(A)$ is the set $rev(X) = \{ rev(u) \mid u \in X \}$.*

Example 1.20
If $u = ab \diamond d$, then $rev(u) = d \diamond ba$.   ⬚

Recursively, the reversal of a partial word is described in the following way:

1. $rev(\varepsilon) = \varepsilon$, and

2. $rev(xa) = a\, rev(x)$

where $x \in A^*$ and $a \in A$.

In a similar fashion, we provide a recursive description of A^*, the set of all words over an alphabet A:

1. $\varepsilon \in A^*$

2. If $x \in A^*$ and $a \in A$, then $xa \in A^*$.

It is often very useful to use mathematical induction in order to prove results related to partial words. Below we provide an example of using induction on the length of a partial word to prove a result related to the reversal of the product of two words.

Example 1.21
Let x, y be words over an alphabet A. Show that

$$rev(xy) = rev(y)\,rev(x)$$

As stated, we prove this by induction on $|y|$. First, suppose $|y| = 0$ or $y = \varepsilon$. Clearly,

$$\begin{aligned} rev(xy) &= rev(x) \\ &= \varepsilon\, rev(x) \\ &= rev(\varepsilon)\, rev(x) \\ &= rev(y)\, rev(x) \end{aligned}$$

Now assume that our result holds for all words y where $|y| = n$ for some nonnegative integer n. According to the process of induction, it remains for us to show that the result holds for words of length $n + 1$.

Let $|y| = n$ and $a \in A$. Then ya is a word of length $n + 1$. Now,

$$\begin{aligned} \mathrm{rev}(x(ya)) &= \mathrm{rev}((xy)a) \\ &= a\,\mathrm{rev}(xy) && \text{by definition} \\ &= a\,\mathrm{rev}(y)\mathrm{rev}(x) && \text{by inductive hypothesis} \\ &= \mathrm{rev}(ya)\mathrm{rev}(x) \end{aligned}$$

Thus, the result holds for all words ya of length $n + 1$, and consequently the result is proved for all words x, y in A^*. ☐

REMARK 1.1   The previous result can be generalized easily to partial words by applying the same argument to the companions of partial words x and y. ☐

1.6   Containment and compatibility

We define equality of partial words in the following way.

DEFINITION 1.8   *The partial words u and v are **equal** if u and v are of equal length (that is, $|u| = |v|$), and*

$$D(u) = D(v) \text{ and } u(i) = v(i) \text{ for all } i \in D(u)(= D(v))$$

For full words, the equality of two words is straightforward, namely, letters in corresponding positions must be equal. However, for partial words containing holes, the notion of equality is only part of the picture. This is because the symbol ⋄ is not an element of our alphabet, but a placeholder symbol for a letter we do not know. So although the partial words $a\diamond b\diamond$ and $a\diamond\diamond b$ are not equal by our definition, they may very well be equal, *if we only had more information*. To sharpen our understanding of this possibility, we introduce and discuss two alternative methods of relating partial words: containment and compatibility.

DEFINITION 1.9   *If u and v are two partial words of equal length, then u is said to be **contained in** v, denoted by $u \subset v$, if all elements in $D(u)$ are in $D(v)$ and $u(i) = v(i)$ for all $i \in D(u)$. We sometimes write $u \sqsubset v$ if $u \subset v$ but $u \neq v$.*

Containment can be restated in the following equivalent way:

For partial words u and v, $u \subset v$ if everything that is known about the letters of u is repeated in v. In this sense, the relation is "one-way," from u to v.

In general, $u \subset v$ does not imply that $v \subset u$.

Example 1.22
Let $u = a \diamond b \diamond$. We can easily compare u to other partial words by writing u above and checking our conditions. For $v_1 = a \diamond \diamond b$, we write,

$$u = a \diamond b \diamond$$
$$\downarrow \quad \not{\downarrow}$$
$$v_1 = a \diamond \diamond b$$

We can now easily see that $D(u) \not\subset D(v_1)$, and therefore $u \not\subset v_1$. For $v_2 = a \diamond ab$, we see

$$u = a \diamond b \diamond$$
$$\downarrow \quad \not{\downarrow}$$
$$v_2 = a \diamond a \, b$$

Because $u(2) \neq v_2(2)$, $u \not\subset v_2$. Lastly, for $v_3 = a \diamond bb$,

$$u = a \diamond b \diamond$$
$$\downarrow \quad \downarrow$$
$$v_3 = a \diamond b \, b$$

and $u \subset v_3$. Notice the fact that $v_3(3)$ is defined implies that $v_3 \not\subset u$. ⬜

We can extend the notion of a word being primitive to a partial word being primitive as follows:

A partial word u is *primitive* if there exists no word v such that $u \subset v^i$ with
$$i \geq 2$$

Example 1.23
The partial word $u = a \diamond ab$ is not primitive, because for $v = ab$, $u \subset v^2$. However, the partial word $a \diamond bb$ is primitive. ⬜

REMARK 1.2   Note that if v is primitive and $v \subset u$, then u is primitive as well. The proof of this fact is left as an exercise. ⬜

Whereas the containment relation may be thought of as a nonsymmetric, "one-way" relation between two partial words, we now define a new, symmetric relation on partial words called compatibility.

DEFINITION 1.10   *The partial words u and v are called* **compatible**, *denoted by $u \uparrow v$, if there exists a partial word w such that $u \subset w$ and $v \subset w$. Equivalently, $u \uparrow v$ if*

$$u(i) = v(i) \text{ for every } i \in D(u) \cap D(v)$$

It is obvious that $u \uparrow v$ implies $v \uparrow u$.

Typically it is easier to test for the compatibility of two partial words by writing them one above the other and applying the second formulation of the definition, that is, if two letters "line up," then they must be equal.

Example 1.24
Let $x = a \diamond b \diamond a \diamond$ and $y = a \diamond \diamond c b b$. We write

$$x = a \diamond b \diamond a \diamond$$
$$y = a \diamond \diamond c \, b \, b$$

and because $x(4) \neq y(4)$, $x \not\uparrow y$. Now, let $u = a \diamond bbc \diamond$ and $v = \diamond bb \diamond c \diamond$. We see that $u \uparrow v$:

$$u = a \diamond b \, b \, c \diamond$$
$$v = \diamond b \, b \diamond c \diamond$$

☐

For compatible words, we can construct a partial word w that contains both u and v such that the domain of w is exactly the union of the domains of u and v. In other words, the letters of w are defined "only when they need to be" in order to contain u and v. For this reason, we call w the *least upper bound* of u and v and denote w as $u \vee v$.

DEFINITION 1.11   *Let u and v be partial words such that $u \uparrow v$. The* **least upper bound** *of u and v is the partial word $u \vee v$, where*

$$u \subset u \vee v \text{ and } v \subset u \vee v, \text{ and}$$
$$D(u \vee v) = D(u) \cup D(v)$$

Example 1.25
Let $u = aba \diamond \diamond a$ and $v = a \diamond \diamond b \diamond a$. Writing them over one another, we see $u \uparrow v$ and also how $u \vee v$ is constructed:

$$u \quad = a \, b \, a \diamond \diamond a$$
$$v \quad = a \diamond \diamond b \diamond a$$
$$u \vee v = a \, b \, a \, b \diamond a$$

☐

For a subset X of $W(A)$, we denote by $C(X)$ the set of all partial words compatible with elements of X. More specifically,

$$C(X) = \{u \mid u \in W(A) \text{ and there exists } v \in X \text{ such that } u \uparrow v\}$$

If $X = \{u\}$, then we denote $C(\{u\})$ simply by $C(u)$. We call a subset X of $W(A)$ *pairwise noncompatible* if no distinct partial words $u, v \in X$ satisfy $u \uparrow v$. In other words, X is pairwise non compatible if for all $u \in X$, $X \cap C(u) = \{u\}$.

The following rules are useful for computing with partial words.

LEMMA 1.1

Let u, v, w, x, y be partial words.

Multiplication: *If $u \uparrow v$ and $x \uparrow y$, then $ux \uparrow vy$.*

Simplification: *If $ux \uparrow vy$ and $|u| = |v|$, then $u \uparrow v$ and $x \uparrow y$.*

Weakening: *If $u \uparrow v$ and $w \subset u$, then $w \uparrow v$.*

We end this section with the following lemma.

LEMMA 1.2

Let u, v, x, y be partial words such that $ux \uparrow vy$.

- *If $|u| \geq |v|$, then there exist pwords w, z such that $u = wz$, $v \uparrow w$, and $y \uparrow zx$.*

- *If $|u| \leq |v|$, then there exist pwords w, z such that $v = wz$, $u \uparrow w$, and $x \uparrow zy$.*

PROOF The proof is left as an exercise. $\Box$

COROLLARY 1.1

Let u, v, x, y be full words. If $ux = vy$ and $|u| \geq |v|$, then $u = vz$ and $y = zx$ for some word z.

Throughout the rest of the book, A denotes a fixed alphabet.

Exercises

1.1 The root of a full word u, denoted by $\sqrt{u}$, is defined as the unique primitive word v such that $u = v^n$ for some positive integer n. What is $\sqrt{u}$ if $u = ababab$? What if $u = ababba$?

1.2 ⬚s Let $u = abba\diamond bacba$. Compute

1. The set of periods of u, $\mathcal{P}(u)$.
2. The set of weak periods of u, $\mathcal{P}'(u)$.
3. A partial word v over the alphabet $\{0,1\}$ satisfying all the following conditions:
   (a) $|v| = |u|$
   (b) $H(v) \subset H(u)$
   (c) $\mathcal{P}(v) = \mathcal{P}(u)$
   (d) $\mathcal{P}'(v) = \mathcal{P}'(u)$

1.3 Let u and v be partial words. Prove that if v is primitive and $v \subset u$, then u is primitive as well.

1.4 ⬚s Let u be a partial word of length p, where p is a prime number. Prove that u is not primitive if and only if $\|\alpha(u)\| \leq 1$.

1.5 Construct a partial word with one hole of length 12 over the alphabet $\{a, b\}$ that is weakly 5-periodic, weakly 8-periodic but not 1-periodic.

1.6 Let u be a word over an alphabet A, and let $v = ua$ for any letter a in A. Prove that $p(u) \leq p(v)$.

1.7 For partial words u and v, does $u \uparrow v$ imply $u \subset v$. Is the converse true?

1.8 Show that if for partial words u, v we have that $u \subset v$, then $\mathcal{P}(v) \subset \mathcal{P}(u)$ and $\mathcal{P}'(v) \subset \mathcal{P}'(u)$.

1.9 Consider the factorization $(u, v) = (abb\diamond bab, bb)$ of $w = abb\diamond babbb$. Is $abb\diamond ba \in C(S(u))$? Is $b \in C(P(v))$?

1.10 ⬚s Prove Lemma 1.2.

1.11 ⬚s A nonempty partial word u is *unbordered* if no nonempty words x, v, w exist such that $u \subset xv$ and $u \subset wx$. Otherwise, it is *bordered*. If u is a nonempty unbordered partial word, then show that $p(u) = |u|$ and consequently, unbordered partial words are primitive.

1.12 Different occurrences of the same unbordered factor u in a partial word w never overlap. True or false?

Challenging exercises

1.13 ⬚s Two partial words u and v are called *conjugate* if there exist partial words x and y such that $u \subset xy$ and $v \subset yx$. Prove that conjugacy on full words is an equivalence relation. Is it an equivalence relation on partial words?

1.14 Let x be a partial word and set $u = \diamond x$. Prove for $1 \leq p \leq |x|$:

- $p \in \mathcal{P}(u)$ if and only if $p \in \mathcal{P}(x)$
- $p \in \mathcal{P}'(u)$ if and only if $p \in \mathcal{P}'(x)$

1.15 Referring to Exercise 1.14, what can be said when $u = ax$ with x a nonempty partial word and $a \in A$?

1.16 $\boxed{\text{s}}$ Construct a partial word u over the alphabet $\{0,1\}$ for which no $a \in \{0,1\}$ exists that satisfies ua is primitive.

1.17 $\boxed{\text{H}}$ Let $u \in W(A)$. For $0 < p \leq |u|$, prove that the following are equivalent:

1. The partial word u is weakly p-periodic.
2. The containments $u \subset xv$ and $u \subset wx$ hold for some partial words x, v, w satisfying $|v| = |w| = p$.

1.18 $\boxed{\text{s}}$ Prove that if u is a nonempty partial word, then there exists a primitive word v and a positive integer n such that $u \subset v^n$. (*Hint:* Use induction on the length of u). Show that uniqueness holds for full words, that is, if u is a nonempty full word, then there exists a unique primitive word v and a unique positive integer n such that $u = v^n$. Does uniqueness hold for partial words?

1.19 $\boxed{\text{s}}$ Let x and y be partial words that are compatible. Show that $(xy \vee yx) \subset (x \vee y)^2$. Is the reverse containment true?

1.20 A nonempty word u is unbordered if $p(u) = |u|$. True or false?

1.21 $\boxed{\text{s}}$ Let u be a nonempty bordered partial word. Let x be a shortest nonempty word satisfying $u \subset xv$ and $u \subset wx$ for some nonempty words v, w. If $|v| \geq |x|$, then show that $p(u) < |u|$. Is this true when $|v| < |x|$?

1.22 Can you find partial words x, y and z not contained in powers of a common word and satisfying $x^m y^n \uparrow z^p$ for some integers $m, n, p \geq 2$.

Programming exercises

1.23 Write a program that discovers if two given partial words u, v of equal length are compatible. If so, then the program outputs the least upper bound of u and v, $u \vee v$.

1.24 Describe an algorithm that computes the minimal weak period of a given partial word. What is the complexity of your algorithm?

1.25 Design an applet that provides an implementation of your algorithm of Exercise 1.24, that is, given as input a partial word u, the applet outputs the minimal weak period of u, $p'(u)$.

1.26 Write a program that when given a finite subset X of $W(A) \setminus \{\varepsilon\}$, outputs $F(C(X))$. Run your program on $X = \{a \diamond b, abbaab\}$.

1.27 Repeat Exercise 1.26 to output $F(C(X^*))$.

Bibliographic notes

The study of the combinatorial properties of strings of symbols from a finite alphabet, also referred to as words, is profoundly connected to numerous fields such as biology, computer science, mathematics, and physics. Lothaire's first book *Combinatorics on Words* appeared in 1983 [106], while recent developments culminated in a second book *Algebraic Combinatorics on Words* which appeared in 2002 [107] and a third book *Applied Combinatorics on Words* in 2005 [108]. Several books that appeared quite recently emphasize connections of combinatorics on words to several research areas. We mention the book of Allouche and Shallit where the emphasis is on automata theory [2], the books of Crochemore and Rytter where the emphasis is on text processing algorithms [58, 60], the book of Gusfield where the emphasis is on algorithms related to biology [84], and finally the book of de Luca and Varrichio where the emphasis is on algebra [65].

The stimulus for recent works on combinatorics of words is the study of molecules such as DNA that play a central role in molecular biology [9, 53, 61, 63, 90, 91, 95]. Partial words appear in comparing genes. Indeed, alignment of two such strings can be viewed as a construction of two partial words that are compatible in a sense that was described in Section 1.6. The study of the combinatorics on partial words was initiated by Berstel and Boasson in their seminal paper [10]. Lemmas 1.1 and 1.2 are from [10]. This study was then pursued by Blanchet-Sadri and co-authors [14, 15, 16, 17, 18, 19, 20, 22, 23, 25, 26, 27, 28, 29, 31, 32, 35, 39, 40, 41, 42] as well as other researchers [103, 104, 130, 131]. Primitive and unbordered partial words were introduced by Blanchet-Sadri [17]. There, some well known properties of primitive and unbordered words are extended to primitive and unbordered partial words including Exercises 1.11, 1.12, 1.18 and 1.21. Conjugate partial words were introduced by Blanchet-Sadri and Luhmann [35] (Exercise 1.13 is proved there). Exercise 1.19 was suggested by David Dakota Blair.

Chapter 2

Combinatorial Properties of Partial Words

In this chapter, we analyze two properties of partial words: conjugacy and commutativity.

2.1 Conjugacy

We start by investigating the case for full words and then extend our results to include partial words.

2.1.1 The equation $xz = zy$

Suppose x, y, and z are words such that $xz = zy$. We are interested to know what the relationships between these words must be. Upon inspection, we observe that z must coincide with x in its first part and also with y in its second part. We illustrate with an example.

Example 2.1
Let $x = abcda$, $y = daabc$, and $z = abc$. Then, it is clear that

$$xz = zy, \text{ because}$$

$$(abcda)(abc) = (abc)(daabc)$$

Note that if $|z|$ is greater than $|x|$ and $|y|$, then x will be a prefix of z and y will be a suffix of z. ⬜

In the following lemma, we expand the idea motivated by the previous example.

LEMMA 2.1
Let x, y, z ($x \neq \varepsilon$ and $y \neq \varepsilon$) be words such that $xz = zy$. Then $x = uv$, $y = vu$, and $z = (uv)^n u$ for some words u, v and integer $n \geq 0$.

PROOF   If $|z| \leq |x|$, then we make use of Corollary 1.1 from the last chapter to show $x = zw$ and $y = wz$ for some word w. Putting $u = z$, $v = w$, and $n = 0$, the result holds.

If $|z| > |x|$, then again by Corollary 1.1 $z = xr$ for some word r. By hypothesis, $xz = zy$, thus

$$x(xr) = (xr)y$$

$$\text{implies } xr = ry$$

Since $x \neq \varepsilon$, $|r| < |z|$ and the desired conclusion follows by induction on $|z|$. The initial case is when $|z| = |x| + 1$, and r is a single letter. Then $xr = ry$, and putting $u = r = a_0$, $v = a_1 \ldots a_{|x|-1}$, and $n = 1$ we have our result. Now, assume that the result holds for all words z, $|z| \leq |x| + k$. Let z' be a word such that $|z'| = |x| + k + 1$ and $xz' = z'y$. Then we have $z' = xr$ and $xr = ry$ where $|r| < |z'|$. In other words, $|r| \leq |x| + k$. By the inductive hypothesis, there exist words u and v and an integer $n \geq 0$ where $x = uv$, $y = vu$, and $r = (uv)^n u$. Therefore, $z' = xr = (uv)^{n+1} u$, and the result holds.   ◻

Example 2.2
Applying Lemma 2.1 to Example 2.1, we see that $u = abc$, $v = da$, and $n = 0$.
◻

We note here that, as a consequence of the previous lemma, the word z is $|x|$-periodic. This fact will become important in the forthcoming extension of conjugacy to partial words.

2.1.2   The equation $xz \uparrow zy$

In this section, we consider the conjugacy property of partial words in accordance with the following definition.

DEFINITION 2.1   *Two partial words x and y are* **conjugate** *if there exist partial words u and v such that $x \subset uv$ and $y \subset vu$.*

Consequently, if the partial words x and y are conjugate, then there exists a partial word z satisfying the conjugacy equation $xz \uparrow zy$. Indeed, by setting $z = u$, we get $xz \subset uvu$ and $zy \subset uvu$.

In the previous section, we investigated the equation $xz = zy$ on words. For partial words, we obtain a similar result via the assumption of $xz \vee zy$ being $|x|$-periodic.

THEOREM 2.1
Let x, y, z be partial words with x, y nonempty. If $xz \uparrow zy$ and $xz \vee zy$ is $|x|$-periodic, then $x \subset uv$, $y \subset vu$, and $z \subset (uv)^n u$ for some words u, v and

integer $n \geq 0$.

Example 2.3

Let $x = \diamond ba$, $y = \diamond b\diamond$, and $z = b\diamond ab\diamond\diamond\diamond$. Then we have

$$
\begin{aligned}
xz &= \diamond\, b\, a\, b \diamond a\, b \diamond \diamond \diamond \diamond \\
zy &= b \diamond a\, b \diamond \diamond \diamond \diamond \diamond b \diamond \\
xz \vee zy &= b\, b\, a\, b \diamond a\, b \diamond \diamond\, b \diamond
\end{aligned}
$$

It is clear that $xz \uparrow zy$ and $xz \vee zy$ is $|x|$-periodic. Putting $u = bb$ and $v = a$, we can verify that the conclusion does indeed hold. □

If z is a full word, then the assumption $xz \uparrow zy$ implies the one of $xz \vee zy$ being $|x|$-periodic and the following corollary holds.

COROLLARY 2.1

Let x, y be nonempty partial words, and let z be a full word. If $xz \uparrow zy$, then $x \subset uv$, $y \subset vu$, and $z \subset (uv)^n u$ for some words u, v and integer $n \geq 0$.

Note that Corollary 2.1 does not necessarily hold if z is not full even if x, y are full as is seen in the following example.

Example 2.4

Let $x = a$, $y = b$, and $z = \diamond bb$. Then $xz = a \diamond bb$ and $zy = \diamond bbb$, and it is clear that $xz \uparrow zy$. However, if there exist full words u and v such that $x \subset uv$, then it must be that $a = uv$. This in turn makes it impossible for $y \subset vu$. We see, therefore, that the requirement that $xz \vee zy$ be $|x|$-periodic is necessary even if both x and y are full. □

First, we investigate the equation $xz \uparrow zy$ on partial words under the missing assumption of $xz \vee zy$ being $|x|$-periodic. The following two results give equivalences for conjugacy.

THEOREM 2.2

Let x, y and z be partial words such that $|x| = |y| > 0$. Then $xz \uparrow zy$ if and only if xzy is weakly $|x|$-periodic.

PROOF By the Division Algorithm, there exist integers m, n such that

$$
|z| = m|x| + n, \qquad 0 \leq n < |x|
$$

Equivalently, we can define m as $\lfloor \frac{|z|}{|x|} \rfloor$ and n as $|z| \bmod |x|$.[1] Then let

---

[1] Recall that for a real number x, $\lfloor x \rfloor$ is the greatest integer less than or equal to x.

$x = u_0v_0, y = v_{m+1}u_{m+2}$ and $z = u_1v_1u_2v_2 \ldots u_mv_mu_{m+1}$ where each u_i has length n and each v_i has length $|x| - n$. We may now align xz and zy one above the other in the following way:

$$
\begin{array}{ccccccccc}
u_0 & v_0 & u_1 & v_1 & \ldots & u_{m-1} & v_{m-1} & u_m & & v_m & u_{m+1} \\
u_1 & v_1 & u_2 & v_2 & \ldots & & u_m & & v_m & u_{m+1} v_{m+1} & u_{m+2}
\end{array}
\qquad (2.1)
$$

Assume $xz \uparrow zy$. Then the partial words in any column in (2.1) are compatible by simplification. Therefore for all i such that $0 \le i \le m+1$, $u_i \uparrow u_{i+1}$ and for all j such that $0 \le j \le m$, $v_j \uparrow v_{j+1}$. Thus $xz \uparrow zy$ implies that xzy is weakly $|x|$-periodic. Conversely, assume xzy is weakly $|x|$-periodic. This implies that $u_iv_i \uparrow u_{i+1}v_{i+1}$ for all i such that $0 \le i \le m$. Note that $u_{m+1}v_{m+1}u_{m+2}$ being weakly $|x|$-periodic, as a result $u_{m+1} \uparrow u_{m+2}$. This shows that $xz \uparrow zy$ which completes the proof. ▯

In the previous theorem and the next, it is helpful to realize that we are factoring the partial words xz and zy into words of length $|x|$ and each of these factors is represented by u_iv_i. The trailing u-factor on each partial word can be thought of as the "remainder term" and indeed that is the case. Aligning these factors and demonstrating their compatibility results in our conclusion of $|x|$-periodicity.

THEOREM 2.3

Let x, y and z be partial words such that $|x| = |y| > 0$. Then the following hold:

1. *If $xz \uparrow zy$, then xz and zy are weakly $|x|$-periodic.*

2. *If xz and zy are weakly $|x|$-periodic and $\lfloor \frac{|z|}{|x|} \rfloor > 0$, then $xz \uparrow zy$.*

PROOF   The proof is similar to that of Theorem 2.2. ▯

Example 2.5

Let $x = ab\diamond d\diamond f$, $y = \diamond\diamond\diamond bc\diamond$, and $z = abcdefab\diamond defabcdefabcdefabcdefab\diamond d$. Figure 2.1 displays the compatibility relation $xz \uparrow zy$ and highlights the factorizations of x, y and z as is done in the proof of Theorem 2.2.[2]

The concatenation xzy is seen to be weakly $|x|$-periodic.

---

[2]This graphic and the other that follows were generated using a C++ applet on one of the author's websites, mentioned in the Website Section at the end of this chapter.

```
↑  ab^d ^f abcd ef ab^d ef abcd ef abcd ef abcd ef ab^d
   abcd ef ab^d ef abcd ef abcd ef abcd ef ab^d ^^ ^bc^
```

FIGURE 2.1: An example of the conjugacy equation.

$$
\begin{array}{llllll}
a & b & \diamond & d & \diamond & f \\
a & b & c & d & e & f \\
a & b & \diamond & d & e & f \\
a & b & c & d & e & f \\
a & b & c & d & e & f \\
a & b & c & d & e & f \\
a & b & \diamond & d & \diamond & \diamond \\
\diamond & b & c & \diamond & &
\end{array}
$$

$\Box$

In Theorem 2.3(2), the assumption $\lfloor \frac{|z|}{|x|} \rfloor > 0$ is necessary. To see this, consider $x = aa$, $y = ba$ and $z = a$. Here, xz and zy are weakly $|x|$-periodic, but $xz \not\uparrow zy$ as $aaa \neq aba$.

Second, we consider solving the system of equations $z \uparrow z'$ and $xz \uparrow z'y$. Note that when $z = z'$, this system reduces to $xz \uparrow zy$. As before, let m be defined as $\lfloor \frac{|z|}{|x|} \rfloor$ and n as $|z| \bmod |x|$. Then let $x = u_0 v_0, y = v_{m+1} u_{m+2}$, $z = u_1 v_1 u_2 v_2 \ldots u_m v_m u_{m+1}$, and $z' = u'_1 v'_1 u'_2 v'_2 \ldots u'_m v'_m u'_{m+1}$ where each u_i, u'_i has length n and each v_i, v'_i has length $|x| - n$. The $|x|$-*pshuffle* and $|x|$-*sshuffle* of xz and $z'y$, denoted by $\mathrm{pshuffle}_{|x|}(xz, z'y)$ and $\mathrm{sshuffle}_{|x|}(xz, z'y)$, are defined as

$$u_0 v_0 u'_1 v'_1 u_1 v_1 u'_2 v'_2 \ldots u_{m-1} v_{m-1} u'_m v'_m u_m v_m u'_{m+1} v_{m+1} u_{m+1}$$

and

$$u_{m+1} u_{m+2}$$

respectively. The term shuffle is intentional, as the pshuffle interleaves the $u_i v_i$ and $u'_i v'_i$ factors from z and z'.

THEOREM 2.4
Let x, y, z and z' be partial words such that $|x| = |y| > 0$ and $|z| = |z'| > 0$. Then $z \uparrow z'$ and $xz \uparrow z'y$ if and only if $\mathrm{pshuffle}_{|x|}(xz, z'y)$ is weakly $|x|$-periodic and $\mathrm{sshuffle}_{|x|}(xz, z'y)$ is $(|z| \bmod |x|)$-periodic.

PROOF We may align z and z' (respectively, xz and $z'y$) one above the other in the following way:

$$
\begin{array}{llllllll}
u_1 & v_1 & u_2 & v_2 & \ldots & u_{m-1} & v_{m-1} & u_m & v_m & u_{m+1} \\
u'_1 & v'_1 & u'_2 & v'_2 & \ldots & u'_{m-1} & v'_{m-1} & u'_m & v'_m & u'_{m+1}
\end{array}
\tag{2.2}
$$

$$
\begin{array}{cccccccccc}
u_0 & v_0 & u_1 & v_1 & \cdots & u_{m-1} & v_{m-1} & u_m & v_m & u_{m+1} \\
u'_1 & v'_1 & u'_2 & v'_2 & \cdots & u'_m & v'_m & u'_{m+1} & v_{m+1} & u_{m+2}
\end{array}
\tag{2.3}
$$

Assume $z \uparrow z'$ and $xz \uparrow z'y$. Then the partial words in any column in (2.2) (respectively, (2.3)) are compatible using the simplification rule. Therefore for all $0 \le i < m$, $u'_{i+1}v'_{i+1} \uparrow u_{i+1}v_{i+1}$ (by 2.2) and $u_i v_i \uparrow u'_{i+1}v'_{i+1}$ (by 2.3). Also, we have $v_m \uparrow v_{m+1}$ and the following sequence of compatibility relations: $u_m \uparrow u'_{m+1}$, $u'_{m+1} \uparrow u_{m+1}$, and $u_{m+1} \uparrow u_{m+2}$. This shows that $\text{pshuffle}_{|x|}(xz, z'y)$ is weakly $|x|$-periodic and that $\text{sshuffle}_{|x|}(xz, z'y)$ is $(|z| \bmod |x|)$-periodic. The converse follows symmetrically. ◻

2.2 Commutativity

As before, we start by investigating the case for full words and then extending our results to include partial words.

2.2.1 The equation $xy = yx$

It is well known that two nonempty words x and y commute if and only if both x and y are powers of a common word, and the proof is straightforward.

LEMMA 2.2

Let x and y be nonempty words. Then $xy = yx$ if and only if there exists a word z such that $x = z^m$ and $y = z^n$ for some integers m, n.

PROOF Suppose $xy = yx$. We will use induction on the length of xy. Since the words are nonempty, we begin with $|xy| = 2$. Because $xy = yx$, it is immediate that $x = y$, and we are done. Now assume that the result is true for all x, y such that $|xy| \le k$ for some positive integer k. Assume $|xy| = k+1$. With the equation $xy = yx$ and our conjugacy result in Lemma 2.1, we have that $x = uv = vu$ and $y = (uv)^l u$ for some words u and v and integer $l \ge 0$. If $u = \varepsilon$ or $v = \varepsilon$, then the result follows. Otherwise recall that $y \ne \varepsilon$, and so $|uv| = |x| < |xy| = k + 1$. By the inductive hypothesis, we conclude that u and v are powers of a common word, z. Consequently, x and y are powers of z, and the result is obtained. The converse statement is obvious. ◻

2.2.2 The equation $xy \uparrow yx$

To extend this characterization of commutativity to partial words, we use the notion of containment. Certainly, if there exist a word z and integers m, n

such that $x \subset z^m$ and $y \subset z^n$, then

$$xy \subset z^{m+n}$$
$$yx \subset z^{n+m}$$

and $xy \uparrow yx$. In addition, the converse holds as well, provided the partial word xy has at most <u>one</u> hole. We state the theorem here without proof, as we will later prove a more general result in Theorem 2.6.

THEOREM 2.5
Let x and y be nonempty partial words such that xy has at most one hole. If $xy \uparrow yx$, then there exists a word z such that $x \subset z^m$ and $y \subset z^n$ for some integers m, n.

However, if xy possesses more than one hole, the situation becomes more subtle. Indeed, it is easy to produce a counterexample when xy contains just one more hole.

Example 2.6
Let $x = \diamond bb$ and $y = abb\diamond$. Then

$$xy = \diamond bbabb\diamond \uparrow abb\diamond\diamond bb = yx$$

Since $\gcd(|x|, |y|) = 1$, if x and y were contained in powers of a common word z, then $|z|$ would be equal to 1, which is not possible for y. ⬚

Definition of (k, l)-special partial word

To extend this theorem to the case when xy has at least two holes, we will need to inspect the structure of the partial word xy more carefully by stepping through a sequence of positions. We select positions motivated by the following lemma, the proof of which is left to the reader.

LEMMA 2.3
Let x, y be nonempty partial words. If there exists a full word z such that $x \subset z^m$ and $y \subset z^n$, then xy is $\gcd(|x|, |y|)$-periodic.

We next develop a criterion based on the contrapositive of this statement to determine whether a partial word with at least two holes can be decomposed into x and y as contained in powers of a common word z. That is, if the pword is not $\gcd(|x|, |y|)$-periodic, then such a decomposition cannot be found. If this occurs, we say xy is (k, l)-*special*, where $k = |x|$ and $l = |y|$. We adopt the convention that $k \leq l$, as we can assume without loss of generality that $|x| \leq |y|$.

For a partial word of length $k + l$, we can test for $\gcd(k, l)$-periodicity by checking sequences of letters that are p positions apart, where $p = \gcd(k, l)$. Note that, because we are only interested in testing for periodicity, the order of checking these positions is irrelevant.

For $0 \leq i < k + l$, we define the sequence of i relative to k, l as $\mathrm{seq}_{k,l}(i) = (i_0, i_1, i_2, \ldots, i_n, i_{n+1})$ where $i_0 = i = i_{n+1}$ and where

For $1 \leq j \leq n$, $i_j \neq i$,

For $1 \leq j \leq n + 1$, i_j is defined as

$$i_j = \begin{cases} i_{j-1} + k & \text{if } i_{j-1} < l \\ i_{j-1} - l & \text{otherwise} \end{cases}$$

Note that $\mathrm{seq}_{k,l}(i)$ is stopped at the first occurrence of i, which defines $n + 1$.[3]

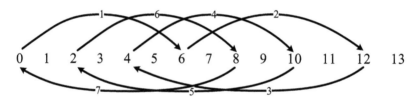

FIGURE 2.2:   The construction of $\mathrm{seq}_{6,8}(0)$.

Example 2.7

If $k = 6$ and $l = 8$, then $\mathrm{seq}_{6,8}(0) = (0, 6, 12, 4, 10, 2, 8, 0)$. The path traversed by this sequence is represented in Figure 2.2. It can be seen that this path selects positions $\gcd(6, 8) = 2$ letters apart, beginning with position 0. To fully verify periodicity, it will be necessary to generate another sequence beginning at $i = 1$, which is $\mathrm{seq}_{6,8}(1) = (1, 7, 13, 5, 11, 3, 9, 1)$. No other sequence is necessary, for if we calculated $\mathrm{seq}_{6,8}(2)$ we simply would obtain a permutation of the first sequence since it already contains the position 2.   ⬜

In general, to fully verify a given pword, $\gcd(k, l)$ sequences are needed corresponding to the positions $0 \leq i < \gcd(k, l)$. Now, we use these sequences

---

[3]Some readers may find the following (equivalent) definition of $\mathrm{seq}_{k,l}(i)$ more intuitive: $i_0 = i$ and for $1 \leq j \leq n + 1$, $i_j = (i_{j-1} + k) \bmod (k + l)$. (Recall that $k + l$ is the length of the partial word being analyzed.) As before, continue the sequence until the first occurrence of i is reached.

to make our definition of (k, l)-special partial word precise in the following manner.

DEFINITION 2.2   *Let k, l be positive integers satisfying $k \leq l$ and let z be a partial word of length $k + l$. We say that z is $(\boldsymbol{k}, \boldsymbol{l})$-special if there exists $0 \leq i < \gcd(k, l)$ such that $seq_{k,l}(i) = (i_0, i_1, i_2, \ldots, i_n, i_{n+1})$ contains (at least) two positions that are holes of z while $z(i_0)z(i_1) \ldots z(i_{n+1})$ is not 1-periodic.*

Notice that in order to show a partial word z is (k, l)-special, it is possible that $\gcd(k, l)$ sequences will need to be calculated. Once a sequence that satisfies the definition is found, then z can be declared (k, l)-special. However, it is necessary to calculate all sequences in order to classify z as not (k, l)-special.

Example 2.8

Let $z = cbca\diamond\diamond cbc\diamond caca$, and let $k = 6$ and $l = 8$ so $|z| = k + l$. We wish to determine if z is $(6, 8)$-special. Find $seq_{6,8}(0) = (0, 6, 12, 4, 10, 2, 8, 0)$ and

$$z(0)\ z(6)\ z(12)\ z(4)\ z(10)\ z(2)\ z(8)\ z(0)$$
$$c\quad c\quad c\quad \diamond\quad c\quad c\quad c\quad c$$

This sequence does not satisfy the definition, and so continue with calculating $seq_{6,8}(1) = (1, 7, 13, 5, 11, 3, 9, 1)$. The corresponding letter sequence is

$$z(1)\ z(7)\ z(13)\ z(5)\ z(11)\ z(3)\ z(9)\ z(1)$$
$$b\quad b\quad a\quad \diamond\quad a\quad a\quad \diamond\quad b$$

Here we have two positions in the sequence which are holes, and the sequence is not 1-periodic. Hence, z is $(6, 8)$-special.   ▯

THEOREM 2.6

Let x, y be nonempty partial words such that $|x| \leq |y|$. If $xy \uparrow yx$ and xy is not $(|x|, |y|)$-special, then there exists a word z such that $x \subset z^m$ and $y \subset z^n$ for some integers m, n.

PROOF   Since $xy \uparrow yx$, there exists a word u such that $xy \subset u$ and $yx \subset u$. Put $|x| = k$ and $|y| = l$. Put $l = mk + r$ where $0 \leq r < k$. Either $r = 0$ or $r > 0$, and for each possibility the proof is split into three cases that refer to a given position i of u. Case 1 refers to $0 \leq i < k$, Case 2 to $k \leq i < l$, and Case 3 to $l \leq i < l + k$ (Cases 1 and 3 are symmetric as is seen by putting $i = l + j$ where $0 \leq j < k$). The following diagram pictures the containments $xy \subset u$ and $yx \subset u$:

$$\begin{array}{c|ccc|ccc|ccc}
xy & x(0) & \dots & x(k-1) & y(0) & \dots & y(l-k-1) & y(l-k) & \dots & y(l-1) \\
yx & y(0) & \dots & y(k-1) & y(k) & \dots & y(l-1) & x(0) & \dots & x(k-1) \\
u & u(0) & \dots & u(k-1) & u(k) & \dots & u(l-1) & u(l) & \dots & u(l+k-1)
\end{array}$$

We prove the result for Case 1 under the assumption that $r > 0$. The other cases follow similarly and are left as exercises for the reader. We consider the cases where $i < r$ and $i \geq r$. If $i < r$, then

$x(i) \subset u(i)$ and $y(i) \subset u(i)$,

$y(i) \subset u(i+k)$ and $y(i+k) \subset u(i+k)$,

$y(i+k) \subset u(i+2k)$ and $y(i+2k) \subset u(i+2k)$,

$y(i+2k) \subset u(i+3k)$ and $y(i+3k) \subset u(i+3k)$,

$\vdots$

$y(i+(m-1)k) \subset u(i+mk)$ and $y(i+mk) \subset u(i+mk)$,

$y(i+mk) \subset u(i+(m+1)k)$ and $x(i+k-r) \subset u(i+(m+1)k)$,

$x(i+k-r) \subset u(i+k-r)$ and $y(i+k-r) \subset u(i+k-r)$,

$y(i+k-r) \subset u(i+2k-r)$ and $y(i+2k-r) \subset u(i+2k-r)$,

$\vdots$

If $i \geq r$, then

$x(i) \subset u(i)$ and $y(i) \subset u(i)$,

$y(i) \subset u(i+k)$ and $y(i+k) \subset u(i+k)$,

$y(i+k) \subset u(i+2k)$ and $y(i+2k) \subset u(i+2k)$,

$y(i+2k) \subset u(i+3k)$ and $y(i+3k) \subset u(i+3k)$,

$\vdots$

$y(i+(m-2)k) \subset u(i+(m-1)k)$ and $y(i+(m-1)k) \subset u(i+(m-1)k)$,

$y(i+(m-1)k) \subset u(i+mk)$ and $x(i-r) \subset u(i+mk)$,

$x(i-r) \subset u(i-r)$ and $y(i-r) \subset u(i-r)$,

$y(i-r) \subset u(i+k-r)$ and $y(i+k-r) \subset u(i+k-r)$,

$\vdots$

If $i < r$, then let $x(i)y(i)y(i+k)\ldots y(i+mk)x(i+k-r)\ldots x(i) = v_i$, and if $i \geq r$, then let $x(i)y(i)y(i+k)\ldots y(i+(m-1)k)x(i-r)\ldots x(i) = v_i$. In either case, we claim that v_i is 1-periodic, say with letter a_i in $A \cup \{\diamond\}$. The claim follows from the above containments in case v_i has less than two holes. For the case where v_i has at least two holes, the claim follows since xy is not (k, l)-special. It turns out that $a_j = a_{j+r} = \cdots$ for $0 \leq j < r$. Let $z = a_0 a_1 \ldots a_{r-1}$. If r divides k, then $x \subset z^{k/r}$ and $y \subset z^{(mk/r)+1}$. If r does not divide k, then z is 1-periodic with letter a say. In this case, $x \subset a^k$ and $y \subset a^l$. ☐

Example 2.9

Given $x = ab\diamond a\diamond\diamond a\diamond b$ and $y = a\diamond babba\diamond\diamond a\diamond b$, the alignment of xy and yx may be observed with the depiction in Figure 2.3. We can check that $xy \uparrow yx$ and also that xy is not $(|x|, |y|)$-special (the latter is left as an exercise). Here $x \subset (abb)^3$ and $y \subset (abb)^4$. ☐

| ab^ | a^^ | a^b | a^b | abb | a^^ | a^b |
| a^b | abb | a^^ | a^b | ab^ | a^^ | a^b |

FIGURE 2.3: An example of the commutativity equation.

Definition of $\{k, l\}$-special partial word

Next, we define the concept of $\{k, l\}$-special partial word as an extension of (k, l)-special partial word and give two lemmas that provide another sufficient condition for two words x and y to commute.

DEFINITION 2.3 *Let k, l be positive integers satisfying $k \leq l$ and let z be a partial word of length $k+l$. We say that z is $\{k, l\}$-special if there exists $0 \leq i < \gcd(k, l)$ such that $\mathrm{seq}_{k,l}(i)$ satisfies the condition of Definition 2.2 or the condition of containing two consecutive positions that are holes of z.*

Restated, z is $\{k, l\}$-special if there exists i such that $\mathrm{seq}_{k,l}(i)$ either

1. has two positions that are holes and is not 1-periodic, OR

2. has two *consecutive* positions that are holes.

By definition, a partial word z that is (k, l)-special is $\{k, l\}$-special. The converse is not true, as is seen in this example.

Example 2.10

Let $k = 6$, $l = 8$, and $z = \diamond babab\diamond\diamond ababab$. Calculating $\text{seq}_{k,l}(0)$, we have

$$z(0)\ z(6)\ z(12)\ z(4)\ z(10)\ z(2)\ z(8)\ z(0)$$
$$\diamond\quad\diamond\quad a\quad\quad a\quad\quad a\quad\quad a\quad\quad a\quad\quad\diamond$$

Since $\text{seq}_{6,8}(0)$ contains the consecutive positions 0 and 6 that are holes of z, z is $\{6, 8\}$-special. However, after calculating $\text{seq}_{6,8}(1)$, we observe that both sequences are 1-periodic, and thus z cannot be $(6, 8)$-special.　　　　　☐

LEMMA 2.4

Let x, y be nonempty words and let z be a partial word with at most one hole. If $z \subset xy$ and $z \subset yx$, then $xy = yx$.

LEMMA 2.5

Let x, y be nonempty words and let z be a non $\{|x|, |y|\}$-special partial word. If $z \subset xy$ and $z \subset yx$, then $xy = yx$.

PROOF　Put $|x| = k$ and $|y| = l$. Without loss of generality, we can assume that $k \leq l$. Put $l = mk + r$ where $0 \leq r < k$. As before, either $r = 0$ or $r > 0$, and for each possibility the proof is split into three cases that refer to a given position i of z. Case 1 treats the situation when $0 \leq i < k$, Case 2 the situation when $k \leq i < l$, and Case 3 when $l \leq i < l + k$ (Cases 1 and 3 are symmetric). The following diagram pictures the inclusions $z \subset xy$ and $z \subset yx$:

$$
\begin{array}{l|lll|lll}
z & z(0) \ldots z(k-1) & z(k) \ldots & z(l-1) & z(l) & \ldots & z(l+k-1) \\
xy & x(0) \ldots x(k-1) & y(0) \ldots & y(l-k-1) & y(l-k) \ldots & & y(l-1) \\
yx & y(0) \ldots y(k-1) & y(k) \ldots & y(l-1) & x(0) & \ldots & x(k-1)
\end{array}
$$

We prove the result for Case 1 for both $r = 0$ and $r > 0$. The other cases follow similarly and are left as exercises for the reader.

We first treat the case where $r = 0$. If $i \in D(z)$, then $z(i) \subset x(i)$ and $z(i) \subset y(i)$ and so $x(i) = y(i)$. If $i \in H(z)$, then we prove that $x(i) = y(i)$ as follows. We have

$z(i) \subset x(i)$ and $z(i) \subset y(i)$,

$z(i + k) \subset y(i)$ and $z(i + k) \subset y(i + k)$,

$z(i + 2k) \subset y(i + k)$ and $z(i + 2k) \subset y(i + 2k)$,

$$\vdots$$

$z(i + (m-1)k) \subset y(i + (m-2)k)$ and $z(i + (m-1)k) \subset y(i + (m-1)k)$,

$z(i + mk) \subset y(i + (m - 1)k)$ and $z(i + mk) \subset x(i)$.

Here $\mathrm{seq}_{k,l}(i) = (i, i+k, \ldots, i+mk, i)$ and $z(i)z(i+k)z(i+2k) \ldots z(i+mk)z(i)$ does not contain consecutive holes and does not contain two holes while not 1-periodic since z is not $\{k, l\}$-special. So $x(i) = y(i + (m-1)k) = y(i + (m-2)k) = \cdots = y(i + k) = y(i)$ (note that $H(z)$ does not contain in particular $i + k, i + mk$).

We now treat the case where $r > 0$. If $i \in D(z)$, then we proceed as in the case where $r = 0$. If $i \in H(z)$, we consider the cases where $i < r$ and $i \geq r$. If $i < r$, then

$z(i) \subset x(i)$ and $z(i) \subset y(i)$,

$z(i + k) \subset y(i)$ and $z(i + k) \subset y(i + k)$,

$z(i + 2k) \subset y(i + k)$ and $z(i + 2k) \subset y(i + 2k)$,

$\vdots$

$z(i + (m - 1)k) \subset y(i + (m - 2)k)$ and $z(i + (m - 1)k) \subset y(i + (m - 1)k)$,

$z(i + mk) \subset y(i + (m - 1)k)$ and $z(i + mk) \subset y(i + mk)$,

$z(i + (m + 1)k) \subset y(i + mk)$ and $z(i + (m + 1)k) \subset x(i + k - r)$,

$z(i + k - r) \subset x(i + k - r)$ and $z(i + k - r) \subset y(i + k - r)$,

$\vdots$

If $i \geq r$, then

$z(i) \subset x(i)$ and $z(i) \subset y(i)$,

$z(i + k) \subset y(i)$ and $z(i + k) \subset y(i + k)$,

$z(i + 2k) \subset y(i + k)$ and $z(i + 2k) \subset y(i + 2k)$,

$\vdots$

$z(i + (m - 1)k) \subset y(i + (m - 2)k)$ and $z(i + (m - 1)k) \subset y(i + (m - 1)k)$,

$z(i + mk) \subset y(i + (m - 1)k)$ and $z(i + mk) \subset x(i - r)$,

$z(i - r) \subset x(i - r)$ and $z(i - r) \subset y(i - r)$,

$\vdots$

Applying the above repeatedly, we can show that $x(i) = y(i)$. More precisely, in the case where $i < r$, $\mathrm{seq}_{k,l}(i) = (i, i+k, \ldots, i+mk, i+(m+1)k, i+k-r, \ldots, i)$ leads to $y(i) = y(i+k) = \cdots = y(i+(m-1)k) = y(i+mk) = x(i+k-r) = \cdots = x(i)$ since z is not $\{k,l\}$-special. Similarly, in the case where $i \geq r$, $\mathrm{seq}_{k,l}(i) = (i, i+k, \ldots, i+(m-1)k, i+mk, i-r, \ldots, i)$ leads to $y(i) = y(i+k) = \cdots = y(i+(m-2)k) = y(i+(m-1)k) = x(i-r) = \cdots = x(i)$.

Note that in Lemma 2.5, the assumption of z being non $\{|x|, |y|\}$-special cannot be replaced by the weaker assumption of z not being $(|x|, |y|)$-special. To see this, consider the partial words $x = ababab$, $y = cbababab$, and $z = \diamond babab\diamond\diamond ababab$ from Example 2.10. Here, $z \subset xy$ and $z \subset yx$, but $xy \neq yx$.

We end this chapter with the concept of a *pairwise nonspecial set of partial words* that is used in later chapters.

DEFINITION 2.4   Let $X \subset W(A)$. Then X is called **pairwise nonspecial** if all $u, v \in X$ of different positive lengths satisfy the following conditions:

- If $|u| < |v|$, then v is non $\{|u|, |v| - |u|\}$-special.
- If $|u| > |v|$, then u is non $\{|v|, |u| - |v|\}$-special.

Note that any subset of $W_1(A)$ is pairwise nonspecial.

Exercises

2.1 Consider the partial words $x = ab\diamond d\diamond f$, $y = q\diamond mno\diamond$ and $z =$

$$abcdefab\diamond defabcdefabcdefabcdef$$
$$ab\diamond d\diamond\diamond\diamond bo\diamond qrm\diamond opqrmnopqrm\diamond op$$

- Show that $xz \uparrow zy$.
- Show that xzy is weakly $|x|$-periodic.

2.2 Referring to Exercise 2.1, display the factorizations of x, y and z as is done in the proof of Theorem 2.2.

2.3 Set $x = a\diamond cd\diamond\diamond$, $y = \diamond def\diamond b$, and

$$z = abc\diamond\diamond\diamond a\diamond\diamond def\diamond\diamond cdefa\diamond$$

Show that $xz \uparrow zy$ and $xz \vee zy$ is $|x|$-periodic. Find words u, v and an integer $n \geq 0$ that satisfy Theorem 2.1.

2.4 If x and y are nonempty conjugate partial words, then there exists a partial word z satisfying the conjugacy equation $xz \uparrow zy$. Moreover, in this case there exist partial words u, v such that $x \subset uv$, $y \subset vu$, and $z \subset (uv)^n u$ for some integer $n \geq 0$. True or false?

2.5 ⬚s If $k = 4$ and $l = 10$, then determine whether the following partial words are $(4, 10)$-special or not?

- $a \diamond baab \diamond aabaa \diamond \diamond$
- $\diamond babab \diamond babab \diamond b$

2.6 Find k, l such that $z = acbca \diamond \diamond cbc \diamond cac$ is (k, l)-special.

2.7 Let $x = ab \diamond a \diamond \diamond a \diamond b$ and $y = a \diamond babba \diamond \diamond a \diamond b$. Show that $xy \uparrow yx$ and that xy is not $(|x|, |y|)$-special. Find a word z and integers m, n such that $x \subset z^m$ and $y \subset z^n$.

2.8 Prove Lemma 2.3.

2.9 Check that in Example 2.9 xy is not $(|x|, |y|)$-special.

2.10 ⬚s Give an example of a partial word that is $\{3, 6\}$-special without being $(3, 6)$-special.

2.11 ⬚s Give partial words u, v, w such that $w \subset uv$, $w \subset vu$ and $uv \neq vu$. Is w $(|u|, |v|)$-special? Is it $\{|u|, |v|\}$-special? Does your answer contradict Lemma 2.5?

2.12 What can be said if u is a full word over the alphabet $\{0, 1\}$ and satisfies $u0 = 0u$? What can be said if $u0 = 1u$?

2.13 What can be said if u is a partial word with one hole over the alphabet $\{0, 1\}$ and satisfies $u0 = 0u$? What can be said if $u0 = 1u$?

2.14 Repeat Exercise 2.13 if u satisfies $u0 \uparrow 0u$ or $u0 \uparrow 1u$.

Challenging exercises

2.15 Prove Theorem 2.1.

2.16 Prove Corollary 2.1.

2.17 ⬚s Let $u, v \in A^+$ and let $z \in W_1(A)$. If $uz \uparrow zv$, then prove that one of the following holds:

1. There exist partial words x, y, x_1, x_2 such that $u = x_1 y$, $v = y x_2$, $x \subset x_1$, $x \subset x_2$, and $z = (x_1 y)^m x (y x_2)^n$ for some integers $m, n \geq 0$.

2. There exist partial words x, y, y_1, y_2 such that $u = xy_1$, $v = y_2x$, $y \subset y_1$, $y \subset y_2$, and $z = (xy_1)^m xy(xy_2)^n x$ for integers $m, n \geq 0$.

2.18 ☐s Let $u, v \in A^+$. Let $z \in W_1(A) \setminus A^+$ and let $z' \in A^+$. If $z \uparrow z'$ and $uz \uparrow z'v$, then prove that one of the following holds:

1. There exist partial words x, y, x_1, x_2 such that $u = x_1y$, $v = yx_2$, $x \sqsubset x_1$, $x \sqsubset x_2$, $z = (x_1y)^m x(yx_2)^n$, and $z' = (x_1y)^m x_1(yx_2)^n$ for some integers $m, n \geq 0$.

2. There exist partial words x, y, y_1, y_2 such that $u = xy_1$, $v = y_2x$, $y \sqsubset y_1$, $y \sqsubset y_2$, $z = (xy_1)^m xy(xy_2)^n x$, and $z' = (xy_1)^{m+1}(xy_2)^n x$ for some integers $m, n \geq 0$.

2.19 Referring to Exercise 2.18, what can be said when $u, v \in A^+$, $z \in A^+$ and $z' \in W_1(A) \setminus A^+$ are such that $z \uparrow z'$ and $uz \uparrow z'v$?

2.20 Generalize Exercise 2.17 to $z \in W_2(A)$.

2.21 ☐s No primitive word u can be an inside factor of uu. True or false?

2.22 ☐s Prove Case 3 of Theorem 2.6.

2.23 Referring to Theorem 2.6, prove the case where $r = 0$.

2.24 Prove Case 2 of Theorem 2.6 when $r > 0$.

2.25 Prove Case 2 of Lemma 2.5.

2.26 Let x, y be nonempty partial words and let u, v be full words such that $x \subset u$ and $y \subset v$. If xy is non $\{|x|, |y|\}$-special and $yx \subset uv$, then $xy \subset vu$.

Programming exercises

2.27 Referring to Exercise 2.21, give pseudo code for an algorithm that tests primitivity on full words. What is the complexity of your algorithm?

2.28 Write a program to find out whether or not two partial words x and y are conjugate. Run your program on the pairs of partial words:

- $x = a\diamond babb\diamond a$ and $y = \diamond b\diamond\diamond aa\diamond\diamond$
- $x = ba\diamond bbbaa$ and $y = a\diamond babb\diamond a$
- $x = ba\diamond bbbaa$ and $y = \diamond b\diamond\diamond aa\diamond\diamond$

2.29 Write a program that when given a pword u and integers k, l such that $k \leq l$, discovers if u is or is not (k, l)-special. Modify your program to let it also discover if u is $\{k, l\}$-special. What is the output for running the program on $u = a \diamond baab \diamond aabaa \diamond \diamond$, $k = 4$ and $l = 10$.

2.30 Starting with your program of Exercise 2.29, write a program that tests whether or not a set X of partial words is pairwise nonspecial. Find a set X that is pairwise nonspecial and one that is not.

2.31 Write a program that takes as inputs four nonempty partial words x, y, z and z' such that $z \uparrow z'$ and $xz \uparrow z'y$, and outputs a factorization of $\text{pshuffle}_{|x|}(xz, z'y)$ and $\text{sshuffle}_{|x|}(xz, z'y)$ according to the proof of Theorem 2.4 and shows that they are weakly $|x|$-periodic and weakly $(|z| \bmod |x|)$-periodic respectively.

Website

A World Wide Web server interface at

$$\texttt{http://www.uncg.edu/mat/research/equations}$$

has been established for automated use of programs related to the equations discussed in this chapter. In particular, one of the programs takes as input three partial words x, y and z such that $|x| = |y|$ and $xz \uparrow zy$, and outputs a factorization of x, y and z and shows that xzy is weakly $|x|$-periodic (this program implements Theorem 2.2). Another program takes as input a set $\{x, y\}$ of two partial words such that $|x| \leq |y|$, $xy \uparrow yx$ and xy is not $(|x|, |y|)$-special, and outputs a partial word z and integers m, n such that $x \subset z^m$ and $y \subset z^n$ (this program implements Theorem 2.6).

Bibliographic notes

The combinatorial properties of conjugacy and commutativity on full words of Sections 2.1.1 and 2.2.1 are discussed in Shyr's book [132]. The study of conjugacy and commutativity on partial words was initiated by Blanchet-Sadri and Luhmann [35]. It is there that the concept of $\{k, l\}$-special partial word was defined. Lemma 2.5, Theorem 2.1, and Corollary 2.1 are from [35]. The property of conjugacy on partial words was further studied by Blanchet-Sadri, Blair and Lewis [20] (Section 2.1.2), and the property of commutativity by Blanchet-Sadri and Anavekar who defined the concept of (k, l)-special partial word [18] (Section 2.2.2). The one-hole case of Theorem 2.5 and Lemma 2.4 are due to Berstel and Boasson [10]. Exercises 2.17, 2.18 and 2.19 are from Blanchet-Sadri and Duncan [29], while Exercise 2.21 is discussed in [51]. Definition 2.4 of pairwise nonspecial set of partial words is from [16].

Part II

PERIODICITY

Chapter 3

Fine and Wilf's Theorem

In this chapter, we discuss the fundamental periodicity result on words due to Fine and Wilf in the context of partial words. Fine and Wilf's result states that any word having periods p and q and length at least $p + q - \gcd(p, q)$ has period $\gcd(p, q)$. Moreover, the bound $p + q - \gcd(p, q)$ is optimal since counterexamples can be provided for words of smaller length. We extend this result to partial words in two ways:

First, we discuss weak periodicity extensions, that is, we consider long enough partial words having *weak* periods p, q and show that under some conditions they also have period $\gcd(p, q)$. We start with partial words with one, two, and three holes, and then generalize the result for partial words with an arbitrary number of holes. The following table describes the number of holes and section numbers where these results are discussed:

Holes	Sections
0–1	3.1
2–3	3.2
arbitrary	3.3, 3.4 and 3.5

Second, we discuss strong periodicity extensions, that is, we consider in Section 3.6 long enough partial words having *strong* periods p, q and show that under some conditions they also have period $\gcd(p, q)$.

3.1 The case of zero or one hole

In this section, we restrict ourselves to partial words with zero or one hole. We omit the proof of the following theorem, because we will prove the general result later in this chapter.

THEOREM 3.1

Let p and q be positive integers.

  1. *(Fine and Wilf) Let u be a word. If u is p-periodic and q-periodic and $|u| \geq p + q - \gcd(p, q)$, then u is $\gcd(p, q)$-periodic.*

2. *Let u be a partial word such that $\|H(u)\| = 1$. If u is weakly p-periodic and weakly q-periodic and $|u| \geq p + q$, then u is $\gcd(p, q)$-periodic.*

The bounds for the minimal length of the partial word u in the above theorem are optimal, that is, the result does not hold for partial words that are weakly p-periodic and weakly q-periodic but of smaller length. In the next examples we present a counterexample for each statement in Theorem 3.1.

Example 3.1

The bound $p + q - \gcd(p, q)$ is optimal in Theorem 3.1(1). For example, using $p = 3, q = 4$ and $p + q - \gcd(p, q) = 6$, the following picture shows that the word *aabaa* of length 5 is 3-periodic and 4-periodic but is not 1-periodic:

$$0\ 1\ 2\ \ \ 3\ 4$$
$$a\ a\ b\ |\ a\ a$$

$$0\ 1\ 2\ 3\ \ \ 4$$
$$a\ a\ b\ a\ |\ a$$

$\square$

Example 3.2

The bound $p + q$ is optimal in Theorem 3.1(2), as can be seen with *aabaa⋄* of length 6 which is weakly 3-periodic and weakly 4-periodic but not 1-periodic:

$$0\ 1\ 2\ \ \ 3\ 4\ 5$$
$$a\ a\ b\ |\ a\ a\ \diamond$$

$$0\ 1\ 2\ 3\ \ \ 4\ 5$$
$$a\ a\ b\ a\ |\ a\ \diamond$$

$\square$

3.2   The case of two or three holes

Theorem 3.1 does not hold for partial words with two holes. For instance, the partial word $u = aabaa\diamond\diamond$ is weakly 3-periodic and weakly 4-periodic and $|u| \geq 3 + 4$ but u is not $\gcd(3, 4)$-periodic.

To extend Theorem 3.1 to partial words with two or three holes (and beyond), we will emulate the process of the last chapter and define a subset of partial words u as $(\|H(u)\|, p, q)$-*special* based on specific criteria. We will then be able to extend our periodicity result to those pwords that are not $(\|H(u)\|, p, q)$-special.

In this section, we limit ourselves to pwords with only two or three holes. We provide definitions for $(2, p, q)$-special and $(3, p, q)$-special for completion, but remind the reader that these definitions will be generalized in future sections.

DEFINITION 3.1 *Let p and q be positive integers satisfying $p < q$. A partial word u is called*

1. $(2, p, q)$-**special** *if at least one of the following holds:*

    (a) $q = 2p$ *and there exists $p \leq i < |u| - 4p$ such that $i+p, i+q \in H(u)$.*

    (b) *There exists $0 \leq i < p$ such that $i + p, i + q \in H(u)$.*

    (c) *There exists $|u| - p \leq i < |u|$ such that $i - p, i - q \in H(u)$.*

2. $(3, p, q)$-**special** *if it is $(2, p, q)$-special or if at least one of the following holds:*

    (a) $q = 3p$ *and there exists $p \leq i < |u| - 5p$ such that $i+p, i+2p, i+3p \in H(u)$ or there exists $p \leq i < |u| - 7p$ such that $i+p, i+3p, i+5p \in H(u)$.*

    (b) *There exists $0 \leq i < p$ such that $i + q, i + 2p, i + p + q \in H(u)$.*

    (c) *There exists $|u| - p \leq i < |u|$ such that $i - q, i - 2p, i - p - q \in H(u)$.*

    (d) *There exists $p \leq i < q$ such that $i - p, i + p, i + q \in H(u)$.*

    (e) *There exists $|u| - q \leq i < |u| - p$ such that $i - p, i + p, i - q \in H(u)$.*

    (f) $2q = 3p$ *and there exists $p \leq i < |u| - 5p$ such that $i + q, i + 2p, i + p + q \in H(u)$.*

If p and q are positive integers satisfying $p < q$ and $\gcd(p, q) = 1$, then for each $n > 0$ we can construct a binary partial word u_n with two holes such that u_n is weakly p- and q-periodic but not $\gcd(p, q)$-periodic. Put

$$u_n = ab^{p-1} \diamond b^{q-p-1} \diamond b^n$$

Writing u_n into p columns, the first hole falls under the first letter a and we see that u_n indeed is weakly p-periodic. Similarly, when u_n is written into q columns, the second hole falls under the first letter a and thus u_n is weakly q-periodic.

The partial word u_n is $(2, p, q)$-special by Definition 3.1(1)(b) as is seen by setting $i = 0$. We can think of a, the letter in position 0, as being "isolated" by the holes in the pword u_n, allowing u_n to not be 1-periodic.[1]

Similarly, the infinite sequence

---

[1] Indeed, we will use this notion of "isolation" to define $(\|H(u)\|, p, q)$-special pwords u for an arbitrary number of holes.

$$\overbrace{\begin{matrix} a & b & ... & b \end{matrix}}^{p \text{ columns}}$$
$$\Diamond \quad b \quad ...$$

$$\overbrace{\begin{matrix} a & b & ... & b & \Diamond & b & ... & b \end{matrix}}^{q \text{ columns}}$$
$$\Diamond \quad b \quad ...$$

FIGURE 3.1:   A $(2, p, q)$-special binary partial word.

$$(ab^{p-1}\Diamond b^{q-p-1}\Diamond b^{n}\Diamond)_{n>0}$$

consists of binary $(3, p, q)$-special partial words with three holes that are weakly p-periodic and weakly q-periodic but not 1-periodic.

We now state the theorem for partial words containing two or three holes.

THEOREM 3.2
Let p and q be positive integers satisfying $p < q$.

1. *Let u be a partial word such that $\|H(u)\| = 2$ and assume that u is not $(2, p, q)$-special. If u is weakly p-periodic and weakly q-periodic and $|u| \geq 2(p+q) - \gcd(p,q)$, then u is $\gcd(p,q)$-periodic.*

2. *Let u be a partial word such that $\|H(u)\| = 3$ and assume that u is not $(3, p, q)$-special. If u is weakly p-periodic and weakly q-periodic and $|u| \geq 2(p+q)$, then u is $\gcd(p,q)$-periodic.*

The bound $2(p+q) - \gcd(p,q)$ turns out to be optimal in Theorem 3.2(1). For instance, the partial word $abaaba\Diamond\Diamond abaaba$ of length 14 is weakly 3-periodic and weakly 5-periodic but is not 1-periodic:

$$\begin{matrix} a & b & a \\ a & b & a \\ \Diamond & \Diamond & a \\ b & a & a \\ b & a \end{matrix}$$

$$\begin{matrix} a & b & a & a & b \\ a & \Diamond & \Diamond & a & b \\ a & a & b & a \end{matrix}$$

A similar result holds for the bound $2(p+q)$ in Theorem 3.2(2) by considering $abaaba\Diamond\Diamond abaaba\Diamond$ where $p = 3, q = 5$ and $2(p+q) = 16$:

$$\begin{matrix} a & b & a \\ a & b & a \\ \Diamond & \Diamond & a \\ b & a & a \\ b & a & \Diamond \end{matrix}$$

$$a \ b \ a \ a \ b$$
$$a \ \diamond \ \diamond \ a \ b$$
$$a \ a \ b \ a \ \diamond$$

3.3 Special partial words

In this section, we give an extension of the notions of $(2, p, q)$- and $(3, p, q)$-special partial words. We first discuss the case where $p = 1$ and then the case where $p > 1$.

3.3.1 $p = 1$

Throughout this section, we fix $p = 1$. Let q be an integer satisfying $q > 1$. Let u be a partial word of length n that is weakly p-periodic and weakly q-periodic. Then u can be represented in the following fashion:

$$
\begin{array}{cccc}
u(0) & u(q) & u(2q) & \cdots \\
u(1) & u(1+q) & u(1+2q) & \cdots \\
\vdots & \vdots & \vdots & \\
u(q-1) & u(2q-1) & u(3q-1) & \cdots
\end{array}
$$

The advantage to this representation is that the columns display the weak p-periodicity of u and the rows display the q-periodicity of u. We can continue this visualization and wrap the array around and sew the last row to the first row so that $u(q-1)$ is sewn to $u(q)$, $u(2q-1)$ is sewn to $u(2q)$, and so on. From this, we get a cylinder for u, and sometimes refer to this as the 3-dimensional representation of u.

Example 3.3
Let $p = 1$ and $q = 5$ for a word u. In Figure 3.2 we graphically show how an array is sewn together to form the cylinder for u. Figure 3.3 shows the cylinder in perspective.[2]   ⬚

We say that $i - p$ (respectively, $i + p$) is immediately *above* (respectively, *below*) i whenever $p \leq i < n$ (respectively, $0 \leq i < n - p$). Similarly, we say that $i - q$ (respectively, $i + q$) is immediately *left* (respectively, *right*) of i whenever $q \leq i < n$ (respectively, $0 \leq i < n - q$). The fact that u is weakly p-periodic implies that if $i, i+p \in D(u)$, then $u(i) = u(i+p)$. Similarly, the fact that u is weakly q-periodic implies that if $i, i+q \in D(u)$, then $u(i) = u(i+q)$.

---

[2]These graphics, and the others that follow, were generated using a Java applet on one of the author's websites, mentioned in the Website Section at the end of this chapter.

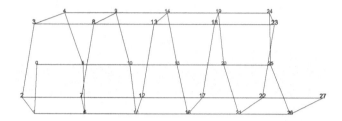

FIGURE 3.2:   A word u with $p = 1$ and $q = 5$.

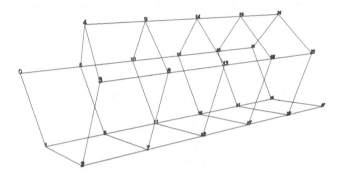

FIGURE 3.3:   Perspective view of a word u with $p = 1$ and $q = 5$.

The following definitions describe three types of isolation that will be acceptable in our definition of special partial word. In each, we have a continuous sequence of holes isolating a subset of defined positions. The type of isolation indicates where the isolation occurs: Type 1 is at the beginning of the partial word, Type 2 is in the interior of the partial word, and Type 3 is at the end of the partial word.

DEFINITION 3.2   *Let S be a nonempty proper subset of $D(u)$. We say that $H(u)$ **1-isolates** S (or that S is 1-isolated by $H(u)$) if the following hold:*

1. **Left** *If $i \in S$ and $i \geq q$, then $i - q \in S$ or $i - q \in H(u)$.*

2. **Right** *If $i \in S$, then $i + q \in S$ or $i + q \in H(u)$.*

3. **Above** *If $i \in S$ and $i \geq p$, then $i - p \in S$ or $i - p \in H(u)$.*

4. **Below** *If $i \in S$, then $i + p \in S$ or $i + p \in H(u)$.*

DEFINITION 3.3   *Let S be a nonempty proper subset of $D(u)$. We say that $H(u)$ **2-isolates** S (or that S is 2-isolated by $H(u)$) if the following hold:*

1. **Left** *If $i \in S$, then $i - q \in S$ or $i - q \in H(u)$.*

2. **Right** *If $i \in S$, then $i + q \in S$ or $i + q \in H(u)$.*

3. **Above** *If $i \in S$, then $i - p \in S$ or $i - p \in H(u)$.*

4. **Below** *If $i \in S$, then $i + p \in S$ or $i + p \in H(u)$.*

DEFINITION 3.4   *Let S be a nonempty proper subset of $D(u)$. We say that $H(u)$ **3-isolates** S (or that S is 3-isolated by $H(u)$) if the following hold:*

1. **Left** *If $i \in S$, then $i - q \in S$ or $i - q \in H(u)$.*

2. **Right** *If $i \in S$ and $i < n - q$, then $i + q \in S$ or $i + q \in H(u)$.*

3. **Above** *If $i \in S$, then $i - p \in S$ or $i - p \in H(u)$.*

4. **Below** *If $i \in S$ and $i < n - p$, then $i + p \in S$ or $i + p \in H(u)$.*

Example 3.4

As a first example, consider the partial word u_1 represented as the 3-dimensional structure below. Here, u_1 is weakly 1-periodic and weakly 5-periodic:

```
  0 5 10 15 20 25 30 35 40 45 50 55 60
```

```
0  c ◇ a  a  a  ◇ d ◇ e ◇ f f ◇
1  c ◇ ◇  a  ◇ h ◇ e ◇ f f ◇ i
2  ◇ ◇ b  ◇  ◇  ◇ e e e ◇ f f ◇
3  a ◇ b  b  ◇ e e e ◇ f f f f
4  a a ◇  ◇  g ◇ e e e ◇ f f
```

The set of positions with letter a is 1-isolated by $H(u_1)$; the set of positions with letter b is 2-isolated by $H(u_1)$; the set of positions with letter c is 1-isolated by $H(u_1)$; the set of positions with letter d is 2-isolated by $H(u_1)$; the set of positions with letter e is 2-isolated by $H(u_1)$; the set of positions with letter f is 3-isolated by $H(u_1)$; the set of positions with letter g is 2-isolated by $H(u_1)$; the set of positions with letter h is 2-isolated by $H(u_1)$; and the set of positions with letter i is 3-isolated by $H(u_1)$. ⬚

FIGURE 3.4:   Entire cylinder for the partial word in Example 3.4.

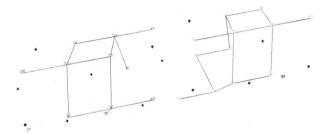

FIGURE 3.5:   Latter part of the cylinder for the pword in Example 3.4.

Example 3.5
As a second example, consider the weakly 1-periodic and weakly 5-periodic partial word u_2 represented as the 3-dimensional structure below. We can see that $D(u_2)$ does not contain a nonempty subset of isolated positions:

0 5 10 15 20 25 30 35 40 45 50 55 60

```
0  a  a  a  a  a  ◇  a  a  a  ◇  a  a  ◇
1  a  ◇  ◇  a  ◇  a  ◇  a  ◇  a  a  ◇  a
2  ◇  ◇  a  a  ◇  a  a  a  a  a  a  a  a
3  a  ◇  a  a  a  a  a  a  ◇  a  a  a  a
4  a  a  ◇  ◇  a  ◇  a  a  a  ◇  a  a
```

⬜

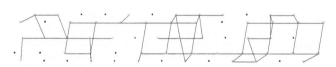

FIGURE 3.6:   Entire cylinder for the partial word in Example 3.5.

DEFINITION 3.5   *Let q be an integer satisfying $q > 1$. For $1 \leq i \leq 3$, the partial word u is called $(\|H(u)\|, 1, q)$-special of Type i if $H(u)$ i-isolates a nonempty proper subset of $D(u)$. The partial word u is called $(\|H(u)\|, 1, q)$-special if u is $(\|H(u)\|, 1, q)$-special of Type i for some $i \in \{1, 2, 3\}$.*

It is a simple matter to check that the above definition extends the notion of $(2, 1, q)$-special and the notion of $(3, 1, q)$-special (as given in Definition 3.1).

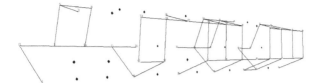

FIGURE 3.7: Perspective view of the cylinder for the partial word in Example 3.5.

Below we present the verification for $(2, 1, q)$-special, and leave the verification of $(3, 1, q)$-special to the reader.

Example 3.6

Definition 3.1(1)(a) corresponds to arrays like the following (with $q = 2$ as per the definition):

$$u(0)\ u(2)\ \cdots\quad u(2m)\quad \diamond\ u(4 + 2m)\ \cdots$$
$$u(1)\ u(3)\ \cdots\ u(1 + 2m)\ \diamond\ u(5 + 2m)\ \cdots$$

or

$$u(0)\ u(2)\ \cdots\quad u(2m)\quad u(2 + 2m)\quad\ \diamond\quad\ u(6 + 2m)\ \cdots$$
$$u(1)\ u(3)\ \cdots\ u(1 + 2m)\quad\ \diamond\quad\ u(5 + 2m)\ u(7 + 2m)\ \cdots$$

For Definition 3.1(1)(b) we have the following array which shows a 1-isolation:

$$u(0)\qquad \diamond\qquad u(2q)\quad \cdots$$
$$\diamond\qquad u(1 + q)\ u(1 + 2q)\ \cdots$$
$$u(2)\quad u(2 + q)\ u(2 + 2q)\ \cdots$$
$$\vdots\qquad \vdots\qquad \vdots$$
$$u(q - 1)\ u(2q - 1)\ u(3q - 1)\ \cdots$$

The symmetrical of the above array demonstrates Definition 3.1(1)(c) and possesses a 3-isolation. ▯

We can also check that the partial word u_1 depicted in Example 3.4 is $(25, 1, 5)$-special, but the partial word u_2 depicted in Example 3.5 is not $(18, 1, 5)$-special.

3.3.2 $p > 1$

Throughout this section, we fix $p > 1$. Let q be an integer satisfying $p < q$. Let u be a partial word of length n that is weakly p-periodic and weakly q-periodic. We illustrate with examples how the positions of u can be represented as a 3-dimensional structure.

In a case where $\gcd(p, q) = 1$ (like $p = 2$ and $q = 5$) we get 1 array:

$$u(0) \quad u(5) \quad u(10) \; u(15) \; \cdots$$
$$u(2) \quad u(7) \quad u(12) \; u(17) \; \cdots$$
$$u(4) \quad u(9) \quad u(14) \; u(19) \; \cdots$$
$$u(1) \quad u(6) \quad u(11) \; u(16) \; u(21) \; \cdots$$
$$u(3) \quad u(8) \quad u(13) \; u(18) \; u(23) \; \cdots$$

If we wrap the array around and sew the last row to the first row so that $u(3)$ is sewn to $u(5)$, $u(8)$ is sewn to $u(10)$, and so on, then we get a cylinder for the positions of u.

In a case where $\gcd(p, q) > 1$ (like $p = 6$ and $q = 8$) we get 2 arrays:

$$u(0) \quad u(8) \quad u(16) \; u(24) \; \cdots$$
$$u(6) \quad u(14) \; u(22) \; u(30) \; \cdots$$
$$u(4) \quad u(12) \; u(20) \; u(28) \; u(36) \; \cdots$$
$$u(2) \; u(10) \; u(18) \; u(26) \; u(34) \; u(42) \; \cdots$$

and

$$u(1) \quad u(9) \quad u(17) \; u(25) \; \cdots$$
$$u(7) \quad u(15) \; u(23) \; u(31) \; \cdots$$
$$u(5) \quad u(13) \; u(21) \; u(29) \; u(37) \; \cdots$$
$$u(3) \; u(11) \; u(19) \; u(27) \; u(35) \; u(43) \; \cdots$$

If we wrap the first array around and sew the last row to the first row so that $u(2)$ is sewn to $u(8)$, $u(10)$ is sewn to $u(16)$, and so on, then we get a cylinder for some of the positions of u. The other positions are in the second array where we wrap around and sew the last row to the first row so that $u(3)$ is sewn to $u(9)$, $u(11)$ is sewn to $u(17)$, and so on.

In general, if $\gcd(p, q) = d$, we get d arrays. In this case, we say that $i - p$ (respectively, $i + p$) is immediately *above* (respectively, *below*) i (within one of the d arrays) whenever $p \leq i < n$ (respectively, $0 \leq i < n - p$). Similarly, we say that $i - q$ (respectively, $i + q$) is immediately *left* (respectively, *right*) of i (within one of the d arrays) whenever $q \leq i < n$ (respectively, $0 \leq i < n - q$). As before, the fact that u is weakly p-periodic implies that if $i, i + p \in D(u)$, then $u(i) = u(i + p)$. Similarly, the fact that u is weakly q-periodic implies that if $i, i + q \in D(u)$, then $u(i) = u(i + q)$.

In what follows, we define $N_j = \{i \mid i \geq 0 \text{ and } i \equiv j \bmod \gcd(p, q)\}$ for $0 \leq j < \gcd(p, q)$. Alternatively, N_j is the set of indices in the jth array.

DEFINITION 3.6   *Let p and q be positive integers satisfying $p < q$. For $1 \leq i \leq 3$, the partial word u is called $(\|H(u)\|, p, q)$-**special of Type i** if there exists $0 \leq j < \gcd(p, q)$ such that $H(u)$ i-isolates a nonempty proper subset of $D(u) \cap N_j$. The partial word u is called $(\|H(u)\|, p, q)$-special if u is $(\|H(u)\|, p, q)$-special of Type i for some $i \in \{1, 2, 3\}$.*

Example 3.7

As a first example, the partial word $u_3 = ababa\diamond\diamond\diamond bab\diamond bb\diamond bbbbbbbbb$ along with its array of indices is shown below. It is $(5, 2, 5)$-special ($p = 2$ and $q = 5$). The set of positions $\{0, 2, 4, 9\}$ is 1-isolated by $H(u_3)$:

0 5 10 15 20	$a \diamond b\ b\ b$
2 7 12 17 22	$a \diamond b\ b\ b$
4 9 14 19	$a\ a \diamond b$
1 6 11 16 21	$b \diamond \diamond b\ b$
3 8 13 18 23	$b\ b\ b\ b\ b$

☐

Example 3.8

As a second example, the partial word

$$u_4 = ababababababab\diamond b\diamond\diamond ababa\diamond ababa\diamond\diamond babababab$$

below is not $(6, 6, 8)$-special. Note that because $\gcd(6, 8) = 2$, u_4 is written as two disjoint arrays:

$a\ a\ a\ a\ a$		$b\ b\ b\ b\ b$
$a \diamond a\ a$		$b \diamond b\ b$
$a \diamond a \diamond a$	and	$b\ b \diamond b\ b$
$a\ a\ a\ a\ a$		$b\ b\ b \diamond b$

☐

3.4 Graphs associated with partial words

Let p and q be positive integers satisfying $p < q$. In this section, we associate to a partial word u that is weakly p-periodic and weakly q-periodic an undirected graph $G_{(p,q)}(u)$. Whether or not u is $(\|H(u)\|, p, q)$-special will be seen from $G_{(p,q)}(u)$.

As explained in Section 3.3, u can be represented as a 3-dimensional structure with $\gcd(p, q)$ disjoint arrays. Each of the $\gcd(p, q)$ arrays of u is associated with a subgraph $G = (V, E)$ of $G_{(p,q)}(u)$ as follows:

V is the subset of $D(u)$ comprising the defined positions of u within the array,

$E = E_1 \cup E_2$ where

$$E_1 = \{\{i, i - p\} \mid i, i - p \in V\},$$

$$E_2 = \{\{i, i - q\} \mid i, i - q \in V\}.$$

For $0 \leq j < \gcd(p, q)$, the subgraph of $G_{(p,q)}(u)$ corresponding to $D(u) \cap N_j$ will be denoted by $G^j_{(p,q)}(u)$. Whenever $\gcd(p, q) = 1$, $G^0_{(p,q)}(u)$ is just $G_{(p,q)}(u)$.

Example 3.9
As a first example, the graph of the partial word u_3 of Example 3.7, $G_{(2,5)}(u_3)$, is shown in Figure 3.8 and is seen to be disconnected. The cylinder for u_3 is also seen to be disconnected in Figure 3.9. □

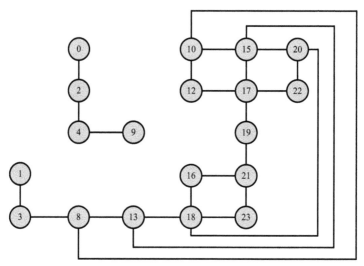

FIGURE 3.8: $G_{(2,5)}(u_3)$

Example 3.10
As a second example, consider the partial word u_4 of Example 3.8. The subgraphs of $G_{(6,8)}(u_4)$ corresponding to the two arrays of u_4, $G^0_{(6,8)}(u_4)$ and $G^1_{(6,8)}(u_4)$, are shown in Figures 3.10 and 3.11 and are seen to be connected. The corresponding cylinders for u_4 are seen in Figures 3.12, 3.13, and 3.14. □

We now define the critical lengths. We consider an even number of holes $2N$ and an odd number of holes $2N + 1$.

DEFINITION 3.7   *Let p and q be positive integers satisfying $p < q$. The*

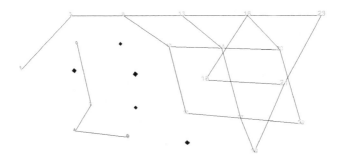

FIGURE 3.9: Cylinder for u_3 in Example 3.9.

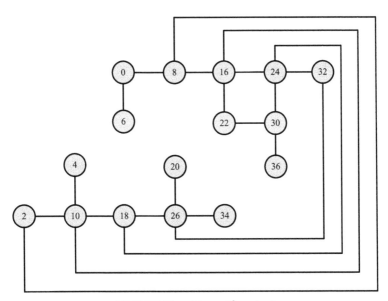

FIGURE 3.10: $G^0_{(6,8)}(u_4)$

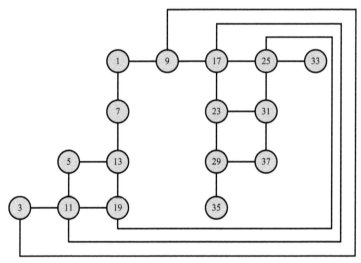

FIGURE 3.11: $G^1_{(6,8)}(u_4)$

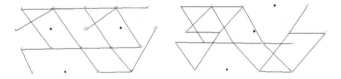

FIGURE 3.12:   Both cylinders for u_4 in Example 3.10.

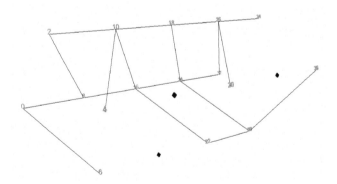

FIGURE 3.13:   Cylinder for $G^0_{(6,8)}(u_4)$ in Example 3.10.

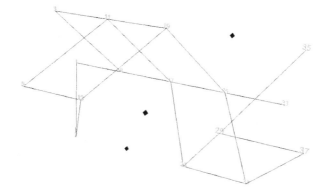

FIGURE 3.14: Cylinder for $G^1_{(6,8)}(u_4)$ in Example 3.10.

critical lengths *for p and q are defined as follows:*

1. $l_{(2N,p,q)} = (N+1)(p+q) - \gcd(p,q)$ *for $N \geq 0$, and*

2. $l_{(2N+1,p,q)} = (N+1)(p+q)$ *for $N \geq 0$.*

In the following lemma, we establish the important connection between (H, p, q)-special partial words and their graphs; namely, if all the graphs for a given partial word are connected, then the word cannot be (H, p, q)-special.

LEMMA 3.1
Let p and q be positive integers satisfying $p < q$, and let H be a positive integer. Let u be a partial word such that $\|H(u)\| = H$ and assume that $|u| \geq l_{(H,p,q)}$. Then u is not (H, p, q)-special if and only if $G^j_{(p,q)}(u)$ is connected for all $0 \leq j < \gcd(p,q)$.

PROOF   We first show that if u is (H, p, q)-special, then there exists $0 \leq j < \gcd(p, q)$ such that $G^j_{(p,q)}(u)$ is not connected. Three cases arise.

Case 1. u is (H, p, q)-special of Type 1.
There exists $0 \leq j < \gcd(p, q)$ such that $H(u)$ 1-isolates a nonempty proper subset S of $D(u) \cap N_j$. The subgraph of $G^j_{(p,q)}(u)$ with vertex set S constitutes a union of components (one component or more). There are therefore at least two components in $G^j_{(p,q)}(u)$ since S is proper.

Case 2. u is (H, p, q)-special of Type 2.
This case is similar to Case 1.

Case 3. u is (H, p, q)-special of Type 3.
This case is similar to Case 1.

We now show that if there exists $0 \leq j < \gcd(p,q)$ such that $G^j_{(p,q)}(u)$ is not connected, then u is (H,p,q)-special (or $H(u)$ isolates a nonempty proper subset of $D(u) \cap N_j$). Consider such a j. Put $p = p' \gcd(p,q)$ and $q = q' \gcd(p,q)$. As before, the partial word u_j is defined by

$$u_j = u(j)u(j + \gcd(p,q))u(j + 2\gcd(p,q))\ldots$$

If $H = 2N$ for some N, u_j is of length at least $(N+1)(p'+q') - 1$; and if $H = 2N+1$ for some N, u_j is of length at least $(N+1)(p'+q')$. In order to simplify the notation, let us denote $G_{(p',q')}(u_j)$ by G^j. Our assumption implies that G^j is not connected.

1. Let $G^j_\diamond$ be the graph constructed for the word u_j, so there are no holes. Then G^j is a subgraph of $G^j_\diamond$ obtained by removing the "hole" vertices.

2. Consider a set of consecutive indices in the domain of u_j, say $i, i + \gcd(p,q), \ldots, i + n\gcd(p,q)$. Call such a set a "domain interval," of length $n + 1$.

3. Every domain interval of length $p' + q'$ is the set of vertices of a cycle in $G^j_\diamond$; that is, there is a closed path in $G^j_\diamond$ which goes through exactly this set of vertices. The point is that a cycle cannot be disconnected by just one point.

4. Suppose C and C' are components of G^j with vertex sets S and S', and suppose neither S nor S' is isolated. Then each domain interval of length $p' + q'$ must contain a point v from S and a point v' from S'.

5. There must be two holes in each domain interval of length $p' + q'$, since otherwise the points v and v' from Item 4 would be connected by a path in the cycle formed by the domain interval.

6. If the number of holes is $2N + 1$ and the length of u_j is at least $(N+1)(p'+q')$ then Item 5 is impossible, since u_j would have $N+1$ pairwise disjoint domain intervals of length $p'+q'$ and Item 5 would then require $2(N+1)$ holes. Similarly, if the number of holes is $2N$ and the length of u_j is at least $(N+1)(p'+q')-1$ then Item 5 is impossible since u_j would have N pairwise disjoint intervals of length $p' + q'$ and one remaining of length $p' + q' - 1$, and so Item 5 would require $2N + 1$ holes.

Note that this proves the lemma in case the number of holes is positive, and in fact Item 3 is essentially the proof in the case of exactly one hole. The case of 0 holes follows from the fact that every domain interval of length $p'+q'-1$ is the set of vertices of a path in $G^j_\diamond$. ∎

3.5 The main result

In this section, we give the main result which extends Theorems 3.1 and 3.2 to an arbitrary number of holes.

LEMMA 3.2
Let p and q be positive integers satisfying $p < q$ and $\gcd(p, q) = 1$. Let u be a partial word that is weakly p-periodic and weakly q-periodic. If $G_{(p,q)}(u)$ is connected, then u is 1-periodic.

PROOF Let i be a fixed position in $D(u)$. If $i' \in D(u)$ and $i' \neq i$, then there is a path in $G_{(p,q)}(u)$ between i' and i. Let $i', i_1, i_2, \ldots, i_n, i$ be such a path. We get $u(i') = u(i_1) = u(i_2) = \cdots = u(i_n) = u(i)$. ▯

THEOREM 3.3
Let p and q be positive integers satisfying $p < q$. Let u be a partial word that is weakly p-periodic and weakly q-periodic. If $G^i_{(p,q)}(u)$ is connected for all $0 \leq i < \gcd(p, q)$, then u is $\gcd(p, q)$-periodic.

PROOF The case where $\gcd(p, q) = 1$ follows by Lemma 3.2. So consider the case where $\gcd(p, q) > 1$. Define for each $0 \leq i < \gcd(p, q)$ the partial word u_i by

$$u_i = u(i)u(i + \gcd(p, q))u(i + 2\gcd(p, q))\ldots$$

Put $p = p' \gcd(p, q)$ and $q = q' \gcd(p, q)$. Each u_i is weakly p'-periodic and weakly q'-periodic. If $G^i_{(p,q)}(u)$ is connected for all i, then $G_{(p',q')}(u_i)$ is connected for all i. Consequently, each u_i is 1-periodic by Lemma 3.2, and u is $\gcd(p, q)$-periodic. ▯

THEOREM 3.4
Let p and q be positive integers satisfying $p < q$, and let H be a positive integer. Let u be a partial word such that $\|H(u)\| = H$ and assume that u is not (H, p, q)-special. If u is weakly p-periodic and weakly q-periodic and $|u| \geq l_{(H,p,q)}$, then u is $\gcd(p, q)$-periodic.

PROOF If u is not (H, p, q)-special and $|u| \geq l_{(H,p,q)}$, then $G^i_{(p,q)}(u)$ is connected for all $0 \leq i < \gcd(p, q)$ by Lemma 3.1. Then u is $\gcd(p, q)$-periodic by Theorem 3.3. ▯

The bound $l_{(2N,p,q)}$ turns out to be optimal for an even number of holes $2N$, and the bound $l_{(2N+1,p,q)}$ optimal for an odd number of holes $2N + 1$. The following builds a sequence of partial words showing this optimality.

DEFINITION 3.8   *Let p and q be positive integers satisfying $1 < p < q$ and $\gcd(p, q) = 1$. Let N be a positive integer.*

1. *The partial word $\boldsymbol{u_{(2N,p,q)}}$ over $\{a, b\}$ of length $l_{(2N,p,q)} - 1$ is defined by*

   (a) $H(u_{(2N,p,q)}) = \{p+q-2, p+q-1, 2(p+q)-2, 2(p+q)-1, \ldots, N(p+q) - 2, N(p+q) - 1\}$.

   (b) *The component of the graph $G_{(p,q)}(u_{(2N,p,q)})$ containing $p - 2$ is colored with letter a.*

   (c) *The component of the graph $G_{(p,q)}(u_{(2N,p,q)})$ containing $p - 1$ is colored with letter b.*

2. *The partial word $\boldsymbol{u_{(2N+1,p,q)}}$ over $\{a, b\}$ of length $l_{(2N+1,p,q)} - 1$ is defined by $u_{(2N+1,p,q)} = u_{(2N,p,q)}\diamond$ so that $H(u_{(2N+1,p,q)}) = H(u_{(2N,p,q)}) \cup \{(N+1)(p+q) - 2\}$.*

The partial word $u_{(2N,p,q)}$ can be thought as two bands of holes $\text{Band}_1 = \{p+q-1, 2(p+q) - 1, \ldots, N(p+q) - 1\}$ and $\text{Band}_2 = \{p+q-2, 2(p+q) - 2, \ldots, N(p+q) - 2\}$ where between the bands the letter is a and outside the bands it is b or vice versa (a similar statement holds for $u_{(2N+1,p,q)}$).

Example 3.11

For example, the partial word $u_{(4,2,5)}$ of length 19 is represented as the 3-dimensional structure

$$
\begin{array}{ccc}
a & \diamond\ b & b \\
a & a\ \diamond & b \\
a & a & a \\
b & \diamond\ a & a \\
b & b\ \diamond & a
\end{array}
$$

It is weakly 2-periodic and weakly 5-periodic but is not 1-periodic (it is not $(4, 2, 5)$-special).   ⬚

Example 3.12

Similarly, the partial word $u_{(5,2,5)}$ of length 20 is represented as the 3-dimensional structure

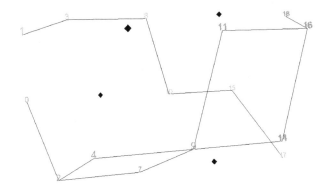

FIGURE 3.15: Cylinder for $u_{(4,2,5)}$ in Example 3.11.

$$a \diamond b \ b$$
$$a \ a \diamond b$$
$$a \ a \ a \diamond$$
$$b \diamond a \ a$$
$$b \ b \diamond a$$

It is weakly 2-periodic and weakly 5-periodic but is not 1-periodic (it is not $(5, 2, 5)$-special). ⬚

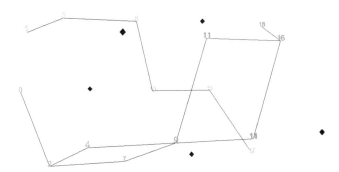

FIGURE 3.16: Cylinder for $u_{(5,2,5)}$ in Example 3.12.

PROPOSITION 3.1

Let p and q be positive integers satisfying $1 < p < q$ and $\gcd(p, q) = 1$. Let H be a positive integer. The partial word $u_{(H,p,q)}$ of length $l_{(H,p,q)} - 1$ is *not (H, p, q)-special, but is weakly p-periodic and weakly q-periodic. However $u_{(H,p,q)}$ is not 1-periodic.*

PROOF   We prove the result when $H = 2N + 1$ for some N (the even case $H = 2N$ is left as an exercise). As stated earlier, the partial word $u_{(2N+1,p,q)}$ of length $(N + 1)(p + q) - 1$ can be thought as two bands of holes $Band_1 = \{p + q - 1, 2(p + q) - 1, \ldots, N(p + q) - 1\}$ and $Band_2 = \{p + q - 2, 2(p + q) - 2, \ldots, N(p + q) - 2, (N + 1)(p + q) - 2\}$. The position $p - 1$ is between the bands and $p - 2$ is outside the bands or vice versa. Let S_1 be the component that contains $p - 1$ and S_2 be the component that contains $p - 2$. The partial word $u_{(2N+1,p,q)}$ is not $(2N + 1, p, q)$-special of Type 2 since neither S_1 nor S_2 is 2-isolated by $H(u_{(2N+1,p,q)})$. To see this, Definition 3.3(1) fails with $i = p - 1$ or $i = p - 2$. To show that $u_{(2N+1,p,q)}$ is not $(2N + 1, p, q)$-special of Type 3, we can use Definition 3.4(1) with $i = p - 1$ or $i = p - 2$. To show that $u_{(2N+1,p,q)}$ is not $(2N + 1, p, q)$-special of Type 1, we can use Definition 3.2(2) with $i = N(p + q) - 1 + q$ or $i = N(p + q) - 2 + q$.   ☐

We end this section with Table 3.1 which summarizes the optimal lengths for Fine and Wilf's weak periodicity extensions.

TABLE 3.1:   Optimal lengths for weak periodicity.

Holes	Lengths
0	$p + q - \gcd(p, q)$
1	$p + q$
2	$2(p + q) - \gcd(p, q)$
3	$2(p + q)$
4	$3(p + q) - \gcd(p, q)$
5	$3(p + q)$
$\vdots$	$\vdots$
$2N$	$(N + 1)(p + q) - \gcd(p, q)$
$2N + 1$	$(N + 1)(p + q)$

3.6   Related results

We now discuss another extension of Fine and Wilf's periodicity result in the context of partial words. The next remark justifies the results of this section.

REMARK 3.1   There exists an integer L (that depends on H, p and q) such that if a partial word u with H holes has (strong) periods p and q and $|u| \geq L$, then u has period $\gcd(p, q)$. $\quad\square$

If $p < q$ are positive integers, then the following result gives a bound $L_{(H,p,q)}$ for H holes when q is large enough.

THEOREM 3.5
Let H be a positive integer, and let p and q be positive integers satisfying $q > x(p, H)$ where

$$x(p, H) = \begin{cases} \left(\frac{H}{2}\right) p & \text{if } H \text{ is even} \\ \left(\frac{H+1}{2}\right) p & \text{if } H \text{ is odd} \end{cases}$$

If a partial word u with H holes is p-periodic and q-periodic and $|u| \geq L_{(H,p,q)}$, then u is $\gcd(p, q)$-periodic where

$$L_{(H,p,q)} = \begin{cases} \left(\frac{H+2}{2}\right) p + q - \gcd(p, q) & \text{if } H \text{ is even} \\ \left(\frac{H+1}{2}\right) p + q & \text{if } H \text{ is odd} \end{cases}$$

PROOF   Set $L = L_{(H,p,q)}$, and suppose that $\gcd(p, q) = 1$. First, let $H = 2N + 1$ for some N. Then we have that $x(p, H) = (N + 1) p$. So $q > x(p, H)$ implies that $q = (N + 1) p + k$ for some $k > 0$. It is enough to show that if $|u| = L$, then u has period 1 because if $|u| > L$, then all factors of u of length L would have period 1, and so u itself would. To see this, suppose $|u| = L + 1$. The prefix of u of length L has periods p and q, and so it has period 1. The same holds for the suffix of u of length L. If u starts or ends with $\diamond$, then the result trivially holds. Otherwise, $u = au'b$ for some u' of length $L - 1$ and some $a, b \in A$. There exists an occurrence of the letter b in u' because $D(u') \neq \emptyset$ by the way L is defined. The equality $b = a$ hence holds. Thus, by induction, any word u of length $\geq L$ satisfying our assumptions is 1-periodic. Now, since $|u| = (N + 1) p + q$ and $q = (N + 1) p + k$, we have that $|u| = (H + 1)p + k = 2q - k$.

Consider the graph of u. Since $|u| = 2q - k$, positions of u within $\{q - k, q - k + 1, \ldots, q - 2, q - 1\}$ have no E_2-edges, and all other elements within $\{0, \ldots, q-k-1\}$ have exactly one E_2-edge. Therefore, the number of positions of u which have exactly one E_2-edge is $|u| - k = (H + 1)p$. Thus, each p-class has exactly $H + 1$ elements with exactly one E_2-edge and all other elements of the p-class have no E_2-edges. In each ith p-class, $N + 1$ elements have E_2-edges with elements in the $((i + q) \bmod p)^{th}$ p-class and $N + 1$ elements have E_2-edges with elements in the $((i-q) \bmod p)^{th}$ p-class. Thus, there are at least

$N + 1$ disjoint cycles in the graph that visit all p-classes and contain all the vertices with E_2-edges. In order to build $N + 1$ such disjoint cycles, pick the smallest vertex v_0 in the $0^{th} = i_0^{th}$ p-class that has not been visited and that has a E_2-edge with an element w_1 of the i_1^{th} p-class. Then visit the vertex w_1 followed by the smallest nonvisited vertex v_1 of that i_1^{th} p-class. Go on like this visiting vertices until you visit w_p in the 0th p-class. Then return to v_0. Such cycle has the form $v_0, w_1, v_1, w_2, v_2, \ldots, w_{p-1}, v_{p-1}, w_p, v_0$. Also, for each such cycle, every element of the graph either belongs to the cycle, or is p-connected to a member of the cycle. There are two types of disconnections possible: one that isolates a set of vertices with elements in different p-classes, and one that isolates a set of vertices within a p-class. Thus in order to disconnect the graph, either all $N + 1$ cycles must be disconnected or all $H + 1$ E_2-edges of a single p-class must be removed. The latter case clearly takes more than H holes, and since two holes are required to disconnect a cycle, we see that at least $H + 1$ holes are required to disconnect the graph in the former case. Thus the graph of u is connected and u is 1-periodic.

Now, let $H = 2N$ for some N. The idea of the proof in this case is similar to that of an odd number of holes. We must disconnect N cycles that each requires two holes to break and one path that requires one hole to break. Hence we require $H + 1$ holes to disconnect the graph of length $L_{(H,p,q)}$.

Suppose $\gcd(p, q) = d \neq 1$. Also suppose that $H = 2N$ for some N (the odd case $H = 2N + 1$ is similar). Thus $|u| = (N + 1)p + q - d$. Consider the set of partial words $u_0, \ldots, u_{d-1}$ where $u_i = u(i)u(i + d)u(i + 2d)\ldots$. Each of these words has periods $\frac{p}{d}$ and $\frac{q}{d}$ which are co-prime. So if each u_i had period 1, then the word u has period d. Each u_i has length $(N + 1)\frac{p}{d} + \frac{q}{d} - 1$ and at most H holes. Thus, by the proof given of this theorem for the case $\gcd(p, q) = 1$, each u_i has period 1, and therefore u is d-periodic. ∎

In the case of no hole, we see that $x(p, 0) = 0$ and the formula presented in Theorem 3.5 agrees with $l_{(0,p,q)} = p + q - \gcd(p, q)$ of Theorem 3.1(1). The case of one hole yields $x(p, 1) = p$ and once again, the formula gives $L_{(1,p,q)} = p + q$ which corresponds to the expression given in Theorem 3.1(2).

If $q > x(p, H)$, then the bound $L_{(H,p,q)}$ is optimal for H holes. The following builds a sequence of partial words showing this optimality.

Let p and q be integers satisfying $1 < p < q$ and $\gcd(p, q) = 1$. Let $W_{0,p,q}$ denote the set of all words of length $p + q - 2$ having periods p and q. We denote by PER_0 the set of all words of maximal length for which Theorem 3.1(1) does not apply, that is,

$$\mathrm{PER}_0 = \bigcup_{\gcd(p,q)=1} W_{0,p,q}$$

The reader is asked to show that $W_{0,p,q}$ contains a unique word w (up to a renaming of letters) such that $\|\alpha(w)\| = 2$, in which case w is a palindrome (or w reads the same forward and backward). It is easy to verify that $w = aabaabaabaa$ when $p = 3$ and $q = 10$.

DEFINITION 3.9   *Let p and q be positive integers satisfying $1 < p \leq x(p, H) < q$ and $\gcd(p, q) = 1$. Let N be a positive integer.*

1. *The partial word $v_{(2N,p,q)}$ over $\{a, b\}$ of length $L_{(2N,p,q)} - 1$ is defined by $v_{(2N,p,q)} = (w[0..p-2)\diamond\diamond)^N w$ where w is the unique palindome over $\{a, b\}$ in $W_{0,p,q}$ of length $p + q - 2$.*

2. *The partial word $v_{(2N+1,p,q)}$ over $\{a, b\}$ of length $L_{(2N+1,p,q)} - 1$ is defined by $v_{(2N+1,p,q)} = v_{(2N,p,q)}\diamond$.*

Example 3.13
For example, the partial word $v_{(4,3,10)} = a\diamond\diamond a\diamond\diamond aabaabaabaa$ has four holes, is 3-periodic and 10-periodic, has length $17 = 3(3) + 10 - \gcd(3, 10) - 1$, but is not 1-periodic.   ▯

PROPOSITION 3.2
Let p and q be positive integers satisfying $1 < p \leq x(p, H) < q$ and $\gcd(p, q) = 1$. Let H be a positive integer. The partial word $v_{(H,p,q)}$ of length $L_{(H,p,q)} - 1$ is p-periodic and q-periodic. However $v_{(H,p,q)}$ is not 1-periodic.

PROOF   We prove the result when $H = 2N$ for some N (the odd case $H = 2N + 1$ is left as an exercise). Set $u = v_{(H,p,q)}$. First, note that since w is not 1-periodic, we also have that u is not 1-periodic. Now, note that w is p-periodic. Also, $w[0..p-2)\diamond\diamond$ has length p and since $w[0..p-2)\diamond\diamond \subset w[0..p)$, we see that u is p-periodic. Since $q > x(p, H) = Np$, w is of length $q+p-2 > Np + p - 2$. In order to show that u is q-periodic, it is enough to show that

$$u[0..Np + p - 2) \uparrow u[|u| - (Np + p - 2)..|u|)$$

Now, $u[0..Np + p - 2) = (w[0..p-2)\diamond\diamond)^N w[0..p-2)$, and

$$u[|u| - (Np + p - 2)..|u|) = w[|w| - (Np + p - 2)..|w|) = w[0..Np + p - 2)$$

since w is a palindrome. Since w is p-periodic, we have $w[0..Np + p - 2) = (w[0..p))^N w[0..p - 2)$ and the desired compatibility relationship follows.   ▯

Table 3.2 summarizes the optimal lengths for Fine and Wilf's extensions for strong periodicity when q is large enough.
We end this chapter with the following result which is left as an exercise for the reader.

THEOREM 3.6
Let $p < q$ be positive integers. If a partial word u with $H > 0$ holes is p-periodic and q-periodic and $|u| \geq (H + 1)p + q - \gcd(p, q)$, then u is $\gcd(p, q)$-periodic.

TABLE 3.2: Optimal lengths for strong periodicity.

Holes	Lengths	Conditions
0	$p + q - \gcd(p, q)$	
1	$p + q$	
2	$2p + q - \gcd(p, q)$	$q > p$
3	$2p + q$	$q > 2p$
4	$3p + q - \gcd(p, q)$	$q > 2p$
5	$3p + q$	$q > 3p$
$\vdots$	$\vdots$	$\vdots$
$2N$	$(N+1)p + q - \gcd(p, q)$	$q > Np$
$2N + 1$	$(N+1)p + q$	$q > (N+1)p$

Exercises

3.1 Consider $p = 3$ and $q = 5$. For $2 \leq i \leq p$, construct a word u_i such that the following three conditions hold:

1. $|u_i| = p + q - i$,
2. $\|\alpha(u_i)\| = i$ where $\alpha(u_i)$ denotes the set of distinct letters in u_i,
3. u_i has periods p and q.

3.2 Using $p = 3$ and $q = 5$, show that the bound $p + q - \gcd(p, q)$ is optimal in Theorem 3.1(1) by providing a counterexample for a word of smaller length. Repeat for the bound $p + q$ in Theorem 3.1(2).

3.3 ⓢ Using Theorem 3.1(2), prove that if u is a partial word with one hole and q is a weak period of u satisfying $|u| \geq p'(u) + q$, then q is a multiple of $p'(u)$. What can be said when u has no hole?

3.4 ⓢ Referring to Theorem 3.1(2), consider the following corollary: "Let p and q be positive integers and u be a partial word such that $\|H(u)\| = 1$. If u is p-periodic and q-periodic and $|u| \geq p + q$, then u is $\gcd(p, q)$-periodic." Is the bound $p + q$ optimal here?

3.5 Check that Definition 3.5 extends the notion of $(3, 1, q)$-special pword as given in Definition 3.1.

3.6 Let p and q be positive integers satisfying $p < q$ and $\gcd(p, q) = 1$. For each $n > 0$, let $v_n = ab^{p-1} \diamond b^{q-p-1} \diamond b^n \diamond$. Show that v_n is a binary $(3, p, q)$-special partial word with three holes according to Definition 3.1 that is weakly p-periodic and weakly q-periodic but that is not 1-periodic.

3.7 ⌐s⌐ Using Definitions 3.2, 3.3 and 3.4, prove the following statements:

1. The $(5, 2, 5)$-special partial word u pictured in

$$
\begin{array}{ccccc}
a & \diamond & b & b & b \\
a & \diamond & b & b & b \\
  & a & a & \diamond & b \\
b & \diamond & \diamond & b & b \\
b & b & b & b & b
\end{array}
$$

   shows an isolation of Type 1.

2. The $(6, 2, 5)$-special partial word v

$$
\begin{array}{ccccc}
b & b & \diamond & b & b \\
b & \diamond & a & \diamond & b \\
b & \diamond & a & \diamond & \\
b & b & b & \diamond & b \\
b & b & b & b & b
\end{array}
$$

   shows a Type 2 isolation.

3. The $(4, 2, 5)$-special partial word w

$$
\begin{array}{ccccc}
b & b & b & b & \diamond \\
b & b & b & \diamond & a \\
b & b & \diamond & a & \\
b & b & b & b & \diamond \\
b & b & b & b & b
\end{array}
$$

   shows a Type 3 isolation.

3.8 Is the partial word

$$u = ababa\diamond\diamond ababa\diamond\diamond ababa$$

   of length 19 $(4, 2, 5)$-special?

3.9 ⌐s⌐ Let $u = abbaab\diamond a\diamond baab$. Build the undirected graph $G_{(4,7)}(u)$. Is u $(2, 4, 7)$-special? Why or why not?

3.10 Using $p = 3$ and $q = 5$, show that the bound $(N+1)(p+q) - \gcd(p, q)$ is optimal for $2N$ holes in Theorem 3.4 by providing a counterexample for a partial word of smaller length. Repeat for the bound $(N + 1)(p + q)$ for $2N + 1$ holes in Theorem 3.4.

Challenging exercises

3.11 Prove Proposition 3.1 for the even case $H = 2N$.

3.12 Prove that $W_{0,p,q}$ contains a unique word u (up to a renaming of letters) such that $\|\alpha(u)\| = 2$. Also prove that u is a palindrome. Give the unique word of length 11 in PER_0 having periods 5 and 8.

3.13 $\boxed{\text{s}}$ Repeat Exercise 3.6 for $v_n = \diamond ab^{p-1}\diamond b^{q-p-1}\diamond b^n$.

3.14 Does 2-isolation imply 1-isolation or 3-isolation? Does 1-isolation or 3-isolation imply 2-isolation? Why or why not?

3.15 Let u, v be nonempty words, let y, z be partial words, and let w be a word satisfying $|w| \geq |u| + |v| - \gcd(|u|, |v|)$. Show that if $wy \subset u^m$ and $wz \subset v^n$ (respectively, $yw \subset u^m$ and $zw \subset v^n$) for some integers m, n, then there exists a word x of length not greater than $\gcd(|u|, |v|)$ such that $u = x^k$ and $v = x^l$ for some integers k, l.

3.16 $\boxed{\text{H}}$ Fixing an arbitrary number of holes, $H \geq 2$, and positive integers p and q satisfying $p < q$ and $\gcd(p, q) = 1$, construct an infinite sequence $(u_n)_{n>0}$ where u_n is a binary (H, p, q)-special partial word with H holes that is weakly p-periodic and weakly q-periodic but not $\gcd(p, q)$-periodic.

3.17 Using the floor function "$\lfloor \ \rfloor$," rewrite $l_{(H,p,q)}$ of Definition 3.7 in one single expression for any $H \geq 0$.

3.18 Let p, q and r be integers satisfying $1 < p < q$, $\gcd(p, q) = 1$, and $0 \leq r < p + q - 1$. For $i \neq q - 1$ and $0 \leq i < p + q - 1$, we define the sequence of i relative to p, q and r as $\mathrm{seq}_{p,q,r}(i) = (i_0, i_1, i_2, \ldots, i_{n-1}, i_n)$ where $i_0 = i$ and

- If $i = r$, then $i_n = q - 1$,
- If $i \neq r$, then $i_n = r$ or $i_n = q - 1$,
- For $1 \leq j < n$, $i_j \notin \{i, r, q - 1\}$,
- For $1 \leq j \leq n$, i_j is defined as

$$i_j = \begin{cases} i_{j-1} + p \text{ if } i_{j-1} < q - 1 \\ i_{j-1} - q \text{ if } i_{j-1} > q - 1 \end{cases}$$

We define $\mathrm{seq}_{p,q,r}(q - 1) = (q - 1)$.

The sequence $\mathrm{seq}_{p,q,r}(i)$ gives a way of visiting elements of $\{0, \ldots, p + q - 2\}$ starting at i. You increase by p as long as possible, then you decrease by q and you start the process again. If you start at $q-1$ or hit $q-1$, you cannot increase by p and you cannot decrease by q and so you stop. If you hit r, you stop. Compute $\text{seq}_{4,11,5}(i)$ for all i. Show that $\text{seq}_{4,11,5}(3)$ is the longest sequence ending with 5 and $\text{seq}_{4,11,5}(5)$ is the longest ending with 10 and all the other sequences are suffixes of these two.

3.19 Show that the sequence $\text{seq}_{p,q,r}(i)$ defined in Exercise 3.18 is always finite.

3.20 ☐ Let p and q be integers satisfying $1 < p < q$ and $\gcd(p,q) = 1$. Let $W_{1,p,q}$ denote the set of all partial words with one hole of length $p+q-1$ having weak periods p and q. We denote by PER_1 the set of all partial words with one hole of maximal length for which Theorem 3.1(2) does not apply, that is,

$$\text{PER}_1 = \bigcup_{\gcd(p,q)=1} W_{1,p,q}$$

Given a singleton set H satisfying $H \subset \{0,\dots,p+q-2\} \setminus \{p-1,q-1\}$, show that $W_{1,p,q}$ contains a unique partial word u (up to a renaming of letters) such that $\|\alpha(u)\| = 2$ and $H(u) = H$. Also, show that if $H = \{p-1\}$ or $H = \{q-1\}$, then $W_{1,p,q}$ contains a unique partial word u such that $\|\alpha(u)\| = 1$ and $H(u) = H$.

3.21 Prove the odd case $H = 2N + 1$ of Proposition 3.2.

3.22 Prove Theorem 3.6.

Programming exercises

3.23 Write a program that receives as input two positive integers p,q satisfying $p < q$ and that for $2 \le i \le p$, constructs a word u_i such that the three conditions of Exercise 3.1 hold.

3.24 Write a program that finds isolation of Type 1, 2 or 3 if present in a given pword u. Run your program on the partial words of Exercise 3.7.

3.25 Design an applet that builds a two-dimensional representation out of a pword based on two of its weak periods.

3.26 Write a program that computes the critical length for weak periods 4, 7, 8, 12 and number of holes 4 and that provides a counterexample of length one less than the critical length.

3.27 Write a program to compute the critical length for strong periods 2, 3 and number of holes 4. Your program should also output all the counterexample pwords (up to a renaming of letters) of length one less than the critical length (not including symmetric cases).

Websites

World Wide Web server interfaces at

`http://www.uncg.edu/mat/research/finewilf`
`http://www.uncg.edu/mat/research/finewilf2`
`http://www.uncg.edu/mat/research/finewilf3`
`http://www.uncg.edu/cmp/research/finewilf4`

have been established for automated use of programs related to generalizations of Fine and Wilf's periodicity result in the framework of partial words.

- The `finewilf` website provides an applet that builds two- and three-dimensional representations out of a partial word based on two of its weak periods. Isolation is visible if it occurs within the pword. In the 2D version, the array is repeated to show where the top of the array connects with the bottom.

- The `finewilf2` website asks the user to list at least two weak periods in ascending order and to enter a nonnegative integer for the number of holes. The applet outputs the critical length for the given weak period set and number of holes as well as a counterexample pword for one less than the critical length if applicable.

- The `finewilf3` and `finewilf4` websites provide applets that compute the critical length for given number of holes and strong periods p, q with p smaller than q and p not dividing q. The applets also build two-dimensional arrays of all counterexample pwords for one less than the critical length (up to renaming of letters and symmetry).

Bibliographic notes

The problem of computing periods in words has important applications in data compression, string searching and pattern matching algorithms. The periodicity result of Fine and Wilf [77] has been generalized in many ways:

Extension to more than two periods are given in [47, 54, 97, 139]; Periodicity is taken to the wide context of Cayley graphs of groups which produce generalizations that involve the concept of a periodicity vector [81]; Other generalizations are produced in the context of labelled trees [80, 123]; Yet, other related results are shown in [3, 4, 5, 6, 7, 8, 48, 75, 78, 109, 116, 117, 122, 135].

Fine and Wilf's result has been generalized to partial words in two ways:

First, any partial word u with H holes and having weak periods p, q and length at least $l_{(H,p,q)}$ has also period $\gcd(p,q)$ provided u is not (H, p, q)-special. This extension was done for one hole by Berstel and Boasson in their seminal paper [10] where the class of $(1, p, q)$-special partial words is empty (the zero-hole case of Theorem 3.1 in Section 3.1 is from Fine and Wilf [77] and the one-hole case from Berstel and Boasson [10]); for two-three holes by Blanchet-Sadri and Hegstrom [32] (Section 3.2); and for an arbitrary number of holes by Blanchet-Sadri [15] (Sections 3.3, 3.4 and 3.5).

Second, any partial word u with H holes and having periods p, q and length at least the so-denoted $L_{(H,p,q)}$ has also period $\gcd(p,q)$. This extension was initiated by Shur and Gamzova in their papers [130, 131] where they proved Theorem 3.6 (Section 3.6). Theorem 3.5 and Proposition 3.2 are from Blanchet-Sadri, Bal and Sisodia [19] (the two-hole case having first been proved by Shur and Gamzova).

Exercise 3.12 is from Choffrut and Karhümaki [51]. It turns out that PER_0 has several characterizations based on quite different concepts [11, 62, 63, 64]. Blanchet-Sadri extended to PER_1 the well known property that states that PER_0 contains a unique word (up to a renaming of letters) that is binary [14] (Exercise 3.20). Exercise 3.3 is from Blanchet-Sadri and Chriscoe [23], Exercise 3.15 from Blanchet-Sadri [17], and Exercises 3.18 and 3.19 from Blanchet-Sadri [14].

Chapter 4

Critical Factorization Theorem

In this chapter, we discuss the fundamental critical factorization theorem in the framework of partial words.

The critical factorization theorem on full words states that given a word w and nonempty words u, v satisfying $w = uv$, the *minimal local period* associated to the factorization (u, v) of w is the length of the shortest repetition (a *square*) *centered* at position $|u| - 1$. It is easy to see that no minimal local period is longer than the minimal (or global) period of the word. The critical factorization theorem shows that *critical factorizations* are unavoidable. Indeed, for any string, there is always a factorization whose minimal local period is equal to the global period of the string.

In other words, we consider a string $a_0 a_1 \ldots a_{n-1}$ and, for any integer i such that $0 \leq i < n - 1$, we look at the shortest repetition centered in this position, that is, we look at the shortest (virtual) suffix of $a_0 a_1 \ldots a_i$ which is also a (virtual) prefix of $a_{i+1} a_{i+2} \ldots a_{n-1}$. The minimal local period at position i is defined as the length of this shortest square. The critical factorization theorem states, roughly speaking, that the global period of $a_0 a_1 \ldots a_{n-1}$ is simply the maximum among all minimal local periods. As an example, consider the word $w = babbaab$ with global period 6. The minimal local periods of w are 2, 3, 1, 6, 1 and 3 which means that the factorization $(babb, aab)$ is critical.

In summary, the following table describes the number of holes and section numbers where the critical factorization theorem is discussed:

Holes	Sections
0	4.2
arbitrary	4.3, 4.4 and 4.5

4.1 Orderings

A binary relation $\preceq$ defined on an arbitrary set S is a subset of $S \times S$. The relation $\preceq$ is called *reflexive* if $u \preceq u$ for all $u \in S$; *antisymmetric* if $u \preceq v$ and $v \preceq u$ imply $u = v$ for all $u, v \in S$; *transitive* if $u \preceq v$ and $v \preceq w$

imply $u \preceq w$ for all $u, v, w \in S$. A reflexive, antisymmetric, and transitive relation $\preceq$ defined on S is called a *partial ordering*. A *total ordering* is a partial ordering for which $u \preceq v$ or $v \preceq u$ holds for all $u, v \in S$. An element u of S (respectively, of a subset $X \subset S$) ordered by $\preceq$ is *maximal* if for all $v \in S$ (respectively, $v \in X$) the condition $u \preceq v$ implies $u = v$. Of course each subset of a totally ordered set has at most one maximal element.

In this section, we define two total orderings of $W(A)$, $\preceq_l$ and $\preceq_r$, and state some lemmas related to them that will be used to prove the main results.

First, let the alphabet A be totally ordered by $\prec$ and let $\diamond \prec a$ for all $a \in A$.

- The first total ordering, denoted by $\prec_l$, is simply the lexicographic ordering related to the fixed total ordering on A and is defined as follows: $u \prec_l v$, if either u is a proper prefix of v, or

$$u = \mathrm{pre}(u, v) \, a \, x$$
$$v = \mathrm{pre}(u, v) \, b \, y$$

with $a, b \in A \cup \{\diamond\}$ satisfying $a \prec_l b$.[1]

- The second total ordering, denoted by $\prec_r$, is obtained from $\prec_l$ by reversing the order of letters in the alphabet, that is, for $a, b \in A$, $a \prec_l b$ if and only if $b \prec_r a$.

LEMMA 4.1
For all partial words u, v, x, y, the following hold:

- $u \prec_l v$ *if and only if* $xu \prec_l xv$,

- $u \prec_r v$ *if and only if* $xu \prec_r xv$,

- $u \prec_l v$ *and* $u \notin P(v)$ *imply* $ux \prec_l vy$,

- $u \prec_r v$ *and* $u \notin P(v)$ *imply* $ux \prec_r vy$.

Now, if u is a partial word on A and $0 \le i < j \le |u|$, then $u[i..j)$ denotes the factor of u satisfying $(u[i..j))_\diamond = u_\diamond(i) \ldots u_\diamond(j-1)$. The *maximal suffix* of u with respect to $\preceq_l$ (respectively, $\preceq_r$) is defined as $u[i..|u|)$ where $0 \le i < |u|$ and where $u[j..|u|) \preceq_l u[i..|u|)$ (respectively, $u[j..|u|) \preceq_r u[i..|u|)$) for all $0 \le j < |u|$.

Example 4.1
If $a \prec_l b$, then the maximal suffix of $ba\diamond bbaab$ with respect to $\preceq_l$ is $bbaab$ and with respect to $\preceq_r$ is aab. Indeed, the nonempty suffixes are ordered as follows with respect to $\preceq_l$:

---

[1] Recall that $\mathrm{pre}(u, v)$ denotes the maximal common prefix of u and v as defined in Chapter 1.

$\diamond bbaab \prec_l a\diamond bbaab \prec_l aab \prec_l ab \prec_l b \prec_l ba\diamond bbaab \prec_l baab \prec_l bbaab$

and as follows with respect to $\preceq_r$:

$\diamond bbaab \prec_r b \prec_r bbaab \prec_r ba\diamond bbaab \prec_r baab \prec_r a\diamond bbaab \prec_r ab \prec_r aab$

□

LEMMA 4.2

Let u, v, w be partial words.

1. *If v is the maximal suffix of $w = uv$ with respect to $\preceq_l$, then no nonempty partial words x, y are such that $y \subset x$, $u = rx$ and $v = ys$ for some pwords r, s.*

2. *If v is the maximal suffix of $w = uv$ with respect to $\preceq_r$, then no nonempty partial words x, y are such that $y \subset x$, $u = rx$ and $v = ys$ for some pwords r, s.*

PROOF We prove Statement 1 (Statement 2 is similar). Let x, y be nonempty partial words satisfying $y \subset x$, $u = rx$ and $v = ys$ for some pwords r, s. Since $w = uv = rxv = rxys$, by the maximality of v, we have $xv \preceq_l v$ and $s \preceq_l v$. Since $v = ys$, these inequalities can be rewritten as $xys \preceq_l ys$ and $s \preceq_l ys$. Now, from the former inequality we obtain that $yys \preceq_l ys$ since $y \subset x$. We then obtain that $ys \preceq_l s$, which together with $s \preceq_l ys$ imply that $s = ys$. Therefore, $y = \varepsilon$ and $x = \varepsilon$ leading to a contradiction. □

LEMMA 4.3

Let u, v, w be partial words.

1. *If v is the maximal suffix of $w = uv$ with respect to $\preceq_l$, then no nonempty partial words x, y, s are such that $y \subset x$, $u = rx$ and $y = vs$ for some pword r.*

2. *If v is the maximal suffix of $w = uv$ with respect to $\preceq_r$, then no nonempty partial words x, y, s are such that $y \subset x$, $u = rx$ and $y = vs$ for some pwords r.*

PROOF We prove Statement 1 (Statement 2 is similar). Let x, y, s be nonempty partial words satisfying $y \subset x$, $u = rx$ and $y = vs$ for some pword r. Here $w = uv = rxv$, and since v is the maximal suffix with respect to $\preceq_l$, we get $xv \preceq_l v$. Since $y \subset x$, we get $yv \preceq_l v$. Replacing y by vs in the latter inequality yields $vsv \preceq_l v$, leading to a contradiction. □

4.2   The zero-hole case

In this section, we discuss the critical factorization theorem on full words. Intuitively, the theorem states that the minimal period $p(w)$ of a word w of length at least two can be locally determined in at least one position of w. This means that there exists a *critical* factorization (u, v) of w with $u, v \neq \varepsilon$ such that $p(w)$ is the minimal local period of w at position $|u| - 1$.

A factorization of a word w being any tuple (u, v) of words such that $w = uv$, a local period of w at position $|u| - 1$ is defined as follows.

DEFINITION 4.1   *Let w be a nonempty word. A positive integer p is called* **a local period of** w **at position** i *if there exist $u, v \in A^+$ and $x \in A^*$ such that $w = uv$, $|u| = i + 1$, $|x| = p$, and such that one of the following conditions holds for some words r, s:*

1. $u = rx$ and $v = xs$ *(internal square),*

2. $x = ru$ and $v = xs$ *(left-external square if $r \neq \varepsilon$),*

3. $u = rx$ and $x = vs$ *(right-external square if $s \neq \varepsilon$),*

4. $x = ru$ and $x = vs$ *(left- and right-external square if $r, s \neq \varepsilon$).*

The minimal local period of w at position i, *denoted by* $p(w, i)$, *is defined as the smallest local period of w at position i.*

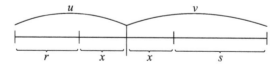

FIGURE 4.1:   Internal square.

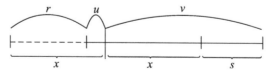

FIGURE 4.2:   Left-external square.

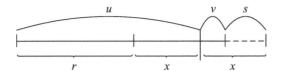

FIGURE 4.3: Right-external square.

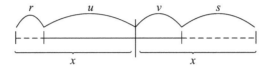

FIGURE 4.4: Left- and right-external square.

Intuitively, around position i, there exists a factor of w having as its minimal period this minimal local period. A factorization (u, v) of w is called *critical* when $u, v \neq \varepsilon$ and $p(w) = p(w, |u| - 1)$. In such case, the position $|u| - 1$ is called a *critical point*. Clearly,

$$1 \leq p(w, i) \leq p(w) \leq |w|$$

Example 4.2
Consider the word $w = babbaab$ with minimal period 6. The minimal local periods of w are: $p(w, 0) = 2$, $p(w, 1) = 3$, $p(w, 2) = 1$, $p(w, 3) = 6$, $p(w, 4) = 1$, and $p(w, 5) = 3$. Here, $p(w) = p(w, 3)$ which means that the factorization $(babb, aab)$ is critical.

i	r	u	v	s	$p(w, i)$	Type of square
0	a	$\underline{b}$	$\underline{a}bbaab$		2	left-external
1	b	$b\underline{a}$	$\underline{bb}aab$		3	left-external
2		$ba\underline{b}$	$\underline{b}aab$		1	internal
3	aa	$ba\underline{bb}$	$\underline{aa}b$	abb	6	left- and right-external
4		$babb\underline{a}$	$\underline{a}b$		1	internal
5		$babb\underline{aa}$	$\underline{b}$	aa	3	right-external

Note that the minimal period of w is simply the maximum among all its minimal local periods. ⬚

The following theorem states that each word of length at least two has at least one critical factorization.

THEOREM 4.1
Let w be a word such that $|w| \geq 2$. Then w has at least one critical factorization (u, v) with $u, v \neq \varepsilon$ and $p(w) = p(w, |u| - 1)$.

While Theorem 4.1 shows the existence of a critical factorization, the following algorithm shows that such a factorization can be found by computing two maximal suffixes of w with respect to the orderings $\preceq_l$ and $\preceq_r$.

ALGORITHM 4.1
The algorithm outputs a critical factorization for a given word w of length at least two.

Step 1: *Compute the maximal suffix of w with respect to $\preceq_l$ (say v) and the maximal suffix of w with respect to $\preceq_r$ (say v').*

Step 2: *Find words u, u' such that $w = uv = u'v'$.*

Step 3: *If $|v| \leq |v'|$, then output (u, v). Otherwise, output (u', v').*

Example 4.3
Returing to Example 4.2, the nonempty suffixes of $w = babbaab$ are ordered as follows (where $a \prec_l b$ and $b \prec_r a$):

$\preceq_l$	$\preceq_r$
aab	b
ab	bbaab
abbaab	babbaab
b	baab
baab	ab
babbaab	abbaab
bbaab	aab

The maximal suffix of w with respect to $\preceq_l$ is $v = bbaab$ and the maximal suffix of w with respect to $\preceq_r$ is $v' = aab$. Here $5 = |v| > |v'| = 3$ which means that the factorization $(u', v') = (babb, aab)$ is critical. ▯

We have omitted the proof of Theorem 4.1 as well as the proof of Algorithm 4.1 because we will prove the general results later in this chapter.

4.3 The main result: First version

In this section, we discuss a first version of the critical factorization theorem for partial words with an arbitrary number of holes. Intuitively, the theorem states that the minimal weak period of a *nonspecial* partial word w of length at least two can be locally determined in at least one position of w. More specifically, if w is nonspecial according to Definition 4.3, then there exists a *critical* factorization (u, v) of w with $u, v \neq \varepsilon$ such that the minimal local period of w at position $|u| - 1$ (as defined below) equals the minimal weak period of w.

DEFINITION 4.2 *Let w be a nonempty partial word. A positive integer p is called **a local period of w at position i** if there exist nonempty partial words u, v, x, y such that $w = uv$, $|u| = i + 1$, $|x| = p$, $x \uparrow y$, and such that one of the following conditions holds for some partial words r, s:*

1. *$u = rx$ and $v = ys$ (internal square),*

2. *$x = ru$ and $v = ys$ (left-external square if $r \neq \varepsilon$),*

3. *$u = rx$ and $y = vs$ (right-external square if $s \neq \varepsilon$),*

4. *$x = ru$ and $y = vs$ (left- and right-external square if $r, s \neq \varepsilon$).*

The minimal local period of w at position i, *denoted by $p(w, i)$, is defined as the smallest local period of w at position i. Clearly,*

$$1 \leq p(w, i) \leq p'(w) \leq |w|$$

and no minimal local period is longer than the minimal weak period.

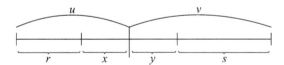

FIGURE 4.5: Internal square.

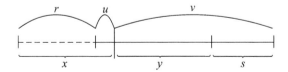

FIGURE 4.6: Left-external square.

A partial word being special is defined as follows.

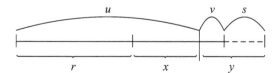

FIGURE 4.7:   Right-external square.

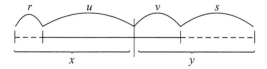

FIGURE 4.8:   Left- and right-external square.

DEFINITION 4.3   *Let w be a partial word such that $p'(w) > 1$. Let v (respectively, v') be the maximal suffix of w with respect to $\preceq_l$ (respectively, $\preceq_r$). Let u, u' be partial words such that $w = uv = u'v'$.*

- *If $|v| \leq |v'|$, then w is called **special** if one of the following holds:*

    1. *$p(w, |u| - 1) < |u|$ and $r \notin C(S(u))$ (as computed according to Definition 4.2).*
    2. *$p(w, |u| - 1) < |v|$ and $s \notin C(P(v))$ (as computed according to Definition 4.2).*

- *If $|v| \geq |v'|$, then w is called **special** if one of the above holds when referring to Definition 4.2 where u is replaced by u' and v by v'.*

*The partial word w is called **nonspecial** otherwise.*

Example 4.4
To illustrate Definition 4.3, first consider $w = aa\diamond\diamond ba\diamond\diamond bb$. The maximal suffixes of w with respect to $\preceq_l$ and $\preceq_r$ are $v = bb$ and $v' = aa\diamond\diamond ba\diamond\diamond bb$ respectively. Here $|v| \leq |v'|$ and $u = aa\diamond\diamond ba\diamond\diamond$. We get that w is special since $1 = p(w, |u| - 1) < |u| = 8$ and $r = aa\diamond\diamond ba\diamond \notin C(S(u))$. Now, consider $w = ab\diamond\diamond a$ with maximal suffixes $v = b\diamond\diamond a$ and $v' = ab\diamond\diamond a$. Again $|v| \leq |v'|$. We have $|u| = 1 \leq 2 = p(w, |u| - 1) < |v| = 4$ but $s = \diamond a \in C(P(v))$, and so w is nonspecial. $\quad\square$

The proof of the following theorem not only shows the existence of a critical factorization for a given nonspecial partial word of length at least two, but also gives an algorithm to compute such a factorization explicitly.

THEOREM 4.2
If the partial word w is nonspecial and satisfies $|w| \geq 2$, then w has at least

one critical factorization. More specifically, if $p'(w) > 1$, then let v denote the maximal suffix of w with respect to $\preceq_l$ and v' the maximal suffix of w with respect to $\preceq_r$. Let u, u' be partial words such that $w = uv = u'v'$. Then the factorization (u, v) is critical when $|v| \leq |v'|$, and the factorization (u', v') is critical when $|v| > |v'|$.

PROOF If $p'(w) = 1$, then

$$w = a_0^{m_0} \diamond a_1^{m_1} \diamond \ldots a_{n-1}^{m_{n-1}} \diamond a_n^{m_n}$$

for some $a_0, a_1, \ldots, a_n \in A$ and integers $m_0, m_1, \ldots, m_n \geq 0$. The result trivially holds in this case. So assume that $p'(w) > 1$ and that $|v| \leq |v'|$ (the case where $p'(w) > 1$ and $|v| > |v'|$ is proved similarly but requires that the orderings $\preceq_l$ and $\preceq_r$ be interchanged). First, assume that $u = \varepsilon$, and thus $w = v$. Since $|v| \leq |v'|$, we also have $w = v'$. Setting $w = az$ for some $a \in A$ and $z \in W(A)$, we argue as follows. If $b \in A$ is a letter in z, then $b \preceq_l a$ and $b \preceq_r a$. Thus, $b = a$ and w is unary. We get $p'(w) = 1$, contradicting our assumption and therefore $u \neq \varepsilon$.

Now, let us denote $p(w, |u| - 1)$ by p. We will use $\beta \, \natural \, \gamma$ as an abbreviation for $\beta \uparrow \gamma$ and $\beta \not\subset \gamma$ and $\gamma \not\subset \beta$ holding simultaneously. The proof is split into four cases that refer to p in relation to $|u|$ and $|v|$. Case 1 refers to $p \geq |u|$ and $p \geq |v|$, Case 2 to $p < |u|$ and $p > |v|$, Case 3 to $p < |u|$ and $p \leq |v|$ and Case 4 to $p \geq |u|$ and $p < |v|$. We prove the result for Cases 1 and 2. The other cases follow similarly and are left as exercises for the reader.

    Case 1. $p \geq |u|$ and $p \geq |v|$

If $p \geq |u|$ and $p \geq |v|$, then Definition 4.2(4) is satisfied. There exist pwords x, y, r, s such that $|x| = p$, $x \uparrow y$, $x = ru$, and $y = vs$. First, if $|r| > |v|$, then $p = |x| = |ru| > |uv| = |w|$, which leads to a contradiction. Similarly, we see that $|s| \leq |u|$ Now, if $|r| \leq |v|$, then we may choose partial words r, s, z, z' such that $v = rz$, $u = z's$, and $z \uparrow z'$. By definition of compatibility, there exists z'' such that $z \subset z''$ and $z' \subset z''$. Thus, $uv = z'srz \subset z''srz''$ showing that $p = |x| = |ru| = |rz's| = |z''sr|$ is a weak period of uv, and so $p'(w) \leq p$. On the other hand, $p'(w) \geq p$. Therefore, $p'(w) = p$ which shows that the factorization (u, v) is critical.

    Case 2. $p < |u|$ and $p > |v|$

If $p < |u|$ and $p > |v|$, then Definition 4.2(3) is satisfied. There exist partial words x, y, r, s, γ such that $|x| = p$, $x \uparrow y$, $u = rx = r\gamma s$, and $y = vs$. If $v \subset \gamma$, then $y \subset x$, and v being the maximal suffix of w with respect to $\preceq_l$, we get a contradiction with Lemma 4.3. If $\gamma \sqsubset v$ or $\gamma \, \natural \, v$, then we consider whether or not $r \in C(S(u))$. If $r \notin C(S(u))$, then w is special by Definition 4.3(1). If $r \in C(S(u))$, then $x'r \uparrow rx$ for some x'. By Theorem 2.2, $u = rx$ is weakly $|x|$-periodic, and so $rxy = rxvs$ is weakly $|x|$-periodic since $x \uparrow y$. Therefore, $p = |x|$ is a weak period of $uv = rxv$ and the result follows as in Case 1.   ☐

REMARK 4.1 In the course of the proof of Theorem 4.2, we showed in addition that if $|v| \leq |v'|$ and the factorization (u, v) is critical, then w is nonspecial, and if $|v| > |v'|$ and the factorization (u', v') is critical, then w is nonspecial. □

Referring to Definition 4.3, assume that $a \prec_l b$ and $b \prec_r a$. We now provide special partial words w with no position i satisfying $p'(w) = p(w, i)$. These examples show why Theorem 4.2 excludes the special partial words.

Example 4.5
For each given pword w, we give answers to the two questions: Is $r \in C(S(u))$ and is $s \in C(P(v))$? We also exhibit u, v, x, y, r and s for the different scenarios.

- Definition 4.3(2) answers "yes" and "no" when $w = aa\diamond b\diamond\diamond\diamond bba$. Here $u = rx$, $v = ys$, $x = \diamond$, $y = b$, $r = aa\diamond b\diamond\diamond$ and $s = ba$.

- Definition 4.3(1) gives "no" and "yes" for $w = baa\diamond bb\diamond$, and computations give $u = rx$, $v = ys$, $x = \diamond$, $y = b$, $r = baa$ and $s = b\diamond$.

- Definition 4.3(2) answers $s \notin C(P(v))$ if $w = ab\diamond a\diamond a$. We can check that $u = a$, $v = ys$, $x = ru$, $y = b\diamond$, $r = b$ and $s = a\diamond a$.

- Definition 4.3(1) gives $r \notin C(S(u))$ for $w = \diamond b\diamond bbabb$, and $u = rx$, $v = bbb$, $x = \diamond bba$, $y = vs$, $r = \diamond b$ and $s = a$.

□

From the proof of Theorem 4.2, we can obtain an algorithm that outputs a critical factorization for a given partial word w with $p'(w) > 1$ and with an arbitrary number of holes of length at least two when w is nonspecial, and that outputs "special" otherwise. The algorithm computes the maximal suffix v of w with respect to $\preceq_l$ and the maximal suffix v' of w with respect to $\preceq_r$. The algorithm finds partial words u, u' such that $w = uv = u'v'$. If $|v| \leq |v'|$, then it computes $p = p(w, |u| - 1)$ and does the following:

1. If $p < |u|$, then it finds partial words x, y, r, s satisfying Definition 4.2. If $r \notin C(S(u))$, then it outputs "special."

2. If $p < |v|$, then it finds partial words x, y, r, s satisfying Definition 4.2. If $s \notin C(P(v))$, then it outputs "special."

3. Otherwise, it outputs (u, v).

If $|v| > |v'|$, then the algorithm computes $p = p(w, |u'| - 1)$ and does the above where u is replaced by u' and v by v'.

Example 4.6

As an example, consider $w = aaab\diamond babb$. Its maximal suffix with respect to $\preceq_l$ (where $a \prec b$) is $v = bb$ and with respect to $\preceq_r$ (where $b \prec a$) is $v' = aaab\diamond babb$. Here $|v| < |v'|$ and the factorization $(aaab\diamond ba, bb)$ is not critical since w is special. Now, if we consider $\text{rev}(w) = bbab\diamond baaa$, its maximal suffix with respect to $\preceq_l$ is $v = bbab\diamond baaa$ and with respect to $\preceq_r$ is $v' = aaa$. Here $|v| > |v'|$ and $\text{rev}(w)$ is nonspecial and so the factorization $(bbab\diamond b, aaa)$ of $\text{rev}(w)$ (which corresponds to the factorization $(aaa, b\diamond babb)$ of w) is critical.

□

The observation in the preceding example on the reversal leads us to improve the algorithm by considering both w and $\text{rev}(w)$.

ALGORITHM 4.2

The algorithm outputs a critical factorization for a given partial word w with $p'(w) > 1$ and $|w| \geq 2$ when w is nonspecial or $\text{rev}(w)$ is nonspecial, and that outputs "special" otherwise.

Step 1: *Compute the maximal suffix v_0 of w with respect to $\preceq_l$ and the maximal suffix v_0' of w with respect to $\preceq_r$. Also compute the maximal suffix v_1 of $\text{rev}(w)$ with respect to $\preceq_l$ and the maximal suffix v_1' of $\text{rev}(w)$ with respect to $\preceq_r$.*

Step 2: *Find partial words u_0, u_0' such that $w = u_0 v_0 = u_0' v_0'$. Also find partial words u_1, u_1' such that $\text{rev}(w) = u_1 v_1 = u_1' v_1'$.*

Step 3: *If $|v_0| \leq |v_0'|$ and $|v_1| \leq |v_1'|$, then compute $p_0 = p(w, |u_0| - 1)$ and $p_1 = p(\text{rev}(w), |u_1| - 1)$.*

Step 4: *If $p_0 \geq p_1$, then do the following:*

1. *If $p_0 < |u_0|$, then find partial words x, y, r, s satisfying Definition 4.2. If $r \notin C(S(u_0))$, then output "special."*

2. *If $p_0 < |v_0|$, then find partial words x, y, r, s satisfying Definition 4.2. If $s \notin C(P(v_0))$, then output "special."*

3. *Otherwise, output (u_0, v_0).*

Step 5: *If $p_0 < p_1$, then do the work of Step 4 with p_1, u_1 and v_1 instead of p_0, u_0 and v_0.*

Step 6: *If $|v_0| > |v_0'|$ (or $|v_1| > |v_1'|$), then do the work of Step 3 with u_0' and v_0' instead of u_0 and v_0 (or do the work of Step 3 with u_1' and v_1' instead of u_1 and v_1). The algorithm may produce (u_0', v_0') unless w is special (or may produce (u_1', v_1') unless $\text{rev}(w)$ is special) (in those cases, output "special").*

4.4   The main result: Second version

In this section, the nonempty suffixes of a given partial word w are ordered as follows according to $\preceq_l$:

$$v_{0,|w|-1} \prec_l v_{0,|w|-2} \prec_l \cdots \prec_l v_{0,0}$$

The factorizations of w called $(u_{0,0}, v_{0,0}), (u_{0,1}, v_{0,1}), \ldots$ result. Similarly, the nonempty suffixes of w are ordered as follows according to $\preceq_r$:

$$v'_{0,|w|-1} \prec_r v'_{0,|w|-2} \prec_r \cdots \prec_r v'_{0,0}$$

The factorizations of w called $(u'_{0,0}, v'_{0,0}), (u'_{0,1}, v'_{0,1}), \ldots$ result. The nonempty suffixes of $\mathrm{rev}(w)$ are ordered as follows:

$$v_{1,|w|-1} \prec_l v_{1,|w|-2} \prec_l \cdots \prec_l v_{1,0}$$
$$v'_{1,|w|-1} \prec_r v'_{1,|w|-2} \prec_r \cdots \prec_r v'_{1,0}$$

The factorizations of $\mathrm{rev}(w)$ called

$$(u_{1,0}, v_{1,0}), (u_{1,1}, v_{1,1}), \ldots, (u'_{1,0}, v'_{1,0}), (u'_{1,1}, v'_{1,1}), \ldots$$

result.

Referring to Definition 4.3, the following table provides examples of special partial words w whose reversals are also special and for which there exists a position i such that $p'(w) = p(w, i)$ or $p'(w) = p(\mathrm{rev}(w), i)$ resulting in a critical factorization (it is assumed that $a \prec_l b$ and $b \prec_r a$):

w	Fact	Crit	Fact	Crit	Fact	Crit
$aaa\diamond\diamond ba$	$(u_{0,0}, v_{0,0})$	no	$(u'_{1,0}, v'_{1,0})$	no	$(u_{1,0}, v_{1,0})$	yes
$abba\diamond abb$	$(u_{0,0}, v_{0,0})$	no	$(u'_{1,0}, v'_{1,0})$	no	$(u_{0,1}, v_{0,1})$	yes
$a\diamond abb\diamond bbbaa$	$(u'_{0,0}, v'_{0,0})$	no	$(u_{1,0}, v_{1,0})$	no	$(u_{0,2}, v_{0,2})$	yes
$a\diamond cbac$	$(u'_{0,0}, v'_{0,0})$	no	$(u'_{1,0}, v'_{1,0})$	no	$(u_{0,2}, v_{0,2})$	yes

The above examples lead us to refine Theorem 4.2. First, we define the concept of an *((k,l))-special* partial word (note that the concept of *special* in Definition 4.3 is equivalent to the concept of $((0,0))$-special in Definition 4.4).

DEFINITION 4.4   *Let w be a partial word such that $p'(w) > 1$, and let k, l be a pair of integers satisfying $0 \leq k, l < |w|$.*

- *If $|v_{0,k}| \leq |v'_{0,l}|$, then w is called **((k, l))-special** if one of the following holds:*

  1. $p(w, |u_{0,k}| - 1) < |u_{0,k}|$ and $r \notin C(S(u_{0,k}))$ (as computed according to Definition 4.2).

2. $p(w, |u_{0,k}| - 1) < |v_{0,k}|$ *and* $s \notin C(P(v_{0,k}))$ *(as computed according to Definition 4.2).*

- If $|v_{0,k}| \geq |v'_{0,l}|$, *then* w *is called* $((k,l))$-**special** *if one of the above holds when referring to Definition 4.2 where* $u_{0,k}$ *is replaced by* $u'_{0,l}$ *and* $v_{0,k}$ *by* $v'_{0,l}$.

The partial word w *is called* $((k,l))$-**nonspecial** *otherwise.*

We now describe the algorithm (based on Theorem 4.3) that outputs a critical factorization for a given partial word w with $p'(w) > 1$ and with an arbitrary number of holes of length at least two when such a factorization exists, and that outputs "no critical factorization exists" otherwise.

ALGORITHM 4.3

Step 1: *Compute the nonempty suffixes of* w *with respect to* $\preceq_l$

$$v_{0,|w|-1} \prec_l \cdots \prec_l v_{0,0}$$

and the nonempty suffixes of w *with respect to* $\preceq_r$

$$v'_{0,|w|-1} \prec_r \cdots \prec_r v'_{0,0}$$

Also compute the nonempty suffixes of $rev(w)$ *with respect to* $\preceq_l$

$$v_{1,|w|-1} \prec_l \cdots \prec_l v_{1,0}$$

and the nonempty suffixes of $rev(w)$ *with respect to* $\preceq_r$

$$v'_{1,|w|-1} \prec_r \cdots \prec_r v'_{1,0}$$

Step 2: *Set* $k_0 = 0$, $l_0 = 0$, $k_1 = 0$, $l_1 = 0$, *and* $mwp = 0$.

Step 3: *If* $k_0 \geq |w| - \|H(w)\|$ *or* $l_0 \geq |w| - \|H(w)\|$ *or* $k_1 \geq |w| - \|H(w)\|$ *or* $l_1 \geq |w| - \|H(w)\|$, *then output "no critical factorization exists".*

Step 4: *If* $v_{0,k_0} \prec_l v'_{0,l_0}$, *then update* l_0 *with* $l_0 + 1$ *and go to Step 3. If* $v'_{0,l_0} \prec_r v_{0,k_0}$, *then update* k_0 *with* $k_0 + 1$ *and go to Step 3. If* $v_{1,k_1} \prec_l v'_{1,l_1}$, *then update* l_1 *with* $l_1 + 1$ *and go to Step 3. If* $v'_{1,l_1} \prec_r v_{1,k_1}$, *then update* k_1 *with* $k_1 + 1$ *and go to Step 3.*

Step 5: *If* $k_0 > 0$ *and* $v'_{0,l_0} = w$, *then update* l_0 *with* $l_0 + 1$ *and go to Step 3. If* $l_0 > 0$ *and* $v_{0,k_0} = w$, *then update* k_0 *with* $k_0 + 1$ *and go to Step 3. If* $k_1 > 0$ *and* $v'_{1,l_1} = rev(w)$, *then update* l_1 *with* $l_1 + 1$ *and go to Step 3. If* $l_1 > 0$ *and* $v_{1,k_1} = rev(w)$, *then update* k_1 *with* $k_1 + 1$ *and go to Step 3.*

Step 6: *Find partial words u_{0,k_0}, u'_{0,l_0} such that $w = u_{0,k_0} v_{0,k_0} = u'_{0,l_0} v'_{0,l_0}$.*
Find partial words u_{1,k_1}, u'_{1,l_1} such that $rev(w) = u_{1,k_1} v_{1,k_1} = u'_{1,l_1} v'_{1,l_1}$.

Step 7: *If $|v_{0,k_0}| \leq |v'_{0,l_0}|$ and $|v_{1,k_1}| \leq |v'_{1,l_1}|$, then compute $p_{0,k_0} = p(w, |u_{0,k_0}| - 1)$ and $p_{1,k_1} = p(rev(w), |u_{1,k_1}| - 1)$.*

Step 8: *If $p_{0,k_0} \leq mwp$, then* move up *which means to update k_0 with $k_0 + 1$ and to go to Step 3. If $p_{1,k_1} \leq mwp$, then* move up *which means to update k_1 with $k_1 + 1$ and to go to Step 3.*

Step 9: *If $p_{0,k_0} \geq p_{1,k_1}$, then update mwp with p_{0,k_0}. Do the following:*

    1. If $p_{0,k_0} < |u_{0,k_0}|$, then find partial words x, y, r, s satisfying Definition 4.2. If $r \notin C(S(u_{0,k_0}))$, then move up *which means update k_0 with $k_0 + 1$ and go to Step 3.*

    2. If $p_{0,k_0} < |v_{0,k_0}|$, then find partial words x, y, r, s satisfying Definition 4.2. If $s \notin C(P(v_{0,k_0}))$, then move up *which means update k_0 with $k_0 + 1$ and go to Step 3.*

    3. Otherwise, output (u_{0,k_0}, v_{0,k_0}).

Step 10: *If $p_{0,k_0} < p_{1,k_1}$, then update mwp with p_{1,k_1} and do the work of Step 9 with p_{1,k_1}, u_{1,k_1} and v_{1,k_1} instead of p_{0,k_0}, u_{0,k_0} and v_{0,k_0}.*

Step 11: *If $|v_{0,k_0}| > |v'_{0,l_0}|$ (or $|v_{1,k_1}| > |v'_{1,l_1}|$), then compute $p_{0,l_0} = p(w, |u'_{0,l_0}| - 1)$ and do the work of Step 8 with p_{0,l_0}, u'_{0,l_0} and v'_{0,l_0} instead of p_{0,k_0}, u_{0,k_0} and v_{0,k_0} (move up here means update l_0 with $l_0 + 1$ and go to Step 3) (or compute $p_{1,l_1} = p(rev(w), |u'_{1,l_1}| - 1)$ and do the work of Step 8 with p_{1,l_1}, u'_{1,l_1} and v'_{1,l_1} instead of p_{1,k_1}, u_{1,k_1} and v_{1,k_1} (move up here means update l_1 with $l_1 + 1$ and go to Step 3)). The algorithm may produce (u'_{0,l_0}, v'_{0,l_0}) unless w is $((k_0, l_0))$-special (or may produce (u'_{1,l_1}, v'_{1,l_1}) unless $rev(w)$ is $((k_1, l_1))$-special) (in those cases, move up).*

We illustrate Algorithm 4.3 with the following example.

Example 4.7

Below are tables for the nonempty suffixes of the partial word $w = a \diamond cbac$ and its reversal $rev(w) = cabc \diamond a$. These suffixes are ordered in two different ways: The first ordering is on the left and is an $\prec_l$-ordering according to the order $\diamond \prec a \prec b \prec c$, and the second is on the right and is an $\prec_r$-ordering where $\diamond \prec c \prec b \prec a$. The tables also contain the indices used by the algorithm, k_0, l_0, k_1, l_1, and the local periods that needed to be calculated in order to compute the critical factorization $(a \diamond c, bac)$. The minimal weak period of w turns out to be equal to 4.

k_0	p_{0,k_0}	v_{0,k_0}	v'_{0,l_0}	p_{0,l_0}	l_0
5		$\diamond cbac$	$\diamond cbac$		5
4		$a\diamond cbac$	c		4
3		ac	$cbac$		3
2	4	bac	bac		2
1	3	c	$a\diamond cbac$		1
0	1	$cbac$	ac	3	0

k_1	p_{1,k_1}	v_{1,k_1}	v'_{1,l_1}	p_{1,l_1}	l_1
5		$\diamond a$	$\diamond a$		5
4		a	$c\diamond a$		4
3		$abc\diamond a$	$cabc\diamond a$		3
2		$bc\diamond a$	$bc\diamond a$		2
1	4	$c\diamond a$	a	1	1
0		$cabc\diamond a$	$abc\diamond a$	3	0

Algorithm 4.3 starts with the pairs

$$(v_{0,0}, v'_{0,0}) = (cbac, ac) \text{ and } (v_{1,0}, v'_{1,0}) = (cabc\diamond a, abc\diamond a)$$

and selects the shortest component of each pair, that is, $v'_{0,0}$ and $v'_{1,0}$. In Step 11, $p_{0,0}$ is computed as 3 and $p_{1,0}$ as 3. Since $p_{0,0} \geq p_{1,0} > mwp = 0$, the factorization $(u'_{0,0}, v'_{0,0}) = (a\diamond cb, ac)$ is chosen and the algorithm discovers that w is $((0,0))$-special according to Definition 4.4. The variable l_0 is then updated to 1 and the pairs

$$(v_{0,0}, v'_{0,1}) = (cbac, a\diamond cbac) \text{ and } (v_{1,0}, v'_{1,0}) = (cabc\diamond a, abc\diamond a)$$

are treated with shortest components $v_{0,0}, v'_{1,0}$ respectively. Now, $p_{0,0}$ is computed as 1 and $p_{1,0}$ as 3. Since $p_{0,0} < p_{1,0} \leq mwp = 3$, k_0 gets updated to 1 and l_1 to 1. Now, the pairs

$$(v_{0,1}, v'_{0,1}) = (c, a\diamond cbac) \text{ and } (v_{1,0}, v'_{1,1}) = (cabc\diamond a, a)$$

are considered and in Step 5, l_0 is updated to 2 since $k_0 = 1 > 0$ and $v'_{0,l_0} = v'_{0,1} = w$. The pairs

$$(v_{0,1}, v'_{0,2}) = (c, bac) \text{ and } (v_{1,0}, v'_{1,1}) = (cabc\diamond a, a)$$

are treated and in Step 5, k_1 is updated to 1 since $l_1 = 1 > 0$ and $v_{1,k_1} = v_{1,0} = rev(w)$. Comes the turn of

$$(v_{0,1}, v'_{0,2}) = (c, bac) \text{ and } (v_{1,1}, v'_{1,1}) = (c\diamond a, a)$$

with shortest components $v_{0,1}$ and $v'_{1,1}$. The algorithm computes $p_{0,1} = 3$ and $p_{1,1} = 1$. Since $p_{1,1} < p_{0,1} \leq mwp = 3$, the indices k_0 and l_1 get updated to 2 and the pairs

$$(v_{0,2}, v'_{0,2}) = (bac, bac) \text{ and } (v_{1,1}, v'_{1,2}) = (c\diamond a, bc\diamond a)$$

are considered with shortest components $v_{0,2}, v_{1,1}$ and with $p_{0,2} = 4, p_{1,1} = 4$ calculated in Step 7. Since $p_{0,2} \geq p_{1,1} > mwp = 3$ leads to an improvement of the number mwp, the algorithm outputs $(u_{0,2}, v_{0,2})$ in Step 9 with $mwp = p_{0,2} = 4$ (here w is $((2,2))$-nonspecial). ▯

We now prove Theorem 4.3.

THEOREM 4.3

1. Let (k_0, l_0) be a pair of nonnegative integers being considered at Step 9 (when $p_{0,k_0} > mwp$ or when $p_{0,l_0} > mwp$). If w is a $((k_0, l_0))$-nonspecial partial word satisfying $|w| \geq 2$ and $p'(w) > 1$, then w has at least one critical factorization. More specifically, the factorization (u_{0,k_0}, v_{0,k_0}) is critical when $|v_{0,k_0}| \leq |v'_{0,l_0}|$, and the factorization (u'_{0,l_0}, v'_{0,l_0}) is critical when $|v_{0,k_0}| > |v'_{0,l_0}|$.

2. Let (k_1, l_1) be a pair of nonnegative integers being considered at Step 10 (when $p_{1,k_1} > mwp$ or when $p_{1,l_1} > mwp$). If $rev(w)$ is a $((k_1, l_1))$-nonspecial partial word satisfying $|w| \geq 2$ and $p'(w) > 1$, then $rev(w)$ has at least one critical factorization. More specifically, the factorization (u_{1,k_1}, v_{1,k_1}) is critical when $|v_{1,k_1}| \leq |v'_{1,l_1}|$, and the factorization (u'_{1,l_1}, v'_{1,l_1}) is critical when $|v_{1,k_1}| > |v'_{1,l_1}|$.

PROOF   We prove Statement 1 (Statement 2 is proved similarly). The pair $(k_0, l_0) = (0,0)$ was treated in Theorem 4.2. So, we may assume that $(k_0, l_0) \neq (0,0)$. We consider the case where $|v_{0,k_0}| \leq |v'_{0,l_0}|$ (the case where $|v_{0,k_0}| > |v'_{0,l_0}|$ is handled similarly but requires that the orderings $\preceq_l$ and $\preceq_r$ be interchanged). Here, $u_{0,k_0} \neq \varepsilon$ unless $v_{0,k_0} = v'_{0,l_0} = w$. In such case, if w begins with $\diamond$, then the algorithm will discover in Step 3 that w has no critical factorization. And if w begins with a for some $a \in A$, then $k_0 < |w| - \|H(w)\|$ and $l_0 < |w| - \|H(w)\|$. In such case, we have $(k_0 > 0$ and $v'_{0,l_0} = w)$ or $(l_0 > 0$ and $v_{0,k_0} = w)$. In the former case, Step 5 will update l_0 with $l_0 + 1$ resulting in the pair $(k_0, l_0 + 1)$ being considered in Step 3; in the latter case, Step 5 will update k_0 with $k_0 + 1$ and $(k_0 + 1, l_0)$ will be considered in Step 3.

We now consider the following cases where p_{0,k_0} denotes $p(w, |u_{0,k_0}| - 1)$. Again, we use $\beta \diamondsuit \gamma$ as an abbreviation for $\beta \uparrow \gamma$, $\beta \not\subset \gamma$ and $\gamma \not\subset \beta$ holding simultaneously. The proof is split into four cases that refer to p_{0,k_0} in relation to $|u_{0,k_0}|$ and $|v_{0,k_0}|$. Case 1 refers to $p_{0,k_0} \geq |u_{0,k_0}|$ and $p_{0,k_0} \geq |v_{0,k_0}|$, Case 2 to $p_{0,k_0} < |u_{0,k_0}|$ and $p_{0,k_0} > |v_{0,k_0}|$, Case 3 to $p_{0,k_0} < |u_{0,k_0}|$ and $p_{0,k_0} \leq |v_{0,k_0}|$, and Case 4 to $p_{0,k_0} \geq |u_{0,k_0}|$ and $p_{0,k_0} < |v_{0,k_0}|$.

We prove the result for Case 1 (the other cases are left as exercises for the reader). Here Definition 4.2(4) is satisfied and there exist pwords x, y, r, s such that $|x| = p_{0,k_0}$, $x \uparrow y$, $x = ru_{0,k_0}$ and $y = v_{0,k_0}s$. First, if $|r| > |v_{0,k_0}|$, then $p_{0,k_0} = |x| = |ru_{0,k_0}| > |u_{0,k_0}v_{0,k_0}| = |w| \geq p'(w)$, which leads to a contradiction. Now, if $|r| \leq |v_{0,k_0}|$, then by Lemma 1.2, there exist r', z such that $v_{0,k_0} = r'z$, $r \uparrow r'$, and $u_{0,k_0} \uparrow zs$. There exists r'' such that $r \subset r''$ and $r' \subset r''$, and there exist z', s' such that $u_{0,k_0} \subset z's'$, $z \subset z'$ and $s \subset s'$. Thus, $u_{0,k_0}v_{0,k_0} \subset z's'r'z'$ showing that $p_{0,k_0} = |z's'r'|$ is a weak period of $u_{0,k_0}v_{0,k_0}$, and $p'(w) \leq p_{0,k_0}$. On the other hand, $p'(w) \geq p_{0,k_0}$. Therefore, $p'(w) = p_{0,k_0}$ which shows that the factorization (u_{0,k_0}, v_{0,k_0}) is critical. □

REMARK 4.2 Referring to the above theorem, the following strengthen Statements 1 and 2:

1. If $|v_{0,k_0}| \leq |v'_{0,l_0}|$ and the factorization (u_{0,k_0}, v_{0,k_0}) is critical, then w is $((k_0, l_0))$-nonspecial, and if $|v_{0,k_0}| > |v'_{0,l_0}|$ and the factorization (u'_{0,l_0}, v'_{0,l_0}) is critical, then w is $((k_0, l_0))$-nonspecial.

2. If $|v_{1,k_1}| \leq |v'_{1,l_1}|$ and the factorization (u_{1,k_1}, v_{1,k_1}) is critical, then $\mathrm{rev}(w)$ is $((k_1, l_1))$-nonspecial, and if $|v_{1,k_1}| > |v'_{1,l_1}|$ and the factorization (u'_{1,l_1}, v'_{1,l_1}) is critical, then $\mathrm{rev}(w)$ is $((k_1, l_1))$-nonspecial.

□

We conclude this section by characterizing the special partial words that admit critical factorizations. If w is such a special partial word satisfying $|v_{0,0}| \leq |v'_{0,0}|$, then $p_{0,0} = p(w, |u_{0,0}| - 1) < p'(w)$. The following theorems give a bound of how far $p_{0,0}$ is from $p'(w)$ and explain why Algorithm 4.3 is faster in average than a trivial algorithm where every position would be tested for critical factorization.

THEOREM 4.4
Let w be a special partial word that admits a critical factorization, and let $v_{0,0}$ (respectively, $v'_{0,0}$) be the maximal suffix of w with respect to $\preceq_l$ (respectively, $\preceq_r$). Let $u_{0,0}, u'_{0,0}$ be partial words such that $w = u_{0,0}v_{0,0} = u'_{0,0}v'_{0,0}$. If w is special according to Definition 4.3(1), then

• *If $|v_{0,0}| \leq |v'_{0,0}|$, then the following hold:*

1. *If $p_{0,0} \leq |v_{0,0}|$, then there exist nonnegative integers m, n, partial words $x_0, \ldots, x_{m+2}, x'_1, \ldots, x'_{m+1}$ of length n, and partial words $y_0, \ldots, y_{m+1}, y'_1, \ldots, y'_m$ of length $p'(w) - p_{0,0} - n$ such that*

    – $x_0 y_0 x'_1 y'_1 x_1 y_1 \ldots x_{m-1} y_{m-1} x'_m y'_m x_m y_m x'_{m+1} y_{m+1} x_{m+1}$ *has a weak period of $p'(w) - p_{0,0}$,*

    – $x_{m+1} \uparrow x_{m+2}$,

    – $p_{0,0} = |x_1 y_1 x_2 y_2 \ldots x_m y_m x_{m+1}| < p_{0,0} + |x_0 y_0| = p'(w)$,

    – $u_{0,0}$ *is a suffix of a weakly $p'(w)$-periodic partial word ending with $x_0 y_0 x_1 y_1 x_2 y_2 \ldots x_m y_m x_{m+1}$,*

 $-$ $v_{0,0}$ is a prefix of a weakly $p'(w)$-periodic partial word starting with $x_1'y_1'x_2'y_2'\ldots x_m'y_m'x_{m+1}'y_{m+1}x_{m+2}$.

2. If $p_{0,0} > |v_{0,0}|$, then let s denote the nonempty suffix of length $p_{0,0} - |v_{0,0}|$ of $u_{0,0}$. Then there exist nonnegative integers m, n and partial words as above except that

  $-$ $p_{0,0} = |x_1y_1x_2y_2\ldots x_my_mx_{m+1}s|$,
  $-$ $u_{0,0}$ is a suffix of a weakly $p'(w)$-periodic partial word ending with $x_0y_0x_1y_1x_2y_2\ldots x_my_mx_{m+1}s$,
  $-$ $v_{0,0} = x_1'y_1'x_2'y_2'\ldots x_m'y_m'x_{m+1}'$.

- If $|v_{0,0}| \geq |v_{0,0}'|$, then the above hold when replacing $u_{0,0}, v_{0,0}$ by $u_{0,0}', v_{0,0}'$ respectively.

PROOF Let $x, y, r \in W(A)\setminus\{\varepsilon\}$ and $s \in W(A)$ be such that $|x| = p_{0,0}$, $x \uparrow y$, $u_{0,0} = rx$, and either $v_{0,0} = ys$ or $y = v_{0,0}s$. We first assume that $v_{0,0} = ys$ (this case is related to Statement 1). Since w admits a critical factorization, there exists $(k_0, l_0) \neq (0,0)$ such that w is $((k_0, l_0))$-nonspecial and either (u_{0,k_0}, v_{0,k_0}) (if $|v_{0,k_0}| \leq |v_{0,l_0}'|$) or (u_{0,l_0}', v_{0,l_0}') (if $|v_{0,k_0}| > |v_{0,l_0}'|$) is critical with minimal local period q (here $p_{0,0} < q = p'(w)$). Let $\alpha, \beta \in W(A) \setminus \{\varepsilon\}$ be such that $\alpha x \uparrow y\beta$, $|\alpha x| = |y\beta| = q$, either $u_{0,0}$ is a suffix of αx or αx is a suffix of $u_{0,0}$, and either $y\beta$ is a prefix of $v_{0,0}$ or $v_{0,0}$ is a prefix of $y\beta$. Let m be defined as $\lfloor \frac{|x|}{|\alpha|} \rfloor$ and n as $|x|(\bmod |\alpha|)$. Then let $\alpha = x_0y_0$, $\beta = y_{m+1}x_{m+2}$, $x = x_1y_1x_2y_2\ldots x_my_mx_{m+1}$, and $y = x_1'y_1'x_2'y_2'\ldots x_m'y_m'x_{m+1}'$ where each x_i, x_i' has length n and each y_i, y_i' has length $|\alpha| - n$. By Theorem 2.2,

$$\text{pshuffle}_{|\alpha|}(\alpha x, y\beta) =$$
$$x_0y_0x_1'y_1'x_1y_1x_2'y_2'\ldots x_{m-1}y_{m-1}x_m'y_m'x_my_mx_{m+1}'y_{m+1}x_{m+1}$$

is weakly $|\alpha|$-periodic and $\text{sshuffle}_{|\alpha|}(\alpha x, y\beta) = x_{m+1}x_{m+2}$ is $|x|(\bmod|\alpha|)$-periodic (which means that $x_{m+1} \uparrow x_{m+2}$) and the result follows. We now assume that $y = v_{0,0}s$ with $s \neq \varepsilon$ (this case is related to Statement 2). Set $x = \gamma s$. Here $\alpha x \uparrow v_{0,0}\beta s$ for some $\alpha, \beta \in W(A) \setminus \{\varepsilon\}$. By simplification, $\alpha\gamma \uparrow v_{0,0}\beta$, and we also have $\gamma \uparrow v_{0,0}$. The result follows similarly as above. $\square$

THEOREM 4.5
Let w be a special partial word that admits a critical factorization, and let $v_{0,0}$ (respectively, $v_{0,0}'$) be the maximal suffix of w with respect to $\preceq_l$ (respectively, $\preceq_r$). Let $u_{0,0}, u_{0,0}'$ be partial words such that $w = u_{0,0}v_{0,0} = u_{0,0}'v_{0,0}'$. If w is special according to Definition 4.3(2), then the following hold:

- *If $|v_{0,0}| \leq |v_{0,0}'|$, then the following hold:*

  1. *If $p_{0,0} \leq |u_{0,0}|$, then there exist nonnegative integers m, n, partial words $x_0, \ldots, x_{m+2}, x_1', \ldots, x_{m+1}'$ of length n, and partial words $y_0, \ldots, y_{m+1}, y_1', \ldots, y_m'$ of length $p'(w) - p_{0,0} - n$ such that*

- $x_0 y_0 x_1' y_1' x_1 y_1 \ldots x_{m-1} y_{m-1} x_m' y_m' x_m y_m x_{m+1}' y_{m+1} x_{m+1}$ *has a weak period of* $p'(w) - p_{0,0}$,
- $x_{m+1} \uparrow x_{m+2}$,
- $p_{0,0} = |x_1' y_1' x_2' y_2' \ldots x_m' y_m' x_{m+1}'| < p_{0,0} + |y_{m+1} x_{m+2}| = p'(w)$,
- $u_{0,0}$ *is a suffix of a weakly* $p'(w)$-*periodic partial word ending with* $x_0 y_0 x_1 y_1 x_2 y_2 \ldots x_m y_m x_{m+1}$,
- $v_{0,0}$ *is a prefix of a weakly* $p'(w)$-*periodic partial word starting with* $x_1' y_1' x_2' y_2' \ldots x_m' y_m' x_{m+1}' y_{m+1} x_{m+2}$.

2. *If* $p_{0,0} > |u_{0,0}|$, *then let* r *denote the nonempty prefix of length* $p_{0,0} - |u_{0,0}|$ *of* $v_{0,0}$. *Then there exist nonnegative integers* m, n *and partial words as above except that*

   - $p_{0,0} = |r x_1' y_1' x_2' y_2' \ldots x_m' y_m' x_{m+1}'|$,
   - $u_{0,0} = x_1 y_1 x_2 y_2 \ldots x_m y_m x_{m+1}$,
   - $v_{0,0}$ *is a prefix of a wealky* $p'(w)$-*periodic partial word starting with* $r x_1' y_1' x_2' y_2' \ldots x_m' y_m' x_{m+1}' y_{m+1} x_{m+2}$.

- *If* $|v_{0,0}| \geq |v_{0,0}'|$, *then the above hold when replacing* $u_{0,0}, v_{0,0}$ *by* $u_{0,0}', v_{0,0}'$ *respectively.*

PROOF Let $x, y, s \in W(A) \setminus \{\varepsilon\}$ and $r \in W(A)$ be such that $|x| = p_{0,0}$, $x \uparrow y$, either $u_{0,0} = rx$ or $x = r u_{0,0}$, $v_{0,0} = ys$, and let (k_0, l_0) and q be as above. Statement 1 is similar to Statement 1 of Theorem 4.4. For Statement 2, let $\alpha, \beta, \gamma \in W(A) \setminus \{\varepsilon\}$ be such that $y = r\gamma$, $r\alpha u_{0,0} \uparrow y\beta$, $|\alpha x| = |y\beta| = q$, and either $y\beta$ is a prefix of $v_{0,0}$ or $v_{0,0}$ is a prefix of $y\beta$. By simplification, $\alpha u_{0,0} \uparrow \gamma\beta$, and we also have $u_{0,0} \uparrow \gamma$. The result follows from Theorem 2.2. □

4.5 Tests

In this chapter, we considered one of the most fundamental results on periodicity of words, namely the critical factorization theorem, and discussed it in the framework of partial words. While the critical factorization theorem on full words, Theorem 4.1, shows that *critical factorizations* are unavoidable, Theorem 4.2 shows that such factorizations can be possibly avoidable for the so-called *special* partial words. Then, Theorem 4.3 refines the class of the special partial words to the class of the so-called *((k,l))-special* partial words. Theorem 4.3's proof leads to an efficient algorithm which, given a partial word with an arbitrary number of holes, outputs "no critical factorization exists" or outputs a critical factorization that gets computed from the lexicographic/reverse lexicographic orderings of the nonempty suffixes of the partial word and its reversal.

Finally, Theorem 4.4 and 4.5 characterize the $((0,0))$-special partial words that admit critical factorizations.

In the testing of the algorithm, it is important to make the distinction between partial words that have a critical factorization and partial words for which no critical factorization exists. In Table 4.1, we provide data concerning partial words without critical factorizations. Tests were run on all partial words with an arbitrary number of holes over a 3-letter alphabet from lengths two to twelve.

TABLE 4.1: Percentage of partial words without critical factorizations.

Length	Number without CFs	Number	Percentage
2	0	16	0.0
3	0	64	0.0
4	24	256	9.375
5	144	1024	14.063
6	816	4096	19.922
7	3852	16384	23.511
8	17376	65536	26.514
9	73962	262144	28.214
10	311460	1048576	29.703
11	1269606	4194304	30.270
12	5115750	16777216	30.492

In the case where a partial word has no critical factorization, Algorithm 4.3 exhaustively searches $|w| - \|H(w)\|$ positions for a factorization. Table 4.2 shows the average values for the indices k_0, l_0, k_1, l_1 after the algorithm completes over the same data set. Also, it shows the average values for these indices when partial words without critical factorizations are ignored.

This data shows that if a partial word has a critical factorization, then Algorithm 4.3 discovers it extremely quickly.

Exercises

4.1 Let A be totally ordered by $a \prec b \prec c$. Order the nonempty suffixes of $u = abc\diamond\diamond cac\diamond$ with respect to $\preceq_l$ and with respect to $\preceq_r$. What are the maximal suffixes of u with respect to $\preceq_l$ and with respect to $\preceq_r$?

4.2 Prove Statement 2 of Lemma 4.2.

4.3 $\boxed{\text{s}}$ Prove Statement 2 of Lemma 4.3.

TABLE 4.2: Average values for the indices k_0, l_0, k_1, l_1.

Length	All partial words				Partial words with CFs			
	k_0	l_0	k_1	l_1	k_0	l_0	k_1	l_1
2	0.0	0.0	0.0	0.0	0.0	0.0	0.0	0.0
3	0.0	0.0	0.0	0.0	0.0	0.0	0.0	0.0
4	0.137	0.180	0.105	0.102	0.0	0.0	0.0	0.0
5	0.352	0.377	0.233	0.212	0.017	0.017	0.010	0.010
6	0.617	0.657	0.453	0.394	0.049	0.049	0.033	0.033
7	0.848	0.910	0.651	0.568	0.083	0.081	0.058	0.058
8	1.093	1.181	0.862	0.763	0.123	0.121	0.091	0.090
9	1.297	1.413	1.050	0.945	0.160	0.158	0.121	0.120
10	1.505	1.650	1.242	1.134	0.196	0.194	0.151	0.150
11	1.676	1.848	1.407	1.301	0.229	0.228	0.180	0.179
12	1.834	2.030	1.562	1.460	0.262	0.261	0.209	0.209

4.4 Use Algorithm 4.1 to find a critical factorization of $w = abbcbac$.

4.5 We call a partial word u a *palindrome* if $u = \text{rev}(u)$. Prove that every full palindrome has at least 2 critical factorizations.

4.6 Classify the square at position 4 of $w = abb \diamond \diamond \diamond cbb$. Is it internal? Left-external? Right-external? Left- and right-external?

4.7 Can you build a pword w that contains squares that are internal, left-external, right-external, and left-and right-external?

4.8 Is the factorization $(abb, c \diamond a \diamond cb)$ of $w = abbc \diamond a \diamond cb$ critical? Why or why not?

4.9 $\boxed{s}$ Compute all minimal local periods of $w = acb \diamond cba$. What is the maximum among all minimal local periods? Does w have a critical factorization?

4.10 $\boxed{s}$ There exist unbordered partial words of length at least two that have no critical factorizations. True or False? Justify your answer.

4.11 $\boxed{s}$ Is $w = ccb \diamond ab \diamond ba$ special? Why or why not?

Challenging exercises

4.12 $\boxed{s}$ Prove, using the definitions, that under some restrictions on u and v, the relations $u \preceq_l v$ and $u \preceq_r v$ together define "u is a prefix of v". In other words, if $u \in A^+$ and $v \in W_1(A) \setminus \{\varepsilon\}$, then prove that both $u \preceq_l v$ and $u \preceq_r v$ if and only if $u \in P(v)$.

4.13 Let u, v be nonempty partial words. Show that both $u \preceq_l v$ and $u \preceq_r v$ if and only if $u \in P(v)$ or there exist pwords x, y and $a \in A$ such that $u = \text{pre}(u, v) \diamond x$ and $v = \text{pre}(u, v) a y$ (the first position where u and v differ is a hole in u).

4.14 Prove Theorem 4.1.

4.15 Prove Algorithm 4.1.

4.16 Prove Cases 3 and 4 of Theorem 4.2.

4.17 Are these partial words special? Do they have critical factorizations?

- $baa \diamond bb$
- $aaa \diamond aabaaa$
- $abb \diamond abba$
- $babbabbab \diamond b$
- $babb \diamond ab$
- $ab \diamond aaba$

4.18 Determine integers k, l for which the following partial words are $((k, l))$-special. Do they have critical factorizations?

- $aab \diamond babbabba$
- $aabba \diamond abbababa$
- $b \diamond baabbaab$
- $baababaa \diamond b$
- $baabbba \diamond baa$
- $aaab \diamond babb$

4.19 ⑤ Run Algorithm 4.3 on input $w = a \diamond cbba$ and discuss as in Example 4.7.

4.20 Repeat Exercise 4.19 for $w = ccb \diamond ab \diamond bbcc \diamond$.

4.21 ⑤ Prove Cases 2, 3 and 4 of Theorem 4.3.

Programming exercises

4.22 Write a program that takes as input a partial word w and a total ordering of $\alpha(w)$, and outputs the nonempty suffixes of w with respect to the two orderings $\preceq_l$ and $\preceq_r$. Run your program on $w = abb \diamond \diamond bba \diamond bcabbc$.

4.23 Design an applet that provides an implementation of Algorithm 4.1, that is, given as input a word w of length at least two, the applet outputs a critical factorization for w.

4.24 Write a program that computes all the minimal local periods of a given pword and that classifies them as internal? left-external? right-external? left- and right-external? Run your program on

- $ccb\diamond ab\diamond ba$
- $abb\diamond\diamond\diamond cbb$

4.25 Give pseudo code for Algorithm 4.2.

4.26 Write a program to determine whether or not a given partial word w is special. Your program should also determine whether or not w has a critical factorization. Run your program on the pwords of Exercise 4.17.

Websites

A World Wide Web server interface at

$$\texttt{http://www.uncg.edu/mat/research/cft2}$$

has been established for automated use of Algorithm 4.3. An earlier version of the algorithm, that works only for one hole, was established at

$$\texttt{http://www.uncg.edu/mat/cft}$$

Bibliographic notes

Several versions of the critical factorization theorem on words exist [49, 51, 68, 69, 70, 106, 107]. Section 4.2 discusses the version which appears in [51].

Algorithm 4.1 is from Crochemore and Perrin who showed that a critical factorization can be found very efficiently from the computation of the maximal suffixes of the word with respect to the two total orderings described in Section 4.1: the lexicographic ordering related to a fixed total ordering on the alphabet $\preceq_l$, and the lexicographic ordering obtained by reversing the order of letters in the alphabet $\preceq_r$ [57]. There exist linear time (in the length of the word) algorithms for such computations [57, 58, 114] (the latter two use the suffix tree construction).

In [29], Blanchet-Sadri and Duncan extended the critical factorization theorem to partial words with one hole. In this case, they called a factorization critical if its minimal local period is equal to the minimal weak period of the partial word. It turned out that for partial words, critical factorizations may be avoidable. They described the class of the special partial words with one hole that possibly avoid critical factorizations. They gave a version of the critical factorization theorem for the nonspecial partial words with one hole. By refining the method based on the maximal suffixes with respect to the lexicographic/reverse lexicographic orderings, they gave a version of the critical factorization theorem for the so-called $((k, l))$-nonspecial partial words with one hole. Their proof led to an efficient algorithm which, given a partial word with one hole, outputs a critical factorization when one exists or outputs "no such factorization exists". Lemmas 4.2 and 4.3 as well as Definition 4.2 are from Blanchet-Sadri and Duncan [29].

In [42], Blanchet-Sadri and Wetzler further investigated the relationship between local and global periodicity of partial words. They extended the critical factorization theorem to partial words with an arbitrary number of holes. They characterized precisely the class of partial words that do not admit critical factorizations. They then developed an efficient algorithm which computes a critical factorization when one exists. Sections 4.3, 4.4 and 4.5 are from Blanchet-Sadri and Wetzler [42].

In [57], a new string matching algorithm was presented, which relies on the critical factorization theorem and which can be viewed as an intermediate between the classical algorithms of Knuth, Morris, and Pratt [98], on the one hand, and Boyer and Moore [44], on the other hand. The algorithm is linear in time and uses constant space as the algorithm of Galil and Seiferas [79]. It presents the advantage of being remarkably simple which consequently makes its analysis possible. The critical factorization theorem has found other important applications which include the design of efficient approximation algorithms for the shortest superstring problem [45, 88, 106].

A periodicity theorem on words, which has strong analogies with the critical factorization theorem, and three applications were derived in [115]. There, the authors improved some results motivated by string matching problems [59, 79]. In particular, they improved the upper bound on the number of comparisons in the text processing of the Galil and Seiferas' time-space optimal string matching algorithm [79]. For other recent developments on the critical factorization theorem and on the study of the local periodic structure of words, we refer the reader to [70, 71, 72].

Chapter 5

Guibas and Odlyzko's Theorem

In this chapter, we discuss a fundamental periodicity result on words due to Guibas and Odlyzko which states that for every word u, there exists a "binary equivalent for u," that is, a binary word v of same length as u that has exactly the same set of periods as u. In summary, the following table describes the number of holes and section numbers where the above mentioned result is discussed:

Holes	Sections
0	5.1
1	5.2 and 5.3

5.1   The zero-hole case

In this section, we restrict ourselves to full words. We first state Guibas and Odlyzko's result.

THEOREM 5.1
For every word u over an alphabet A, there exists a word v of length $|u|$ over the alphabet $\{0, 1\}$ such that $\mathcal{P}(v) = \mathcal{P}(u)$.

Example 5.1
If $u = abacbaba$, then the set of periods of u is $\mathcal{P}(u) = \{5, 7, 8\}$. It is easy to see that $v = 01011010$ satisfies the desired properties in the theorem. Note that the existence of a binary equivalent for u is not unique here since 01010010 does the job as well. ▯

We omit the proof of Theorem 5.1 because we will prove a more general result in the next section. We mention though that an elementary short constructive proof exists that is based on a few properties of words. We start with the following property that relates to primitivity.

LEMMA 5.1
Let u be a word over the alphabet $\{0,1\}$. Then $u0$ or $u1$ is primitive.

Example 5.2
Considering the word 01010, if we append 1 we get the word $(01)^3$ that is clearly nonprimitive, but if we append 0 we get the primitive word 010100. ☐

The next three properties the theorem is based on, stated in Lemmas 5.2, 5.3 and 5.4, can be illustated with the word u equal to

$$abacbabacbabacbaba$$

of length $|u| = 18$ and set of periods $\mathcal{P}(u) = \{5, 10, 15, 17, 18\}$. Factorizing u into blocks of length $p(u) = 5$ gives

$$(abacb)(abacb)(abacb)aba$$

with a leftover block of length 3, and continuing further we get

$$(aba(cb))(aba(cb))(aba(cb))aba$$

Setting $v = aba$ and $w = cb$, we can rewrite u as $(vw)^k v$ with $k = 3$. Note that the periods q of u satisfying $q \le |u| - p(u)$ are 5 and 10 which are multiples of $p(u)$, and the ones that satisfy $|u| - p(u) < q < |u|$ are 15 and 17. If we consider 17 for example, it can be written as $q = 17 = (3-1)5 + 7 = (k-1)p(u) + r$ with $|v| = 3 < 7 < 8 = |v| + p(u)$. Note that $r = 7 \in \mathcal{P}(abacbaba) = \mathcal{P}(vwv)$. A similar statement can be said about the period 15.

LEMMA 5.2
Let u be a word over an alphabet A. If q is a period of u satisfying $q \le |u| - p(u)$, then q is a multiple of $p(u)$.

LEMMA 5.3
Let u be a word over an alphabet A with minimal period $p(u)$. Then there are words v, w (possibly $v = \varepsilon$) and a positive integer k such that $u = (vw)^k v$, $w \ne \varepsilon$ and $p(u) = |vw|$.

LEMMA 5.4
Let u be as in Lemma 5.3 with $k > 1$, and let q be such that $|u| - p(u) < q < |u|$. Put $q = (k-1)p(u) + r$ where $|v| < r < |v| + p(u)$. Then $q \in \mathcal{P}(u)$ if and only if $r \in \mathcal{P}(vwv)$.

The above mentioned properties lead to an algorithm that computes a binary equivalent of any given input.

ALGORITHM 5.1

Given as input a word u over an alphabet A, the algorithm computes a word $Bin(u)$ of length $|u|$ over the alphabet $\{0,1\}$ such that $\mathcal{P}(Bin(u)) = \mathcal{P}(u)$.

Find the minimal period $p(u)$ of u.

Step 1: *If $p(u) = |u|$, then output $Bin(u) = 01^{|u|-1}$.*

Step 2: *If $p(u) \neq |u|$, then find words v, w and a positive integer k such that $u = (vw)^k v$, $w \neq \varepsilon$ and $p(u) = |vw|$.*

- *If $k = 1$, then compute $Bin(v)$, find c in the alphabet $\{0,1\}$ such that $Bin(v)1^{|w|-1}c$ is primitive, and output*

$$Bin(u) = Bin(v)1^{|w|-1}c\,Bin(v)$$

- *If $k > 1$, then compute $Bin(vwv) = v'w'v'$ where $|v'| = |v|$ and $|w'| = |w|$ and output*

$$Bin(u) = (v'w')^k v'$$

REMARK 5.1   Note that $Bin(\varepsilon) = \varepsilon$, and if $u \neq \varepsilon$ then $Bin(u)$ begins with 0. ⬚

We give an example.

Example 5.3

Returning to the word $u = abacbabacbabacbaba$, the following depicts the path pursued by the algorithm on u:

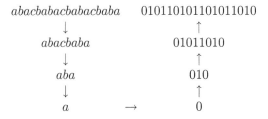

Computations show that

$$Bin(abacbaba) = Bin(aba(cb)aba) = 010(11)010$$

Both *abacbaba* and 01011010 have the periods $5, 7$ and 8 as noticed earlier. Another possible output for *abacbaba* is 01010010 since both 01011 and 01010 are primitive. Now, both $u = (aba(cb))^3 aba$ and $Bin(u) = (010(11))^3 010$ have the periods $5, 10, 15, 17$ and 18. Another possible output for $Bin(u)$ is $(010(10))^3 010$. ⬚

The time complexity of Algorithm 5.1 is stated in the following theorem.

THEOREM 5.2
Given a word u over an alphabet A, a word $Bin(u)$ over the alphabet $\{0,1\}$ with the same length and the same periods of u can be computed in linear time.

5.2 The main result

In this section, we extend Theorem 5.1 to partial words with one hole. We prove that for every partial word u with one hole over an alphabet A, there exists a partial word v of length $|u|$ over the alphabet $\{0,1\}$ such that $H(v) \subset H(u)$, $\mathcal{P}(v) = \mathcal{P}(u)$ and $\mathcal{P}'(v) = \mathcal{P}'(u)$ (Theorem 5.3).

We first define a construction of a word of length n from a given word u of length n over the alphabet $A \cup \{\diamond\}$. Let S be a subset of $\{0, \ldots, n-1\}$ and $a \in A \cup \{\diamond\}$. We define the word $u(S, a)$ as follows:

$$u(S, a)(i) = \begin{cases} u(i) & \text{if } i \notin S \\ a & \text{otherwise} \end{cases}$$

More specifically, $u(S, a)$ is built by replacing all the positions in S by "a."

Example 5.4
Consider the word $u = abb\diamond cbba$ over the alphabet $\{a, b, c, \diamond\}$. We can see that $u(\{0, 3, 4\}, a) = abbaabba$. ▯

If S is the singleton set $\{s\}$, then we will sometimes abbreviate $u(S, a)$ by $u(s, a)$. Throughout this chapter, $\bar{0}$ will denote 1 and $\bar{1}$ will denote 0.

A first step towards our goal is to extend to partial words with one hole the properties of words and periods of Section 5.1. They are Lemmas 5.5, 5.6, 5.7, 5.8 and 5.9 that follow.

LEMMA 5.5
Let u be a partial word with one hole over the alphabet $\{0,1\}$ which is not of the form $x\diamond x$ for any x. Then $u0$ or $u1$ is primitive.

PROOF   Assume that $u0 \subset v^k$, $u1 \subset w^l$ for some primitive words v, w and integers $k, l \geq 2$. Both $|v|$ and $|w|$ are periods of u, and, since $k, l \geq 2$, $|u| = k|v| - 1 = l|w| - 1 \geq 2\max\{|v|, |w|\} - 1 \geq |v| + |w| - 1$.

Case 1. $|u| = |v| + |w| - 1$

Here $|v| = |w|$ and $k = l = 2$. Since v ends with 0 and w with 1, put $v = y0$ and $w = z1$ with $|y| = |z|$. We get $u \subset y0y$ and $u \subset z1z$. We conclude that $u = x \diamond x$ where $x = y = z$, a contradiction.

Case 2. $|u| > |v| + |w| - 1$

By Theorem 3.1, u is also $\gcd(|v|, |w|)$-periodic. However, $\gcd(|v|, |w|)$ divides $|v|$ and $|w|$, and so $u \subset x^m$ with some word x satisfying $|x| = \gcd(|v|, |w|)$ and some integer m. Since v ends with 0 and w with 1, we get that x ends with 0 and 1, a contradiction. ▯

The following lemma gives the structure of the set of weak periods of a partial word with one hole.

LEMMA 5.6
Let u be a partial word with one hole over an alphabet A. If q is a weak period of u satisfying $q \leq |u| - p'(u)$, then q is a multiple of $p'(u)$.

PROOF   See Exercise 3.3. ▯

The following lemma factorizes a partial word with one hole.

LEMMA 5.7
Let u be a partial word with one hole over an alphabet A with minimal weak period $p'(u)$. Then one of the following holds:

1. *There is a positive integer k and there are partial words $v, w_1, w_2, \ldots, w_k$ (possibly $v = \varepsilon$) such that*

$$u = vw_1vw_2 \ldots vw_kv$$

where $p'(u) = |vw_1| = |vw_2| = \cdots = |vw_k|$ and where there exists $1 \leq i \leq k$ such that $w_i = x \diamond y$, $w_j = xay$ if $j < i$, and $w_j = xby$ if $j > i$ for some $a, b \in A$ and $x, y \in A^*$.

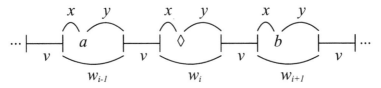

FIGURE 5.1:   Type 1 factorization.

2. *There is a positive integer k and there are partial words $w, v_1, v_2, \ldots, v_{k+1}$ such that*

$$u = v_1 w v_2 w \ldots v_k w v_{k+1}$$

where $p'(u) = |v_1 w| = |v_2 w| = \cdots = |v_k w| = |v_{k+1} w|$, $w \neq \varepsilon$, and where there exists $1 \leq i \leq k+1$ such that $v_i = x \diamond y$, $v_j = xay$ if $j < i$, and $v_j = xby$ if $j > i$ for some $a, b \in A$ and $x, y \in A^*$.

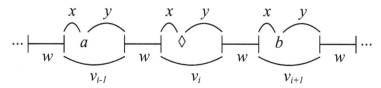

FIGURE 5.2:   Type 2 factorization.

PROOF   Let u be a partial word with one hole over A with minimal weak period $p'(u)$. Then $|u| = kp'(u) + r$ where $0 \leq r < p'(u)$. Put $u = v_1 w_1 v_2 w_2 \ldots v_k w_k v_{k+1}$ where $|v_1 w_1| = |v_2 w_2| = \cdots = |v_k w_k| = p'(u)$ and $|v_1| = |v_2| = \cdots = |v_k| = |v_{k+1}| = r$. Since $p'(u)$ is a weak period of u, $v_i w_i \uparrow v_{i+1} w_{i+1}$ for all $1 \leq i < k$ and $v_k \uparrow v_{k+1}$. Two cases arise.

Case 1. There exists $1 \leq i \leq k$ such that the hole is in w_i.

In this case, $v_1 = v_2 = \cdots = v_k = v_{k+1} = v$ for some possibly empty v. Here we get the situation described in Statement 1.

Case 2. There exists $1 \leq i \leq k+1$ such that the hole is in v_i.

In this case, $w_1 = w_2 = \cdots = w_k = w$ for some nonempty w (if w is empty, then $r = |v_{k+1}| = |v_k| = p'(u)$, a contradiction). Note that $k \geq 1$ (otherwise, $u = v_{k+1}$ and u has weak period $|v_{k+1}| < p'(u)$ contradicting the fact that $p'(u)$ is the minimal weak period of u). Here we get the situation described in Statement 2.   ▯

We illustrate Lemma 5.7 with the following examples.

Example 5.5

If $u = abcdab\diamond dabfd$, then we get the factorization

$$(ab\underline{c}d) \ (ab\diamond d) \ (ab\underline{f}d)$$
$$\quad w_1 \qquad w_2 \qquad w_3$$

where $v = \varepsilon$, $k = 3$, $i = 2$, $x = ab$ and $y = d$. The underlined "c" is the "a" mentioned in Lemma 5.7, while the underlined "f" is the "b" mentioned there. Now,

$$abcdab\underline{f}dab\underline{f}dabcdabcdab\diamond dab\underline{f}dabcdabcdabcdab\underline{f}d$$

can be factorized as

$$\underbrace{(abcdab\underline{f}dabfd)}_{v_1}\ \underbrace{(abcd)}_{w}\ \underbrace{(abcdab\diamond dabfd)}_{v_2}\ \underbrace{(abcd)}_{w}\ \underbrace{(abcdab\underline{c}dabfd)}_{v_3}$$

where $k = 2$, $i = 2$, $x = abcdab$ and $y = dabfd$. $\qquad\Box$

LEMMA 5.8

Let u be as in Lemma 5.7(1) with $k > 1$, and let q be such that $|u| - p'(u) < q < |u|$. Put $q = (k-1)p'(u) + r$ where $|v| < r < |v| + p'(u)$. Also put $H(vw_iv) = \{h\}$. Then $q \in \mathcal{P}(u)$ if and only if $q \in \mathcal{P}'(u)$. Moreover, $q \in \mathcal{P}'(u)$ if and only if the following three conditions hold:

1. *$r \in \mathcal{P}'(vw_iv)$.*

2. *If $i \neq 1$ and $h + r < |v| + p'(u)$, then $(vw_iv)(h + r) = a$.*

3. *If $i \neq k$ and $r \leq h$, then $(vw_iv)(h - r) = b$.*

PROOF For any $0 \leq j < |u| - q = p'(u) + |v| - r$, we have $u(j) = (vw_1v)(j)$ and $u(j+q) = (vw_kv)(j+r)$. Hence $u(j) = u(j+q)$ if and only if $(vw_1v)(j) = (vw_kv)(j+r)$. The latter implies that $q \in \mathcal{P}'(u)$ if and only if Conditions 1–3 hold. To see this, first let us assume that $q \in \mathcal{P}'(u)$ and let $j, j + r \in D(vw_iv)$. We have $j \in D(vw_1v)$ and $j + r \in D(vw_kv)$ and so $j, j + q \in D(u)$. We get $u(j) = u(j+q)$ and so $(vw_iv)(j) = (vw_1v)(j) = (vw_kv)(j+r) = (vw_iv)(j+r)$ showing that Condition 1 holds. To see that Condition 2 holds, note that $h \in D(u)$ and $h + q \in D(u)$. We have $(vw_iv)(h + r) = (vw_kv)(h + r) = u(h + q) = u(h) = (vw_1v)(h) = a$. To see that Condition 3 holds, note that $h - r \in D(u)$ and $h - r + q \in D(u)$. We have $(vw_iv)(h-r) = (vw_1v)(h-r) = u(h - r) = u(h - r + q) = (vw_kv)(h) = b$.

Now, let us show that if Conditions 1–3 hold, then $q \in \mathcal{P}'(u)$. Let $j, j + q \in D(u)$. We get $j \in D(vw_1v)$ and $j + r \in D(vw_kv)$. If $j \notin \{h, h - r\}$, then $j \in D(vw_iv)$ and $j + r \in D(vw_iv)$. In this case, $(vw_1v)(j) = (vw_iv)(j) = (vw_iv)(j+r) = (vw_kv)(j+r)$ since Condition 1 holds, and so $u(j) = u(j+q)$. If $j = h$, then $i \neq 1$ and $j + r \in D(vw_iv)$. In this case, $u(j) = (vw_1v)(j) = a = (vw_iv)(j + r) = (vw_kv)(j + r) = u(j + q)$ since Condition 2 holds. If $j = h - r$, then $i \neq k$ and $j \in D(vw_iv)$. In this case, $u(j) = (vw_1v)(j) = (vw_iv)(j) = b = (vw_kv)(j + r) = u(j + q)$ since Condition 3 holds. $\qquad\Box$

LEMMA 5.9

Let u be as in Lemma 5.7(2) with $k > 1$, and let q be such that $|u| - p'(u) < q < |u|$. Put $q = (k-1)p'(u) + r$ where $|v_i| < r < |v_i| + p'(u)$. Then $q \in \mathcal{P}(u)$ if and only if $q \in \mathcal{P}'(u)$.

1. If $i \neq k+1$ and $H(v_i) = \{h\}$, then $q \in \mathcal{P}'(u)$ if and only if the following two conditions hold:

    (a) $r \in \mathcal{P}'(v_i w v_{i+1})$.

    (b) If $i \neq 1$ and $h + r < |v_i| + p'(u)$, then $(v_i w v_{i+1})(h + r) = a$.

2. If $i \neq 1$ and $H(v_{i-1} w v_i) = \{h\}$, then $q \in \mathcal{P}'(u)$ if and only if the following two conditions hold:

    (a) $r \in \mathcal{P}'(v_{i-1} w v_i)$.

    (b) If $i \neq k+1$ and $r \leq h$, then $(v_{i-1} w v_i)(h - r) = b$.

PROOF   For any $0 \leq j < |u| - q = p'(u) + |v_i| - r$, we have $u(j) = (v_1 w v_2)(j)$ and $u(j+q) = (v_k w v_{k+1})(j+r)$. Hence $u(j) = u(j+q)$ if and only if $(v_1 w v_2)(j) = (v_k w v_{k+1})(j + r)$. The proof is similar to that of Lemma 5.8.
□

Lemma 5.9 is pictured in Figures 5.3 and 5.4.

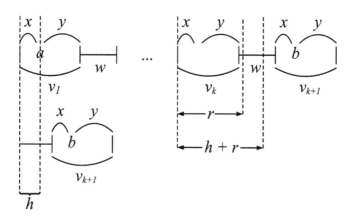

FIGURE 5.3:   The case of Lemma 5.9(1).

Algorithm 5.2, that will be described fully in Section 5.3, works as follows: Let A be an alphabet not containing the special symbol □. Given as input a partial word u with one hole over A where $H(u) = \{h\}$, Algorithm 5.2 computes a triple $T(u) = [\text{Bin}'(u), \alpha_u, \beta_u]$, where $\text{Bin}'(u)$ is a partial word of

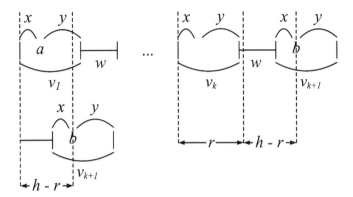

FIGURE 5.4: The case of Lemma 5.9(2).

length $|u|$ over the alphabet $\{0, 1\}$ such that $\mathrm{Bin}'(u)$ does not begin with 1, $H(\mathrm{Bin}'(u)) \subset \{h\}$, where $\mathcal{P}(\mathrm{Bin}'(u)) = \mathcal{P}(u)$ and $\mathcal{P}'(\mathrm{Bin}'(u)) = \mathcal{P}'(u)$, and where

$$\alpha_u = \begin{cases} \square & \text{if } h - p'(u) < 0 \\ u(h - p'(u)) & \text{otherwise} \end{cases}$$

and

$$\beta_u = \begin{cases} \square & \text{if } h + p'(u) \geq |u| \\ u(h + p'(u)) & \text{otherwise} \end{cases}$$

In particular, $T(\diamond) = [0, \square, \square]$, and if $a \in A$ and $k > 1$, then $T(\diamond a^{k-1}) = [0^k, \square, a]$. Moreover, if $\mathcal{P}(u) \neq \mathcal{P}'(u)$, then $H(\mathrm{Bin}'(u)) = \{h\}$ and $\alpha_u = u(h - p'(u)) \neq u(h + p'(u)) = \beta_u$. Also, if $\alpha_u \neq \square$ and $\beta_u \neq \square$, then $H(\mathrm{Bin}'(u)) = \{h\}$.

In summary, the algorithm works as follows:

```
find the minimal weak period p'(u) of u
if p'(u) = |u| then output T(u) = [01^{|u|-1}, □, □]
if p'(u) ≠ |u| then find pwords that satisfy Lemma 5.7
    1.  if the pwords found satisfy Lemma 5.7(1) then
        (a) k = 1
        (b) if k > 1 then compute T(vw_i v) = [Bin'(vw_i v), α, β]
            (i) if i = 1 then
                A. β = □
                B. α = □ and β ≠ □
                C. α ≠ □ and β ≠ □
            (ii) if i = k then
                A. α = □
```

     B. $\alpha \neq \square$ and $\beta = \square$
     C. $\alpha \neq \square$ and $\beta \neq \square$
    (iii) $1 < i < k$ and $a = b$
    (iv) if $1 < i < k$ and $a \neq b$ then
       A. $\alpha \neq \square$ and $\beta = \square$
       B. $\beta \neq \square$ and $x = \varepsilon$
       C. $(\alpha = \square$ and $\beta = \square)$ or $(\beta \neq \square$ and $x \neq \varepsilon)$
2. if the pwords found satisfy Lemma 5.7(2) then
   (a) if $k = 1$ then
     (i) if $v_1 = x \diamond y$ and $v_2 = xby$ then
      compute $T(v_1) = [\text{Bin}'(v_1), \alpha, \beta]$
        A. $\beta = \square$
        B. $\beta \neq \square$
     (ii) if $v_1 = xay$ and $v_2 = x \diamond y$ then
      compute $T(v_2) = [\text{Bin}'(v_2), \alpha, \beta]$
        A. $\alpha = \square$
        B. $\alpha \neq \square$
   (b) if $k > 1$ then
     (i) $i = 1$
     (ii) $i = k + 1$
     (iii) $1 < i < k + 1$ and $a = b$
     (iv) if $1 < i < k + 1$ and $a \neq b$ then
      compute $T(v_i) = [\text{Bin}'(v_i), \alpha, \beta]$
        A.
        B.
        C.
        D.
        E.
        F.

We will prove a series of lemmas that handle different cases of the algorithm, and illustrate them with a few examples. We first concentrate on pwords that satisfy Lemma 5.7(1). Lemma 5.10 deals with $k = 1$ and Lemmas 5.11, 5.12, 5.13 and 5.14 with $k > 1$.

LEMMA 5.10 (Item 1(a))
Let u be as in Lemma 5.7(1) with $k = 1$. Assume that $\text{Bin}(v)$ begins with 0. For $c \in \{0, 1\}$ such that $\text{Bin}(v)1^{|w_1|-1}c$ is primitive, $\mathcal{P}'(u') = \mathcal{P}(u') = \mathcal{P}(u) = \mathcal{P}'(u)$ for the binary word $u' = \text{Bin}(v)1^{|w_1|-1}c\,\text{Bin}(v)$.

PROOF   Put $w_1 = w$. Here u' is a full word and so $\mathcal{P}'(u') = \mathcal{P}(u')$. Also $\mathcal{P}'(u) = \mathcal{P}(u)$ holds since every weak period of u is greater than or equal to $p'(u) = |vw|$. Clearly, $\mathcal{P}(u) \subset \mathcal{P}(u')$, since $\mathcal{P}(\text{Bin}(v)) = \mathcal{P}(v)$ and all periods

q of u satisfy $q \geq p(u) \geq p'(u) = |vw| = |\text{Bin}(v)1^{|w|-1}c|$. Assume then that there exists $q \in \mathcal{P}(u') \setminus \mathcal{P}(u)$ and also that q is minimal with this property. Either $q < |\text{Bin}(v)|$ or $|\text{Bin}(v)| + |w| - 1 \leq q < |u|$, since $\text{Bin}(v)$ does not begin with 1.

If $q < |\text{Bin}(v)|$, then, by the minimality of q, q is the minimal period of u', and Lemma 5.2 implies that $p'(u)$ is a multiple of q, and so $\text{Bin}(v)1^{|w|-1}c$ is not primitive, a contradiction. If $q = |\text{Bin}(v)| + |w| - 1$, then $c = 0$. In this case, if $|w| > 1$, we get $\text{Bin}(v)1 = 0\text{Bin}(v)$, which is impossible, and if $|w| = 1$, we get that $\text{Bin}(v)$ consists of 0's only and $\text{Bin}(v)1^{|w|-1}c = \text{Bin}(v)0$ is not primitive. Therefore $q > |\text{Bin}(v)| + |w| - 1$, and $q > p'(u) = |vw|$ since $p'(u) \notin \mathcal{P}(u') \setminus \mathcal{P}(u)$. Put $q = p'(u) + r$ where $r > 0$. Then r is a period of $\text{Bin}(v)$ and hence of v. But this implies $q \in \mathcal{P}(u)$, a contradiction. ☐

To illustrate Lemma 5.10, we consider the following example.

Example 5.6
Let $u = acac\diamond cbaca$ with $\mathcal{P}(u) = \mathcal{P}'(u) = \{7, 9, 10\}$. We can decompose u as in Lemma 5.7(1) obtaining the factors $v = aca$ and $w_1 = c\diamond cb$. Since $\text{Bin}(v) = \text{Bin}(aca) = 010$, both $c = 0$ and $c = 1$ make $\text{Bin}(v)1^{|w_1|-1}c$ primitive. Thus 0101110010 and 0101111010 have the same periods and weak periods as u. ☐

The following remark will be useful for understanding the next four lemmas.

REMARK 5.2 If u and $q \in \mathcal{P}(u)$ satisfy the assumptions of Lemma 5.8 and $T(vw_iv) = [\text{Bin}'(vw_iv), \alpha, \beta]$, then $r \in \mathcal{P}'(vw_iv)$ and the following hold:

- First, if $h + r < |vw_iv|$, then $h + p'(vw_iv) \leq h + r < |vw_iv|$ and so $\beta \neq \square$. Moreover, $y \neq \varepsilon$ or $v \neq \varepsilon$ (otherwise,

$$|x| + r = |vx| + r = h + r < |vw_iv| = |vx\diamond yv| = |x\diamond| = |x| + 1$$

which leads to a contradiction with the fact that $r > 0$).

- Second, if $r \leq h$, then $p'(vw_iv) \leq r \leq h$ and so $h - p'(vw_iv) \geq 0$ and $\alpha \neq \square$. Moreover, $x \neq \varepsilon$ (otherwise, we get the contradiction $|v| < r \leq h = |vx| = |v|$).

- Third, if $h + r < |vw_iv|$ and $r > h$, then $|x| < |y|$ (otherwise, $h + r > h + h = |vx| + |vx| \geq |vxyv|$ and so $h + r \geq |vx\diamond yv| = |vw_iv|$ a contradiction).

- Fourth, if $h + r \geq |vw_iv|$ and $r \leq h$, then $|x| > |y|$ (otherwise, $|vw_iv| \leq h + r \leq h + h = |vx| + |vx| \leq |vxyv| < |vw_iv|$ a contradiction).

- Fifth, if $h + r < |vw_iv|$ and $\alpha \neq \square$, then $|vw_iv| \geq p'(vw_iv) + r$ (otherwise, $|vw_iv| < p'(vw_iv) + r \leq h + r < |vw_iv|$ a contradiction).

- Sixth, if $r \leq h$ and $\beta \neq \square$, then $|vw_iv| \geq p'(vw_iv) + r$ (otherwise, $|vw_iv| < p'(vw_iv) + r \leq p'(vw_iv) + h$ implying $\beta = \square$ a contradiction).

- Seventh, if $\alpha = \square$ and $\beta \neq \square$, then $|x| < |y|$ (otherwise, $h + p'(vw_iv) > h + h = |vx| + |vx| \geq |vxyv|$ which implies $h + p'(vw_iv) \geq |vw_iv|$ and thus $\beta = \square$ a contradiction).

- Eight, if $\alpha \neq \square$ and $\beta = \square$, then $|x| > |y|$ (otherwise, $h + p'(vw_iv) \leq h + h = |vx| + |vx| \leq |vxyv| < |vw_iv|$ and so $\beta \neq \square$ a contradiction).

$$\square$$

The next four lemmas refer to the following binary values whenever they exist:

$$d_1 = \overline{\mathrm{Bin}'(vw_iv)(h - p'(vw_iv))} \tag{5.1}$$
$$d_2 = \mathrm{Bin}'(vw_iv)(h - p'(vw_iv)) \tag{5.2}$$
$$d_3 = \mathrm{Bin}'(vw_iv)(h + p'(vw_iv)) \tag{5.3}$$
$$d_4 = \overline{\mathrm{Bin}'(vw_iv)(h + p'(vw_iv))} \tag{5.4}$$

They also refer to "T" which means "True" and "F" which means "False."

LEMMA 5.11 (Item 1(b)(i))
Let u be as in Lemma 5.7(1) with $k > 1$. If $T(vw_iv) = [Bin'(vw_iv), \alpha, \beta]$ with $H(Bin'(vw_iv)) \subset H(vw_iv) = \{h\}$ and $i = 1$, then the following hold:

A. *If $\beta = \square$, then put $Bin(vw_kv) = v'w'v'$ where $|v'| = |v|$ and $|w'| = |w_k|$. Then $\mathcal{P}'(u') = \mathcal{P}(u') = \mathcal{P}(u) = \mathcal{P}'(u)$ for the binary word*

$$u' = (v'w')^k v'$$

B. *If $\alpha = \square$ and $\beta \neq \square$, then $|x| < |y|$ and put $Bin'(vw_iv) = v'w'v'$ where $|v'| = |v|$ and $|w'| = |w_i|$. Then $\mathcal{P}'(u') = \mathcal{P}(u') = \mathcal{P}(u) = \mathcal{P}'(u)$ for the binary partial word*

$$u' = (v'w')(h, \diamond)((v'w')(h, d_4))^{k-1} v'$$

C. *Otherwise, put $Bin'(vw_iv) = v'w'v'$ where $|v'| = |v|$ and $|w'| = |w_i|$. Then $\mathcal{P}'(u') = \mathcal{P}(u') = \mathcal{P}(u) = \mathcal{P}'(u)$ for the binary partial word*

$$u' = v'w'((v'w')(h, \bar{d}))^{k-1} v'$$

where d is defined by the following table when $\alpha = \beta$:

$b = \alpha$	$x = \varepsilon$	$y = \varepsilon$	$v = \varepsilon$	$\|x\| < \|y\|$	$\|x\| = \|y\|$	$\|x\| > \|y\|$
T	T	T	T	d_1	d_1	d_1
T	T	T	F	d_1	d_1	d_1
T	T	F	T	d_1	d_1	d_1
T	T	F	F	d_1	d_1	d_1
T	F	T	T	d_1	d_1	d_1
T	F	T	F	d_1	d_1	d_1
T	F	F	T			
T	F	F	F	d_1		d_1
F	T	T	T	d_2	d_2	d_2
F	T	T	F	d_2	d_2	d_2
F	T	F	T	d_2	d_2	d_2
F	T	F	F	d_2	d_2	d_2
F	F	T	T	d_2	d_2	d_2
F	F	T	F	d_2	d_2	d_2
F	F	F	T	d_2	d_2	d_2
F	F	F	F			

and by the following table when $\alpha \neq \beta$:

$b = \alpha$	$x = \varepsilon$	$y = \varepsilon$	$v = \varepsilon$	$\|x\| < \|y\|$	$\|x\| = \|y\|$	$\|x\| > \|y\|$
T	T	T	T	d_1	d_1	d_1
T	T	T	F	d_1	d_1	d_1
T	T	F	T	d_1	d_1	d_1
T	T	F	F	d_1	d_1	d_1
T	F	T	T	d_1	d_1	d_1
T	F	T	F	d_1	d_1	d_1
T	F	F	T	d_1	d_1	d_1
T	F	F	F	d_1		d_1
F	T	T	T	d_2	d_2	d_2
F	T	T	F	d_2	d_2	d_2
F	T	F	T	d_2	d_2	d_2
F	T	F	F	d_2	d_2	d_2
F	F	T	T	d_2	d_2	d_2
F	F	T	F	d_2	d_2	d_2
F	F	F	T	d_2	d_2	d_2
F	F	F	F	d_2	d_2	d_2

unless an entry is empty in which case

$$u' = rev(\,Bin'(\,rev(u)))$$

PROOF   First, let us show that $\mathcal{P}'(u) = \mathcal{P}(u)$. The inclusion $\mathcal{P}(u) \subset \mathcal{P}'(u)$ clearly holds. So let $q \in \mathcal{P}'(u)$. If $q \leq \|u\| - p'(u)$, then q is a multiple of $p'(u)$ by

Lemma 5.6. In this case, since $p'(u) \in \mathcal{P}(u)$, also $q \in \mathcal{P}(u)$. If $q > |u| - p'(u)$, then clearly $q \in \mathcal{P}(u)$.

Now, let us show that $\mathcal{P}(u') = \mathcal{P}(u)$. Obviously, $|u| \in \mathcal{P}(u)$ and $|u| \in \mathcal{P}(u')$.

First, consider q with $q \leq |u| - p'(u)$. If $q \in \mathcal{P}(u)$, then Lemma 5.6 gives that q is a multiple of $p'(u)$, and therefore $q \in \{p'(u), 2p'(u), \ldots, (k-1)p'(u)\}$. We get $q \in \mathcal{P}(u')$ since $p'(u) \in \mathcal{P}(u')$. On the other hand, assume that $q \in \mathcal{P}(u')$. Now, $|u'| = |u| \geq p'(u) + q$, and thus, by Theorem 3.1, $\gcd(p'(u), q) \in \mathcal{P}(u')$. For Statement A, $\gcd(p'(u), q)$ is a period of $v'w'v'$ and hence of vw_kv. So $\gcd(p'(u), q) \in \mathcal{P}(u)$ and since q is a multiple of $\gcd(p'(u), q)$, we also get $q \in \mathcal{P}(u)$. For Statement C, $v'w'v'$ has a hole since $\alpha \neq \square$ and $\beta \neq \square$, and for the nonempty entries $\gcd(p'(u), q)$ is a period of $(v'w')(h, \bar{d})v'$. If $b = \alpha$, then $b = (vw_iv)(h - p'(vw_iv))$ and so $\bar{d} = \bar{d}_1 = \text{Bin}'(vw_iv)(h - p'(vw_iv))$. In this case, $\gcd(p'(u), q)$ is a period of vw_kv. So $\gcd(p'(u), q) \in \mathcal{P}(u)$ and since q is a multiple of $\gcd(p'(u), q)$, we also get $q \in \mathcal{P}(u)$. The case where $b \neq \alpha$ follows similarly.

Second, consider q with $|u| - p'(u) < q < |u|$, and put $q = (k-1)p'(u) + r$ where $|v| < r < p'(u) + |v|$. For Statement A, $q \in \mathcal{P}(u)$ if and only if $r \in \mathcal{P}(vw_kv)$ if and only if $r \in \mathcal{P}(v'w'v')$ if and only if $q \in \mathcal{P}(u')$. For Statements B and C, $\beta \neq \square$ and if $q \in \mathcal{P}(u)$, then the conditions of Lemma 5.8(1,3) hold. Here $r \in \mathcal{P}'(vw_iv)$ and if $r \leq h$, then $(vw_iv)(h - r) = b$.

Case 1. $r \leq h$

By Remark 5.2(2,6), $\alpha \neq \square$ and $x \neq \varepsilon$ and $|vw_iv| \geq p'(vw_iv) + r$. By Lemma 5.6, r is a multiple of $p'(vw_iv)$ and so $b = (vw_iv)(h - r) = (vw_iv)(h - p'(vw_iv)) = \alpha$. For the nonempty entries, we get $(v'w'v')(h - r) = \bar{d}$ since $d = d_1$, and $q \in \mathcal{P}(u')$ by Lemma 5.8.

Case 2. $r > h$

For Statement B, we have $r \in \mathcal{P}(v'w'v')$ and thus $r \in \mathcal{P}'((v'w')(h, \diamond)v')$. For Statement C, we have $r \in \mathcal{P}'(v'w'v')$, and in either case $q \in \mathcal{P}(u')$ by Lemma 5.8.

The cases $r \leq h$ and $r > h$ are handled similarly as above in order to show that if $q \in \mathcal{P}(u')$ then $q \in \mathcal{P}(u)$.

Last, let us show that $\mathcal{P}'(u') = \mathcal{P}'(u)$. Obviously, $|u| \in \mathcal{P}'(u)$ and $|u| \in \mathcal{P}'(u')$. Note that $p'(u) = |vw_1| = \cdots = |vw_k| = |v'w'|$ and so $p'(u) \in \mathcal{P}'(u)$ and $p'(u) \in \mathcal{P}'(u')$.

Consider q with $q \leq |u| - p'(u)$. If $q \in \mathcal{P}'(u)$, then Lemma 5.6 gives that q is a multiple of $p'(u)$, and therefore $q \in \{p'(u), 2p'(u), \ldots, (k-1)p'(u)\}$. We get $q \in \mathcal{P}'(u')$. On the other hand, if $q \in \mathcal{P}'(u')$, then $|u'| = |u| \geq p'(u) + q$, and thus, by Theorem 3.1, $\gcd(p'(u), q) \in \mathcal{P}(u')$. Since $\mathcal{P}(u') = \mathcal{P}(u) \subset \mathcal{P}'(u)$, we get that $\gcd(p'(u), q) \in \mathcal{P}'(u)$. By the minimality of $p'(u)$, we have $\gcd(p'(u), q) = p'(u)$, and therefore $p'(u)$ divides q. We get $q \in \mathcal{P}'(u)$.

Now, consider q with $|u| - p'(u) < q < |u|$, and put $q = (k-1)p'(u) + r$ where $|v| < r < p'(u) + |v|$. We show that $\mathcal{P}'(u) \subset \mathcal{P}'(u')$ (the inclusion $\mathcal{P}'(u') \subset \mathcal{P}'(u)$ is proved similarly). If $q \in \mathcal{P}'(u)$, then $q \in \mathcal{P}(u)$. Since $\mathcal{P}(u) = \mathcal{P}(u')$, we get that $q \in \mathcal{P}(u')$ and hence $q \in \mathcal{P}'(u')$. ☐

Example 5.7
Consider $u = a\diamond cabc$ that is factorized as

$$\underset{w_1}{(a\diamond c)} \; \underset{w_2}{(abc)}$$

according to Lemma 5.7(1) with $v = \varepsilon$, $k = 2 > 1$, $i = 1$, $x = a$ and $y = c$. Here

$$T(vw_iv) = [\text{Bin}'(vw_iv), \alpha, \beta] = [0\diamond 1, a, c]$$

and we have $\alpha = a \neq \square$ and $\beta = c \neq \square$, thus the computation of u' falls into Lemma 5.11(C). The "b" value is b which is not equal to α, $x \neq \varepsilon$, $y \neq \varepsilon$, $v = \varepsilon$ and $|x| = |y|$. Moreover, setting $\text{Bin}'(vw_iv) = v'w'w'$ where $|v'| = |v|$ and $|w'| = |w_i|$ implies that $v' = \varepsilon$ and $w' = 0\diamond 1$. In addition, $\alpha \neq \beta$ and $H(vw_iv) = \{h\} = \{1\}$. The value d is then

$$d_2 = \text{Bin}'(vw_iv)(h - p'(vw_iv)) = (0\diamond 1)(1 - 1) = 0$$

In this case,

$$u' = v'w'((v'w')(h, \bar{0}))^{2-1}v' = 0\diamond 1011$$

We can check that both u and u' have periods 3, 6 and weak periods 3, 6. ▯

LEMMA 5.12 (Item 1(b)(ii))
Let u be as in Lemma 5.7(1) with $k > 1$. If $T(vw_iv) = [\text{Bin}'(vw_iv), \alpha, \beta]$ with $H(\text{Bin}'(vw_iv)) \subset H(vw_iv) = \{h\}$ and $i = k$, then the following hold:

A. *If $\alpha = \square$, then put $\text{Bin}(vw_1v) = v'w'v'$ where $|v'| = |v|$ and $|w'| = |w_1|$. Then $\mathcal{P}'(u') = \mathcal{P}(u') = \mathcal{P}(u) = \mathcal{P}'(u)$ for the binary word*

$$u' = (v'w')^k v'$$

B. *If $\alpha \neq \square$ and $\beta = \square$, then $|x| > |y|$ and put $\text{Bin}'(vw_iv) = v'w'v'$ where $|v'| = |v|$ and $|w'| = |w_i|$. Then $\mathcal{P}'(u') = \mathcal{P}(u') = \mathcal{P}(u) = \mathcal{P}'(u)$ for the binary partial word u' defined as follows. If $v = \varepsilon$ and $y \neq \varepsilon$, then*

$$u' = \text{rev}(\text{Bin}'(\text{rev}(u)))$$

Otherwise

$$u' = ((v'w')(h, d_1))^{k-1}(v'w')(h, \diamond)v'$$

C. *Otherwise, put $\text{Bin}'(vw_iv) = v'w'v'$ where $|v'| = |v|$ and $|w'| = |w_i|$. Then $\mathcal{P}'(u') = \mathcal{P}(u') = \mathcal{P}(u) = \mathcal{P}'(u)$ for the binary partial word*

$$u' = ((v'w')(h, d))^{k-1}v'w'v'$$

where

$$d = \begin{cases} d_3 \ \textit{if } a = \beta \\ d_4 \ \textit{otherwise} \end{cases}$$

PROOF   The proof is similar to that of Lemma 5.11.   ⬜

LEMMA 5.13 (Item 1(b)(iii))
Let u be as in Lemma 5.7(1) with $k > 1$. If $T(vw_iv) = [\,Bin'(vw_iv), \alpha, \beta\,]$ with $H(Bin'(vw_iv)) \subset H(vw_iv) = \{h\}$ and $1 < i < k$ and $a = b$, then put $Bin(vw_1v) = v'w'v'$ where $|v'| = |v|$ and $|w'| = |w_1|$. Then $\mathcal{P}'(u') = \mathcal{P}(u') = \mathcal{P}(u) = \mathcal{P}'(u)$ for the binary partial word

$$u' = (v'w')^{i-1}(v'w')(h, \diamond)(v'w')^{k-i}v'$$

PROOF   The equality $\mathcal{P}'(u) = \mathcal{P}(u)$ is proved as in Lemma 5.11. For the equality $\mathcal{P}(u') = \mathcal{P}(u)$, the case where $q \leq |u| - p'(u)$ is handled similarly as in Lemma 5.12(A). As for the case where $|u| - p'(u) < q < |u|$, $q \in \mathcal{P}(u)$ if and only if $r \in \mathcal{P}(vw_1v)$ if and only if $r \in \mathcal{P}(v'w'v')$ if and only if $q \in \mathcal{P}(u')$. Finally, the equality $\mathcal{P}'(u') = \mathcal{P}'(u)$ is proved as in Lemma 5.11.   ⬜

Figures 5.5 and 5.6 will be useful for understanding Lemma 5.14. In Figure 5.5, if $r > h$, then r is seen to be a period of vw_1v, while in Figure 5.6, if $h + r \geq |vw_iv|$, then r is seen to be a period of vw_kv.

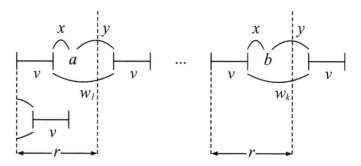

FIGURE 5.5:   The case when $r > h$.

LEMMA 5.14 (Item 1(b)(iv))
Let u be as in Lemma 5.7(1) with $k > 1$. If $T(vw_iv) = [\,Bin'(vw_iv), \alpha, \beta\,]$ with $H(Bin'(vw_iv)) \subset H(vw_iv) = \{h\}$ and $1 < i < k$ and $a \neq b$, then the following hold:

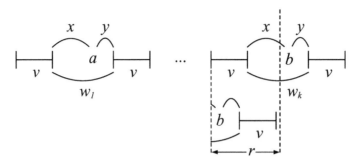

FIGURE 5.6: The case when $h + r \geq |vw_i v|$.

A. If $\alpha \neq \square$ and $\beta = \square$, then <u>put $Bin(vw_k v) = v'w'v'$</u> where $|v'| = |v|$ and $|w'| = |w_k|$, and put $d = \overline{(v'w')(h)}$. Then $\mathcal{P}(u') = \mathcal{P}(u)$ and $\mathcal{P}'(u') = \mathcal{P}'(u)$ for the binary partial word

$$u' = ((v'w')(h,d))^{i-1}(v'w')(h,\diamond)((v'w')(h,\bar{d}))^{k-i}v'$$

B. If $\beta \neq \square$ and $x = \varepsilon$, then put $Bin(vw_1 v) = v'w'v'$ where $|v'| = |v|$ and $|w'| = |w_1|$, and put $d = (v'w')(h)$. Then $\mathcal{P}(u') = \mathcal{P}(u)$ and $\mathcal{P}'(u') = \mathcal{P}'(u)$ for the binary partial word

$$u' = ((v'w')(h,d))^{i-1}(v'w')(h,\diamond)((v'w')(h,\bar{d}))^{k-i}v'$$

C. Otherwise, put $Bin'(vw_i v) = v'w'v'$ where $|v'| = |v|$ and $|w'| = |w_i|$. Then $\mathcal{P}(u') = \mathcal{P}(u)$ and $\mathcal{P}'(u') = \mathcal{P}'(u)$ for the binary partial word

$$u' = ((v'w')(h,d))^{i-1}(v'w')(h,\diamond)((v'w')(h,\bar{d}))^{k-i}v'$$

where $d = 0$ unless $\beta \neq \square$ and $x \neq \varepsilon$.

In the case where $\beta \neq \square$ and $x \neq \varepsilon$, d is defined by the following table if $\alpha = \square$:

$a = \beta$	$|x| < |y|$
T	d_3
F	d_4

and by the following table if $\alpha \neq \square$:

$a=\beta$	$b=\alpha$	$y=\varepsilon$	$v=\varepsilon$	$|x|<|y|$	$|x|=|y|$	$|x|>|y|$
T	T	T	T	d_1	d_1	d_1
T	T	T	F	d_1	d_1	d_1
T	T	F	T	d_1	d_1	d_1
T	T	F	F	d_1	d_1	d_1
T	F	T	T	d_2	d_2	d_2
T	F	T	F	d_2	d_2	d_2
T	F	F	T	d_2	d_2	d_2
T	F	F	F	d_2	d_2	d_2
F	T	T	T	d_2	d_2	d_2
F	T	T	F	d_1	d_1	d_1
F	T	F	T			
F	T	F	F	d_1	d_1	d_1
F	F	T	T	d_2	d_2	d_2
F	F	T	F	d_2	d_2	d_2
F	F	F	T		d_2	d_2
F	F	F	F	d_2	d_2	d_2

unless an entry is empty in which case

$$u' = rev(\,Bin'(\,rev(u)))$$

Example 5.8

The pword $u = abcdab\diamond dabfd$ illustrates Item 1(b)(iv)C of Lemma 5.14.

- The partial words found satisfy Lemma 5.7(1) with $1 < i < k$ and $a \neq b$. Indeed,

$$u = (ab\underline{c}d)(ab\diamond d)(ab\underline{f}d) = w_1 w_2 w_3$$

and the partial words found satisfy Lemma 5.7(1) with $v = \varepsilon$, $k = 3$, $i = 2$, the "a" value c is distinct from the "b" value f.

- And $T(vw_iv) = [Bin'(vw_iv), \alpha, \beta]$ is such that $\alpha = \Box$ and $\beta = \Box$. Indeed, $T(w_2) = [0111, \Box, \Box]$.

In this case,

1. Factorize $Bin'(vw_iv)$ as $v'w'v'$ where $|v'| = |v|$ and $|w'| = |w_i|$. Here $Bin'(w_2) = (\varepsilon)(0111)(\varepsilon)$.

2. Compute h and d. Here $h = 2$ and $d = 0$.

3. Output

$$
\begin{aligned}
u' &= ((v'w')(h, d))^{i-1}(v'w')(h, \diamond)((v'w')(h, \bar{d}))^{k-i}v' \\
&= (0111)(h, 0)(0111)(h, \diamond)(0111)(h, \bar{0}) \\
&= (01\underline{0}1)(01\diamond1)(01\underline{1}1)
\end{aligned}
$$

Both u and u' have only the period 12 and the weak periods 4, 12.

Now, we will concentrate on pwords that satisfy Lemma 5.7(2).

LEMMA 5.15 (Item 2(a)(i)A)
Let u be as in Lemma 5.7(2) with $k = 1$. Assume that $v_1 = x \diamond y$ and $v_2 = xby$, and that $T(v_1) = [Bin'(v_1), \alpha, \beta]$. Also assume that $\beta = \square$. If $c \in \{0,1\}$ is such that $Bin(v_2)1^{|w|-1}c$ is primitive, then $\mathcal{P}'(u') = \mathcal{P}(u') = \mathcal{P}(u) = \mathcal{P}'(u)$ for the binary word

$$u' = Bin(v_2)1^{|w|-1}c\,Bin(v_2)$$

PROOF The proof is very similar to that of Lemma 5.10 and is left as an exercise for the reader.

LEMMA 5.16 (Item 2(a)(i)B)
Let u be as in Lemma 5.7(2) with $k = 1$. Assume that $v_1 = x \diamond y$ and $v_2 = xby$. Assume that $T(v_1) = [Bin'(v_1), \alpha, \beta]$ with $H(Bin'(v_1)) \subset H(v_1) = \{h\}$. Also assume that $\beta \neq \square$. Define d as follows:

$$d = \begin{cases} \overline{Bin'(v_1)(h - p'(v_1))} & \text{if } \alpha \neq \square \text{ and } b = \alpha \\ Bin'(v_1)(h - p'(v_1)) & \text{if } \alpha \neq \square \text{ and } b \neq \alpha \\ 1 & \text{otherwise} \end{cases}$$

1. *If $Bin'(v_1) = 0^{|x|} \diamond 1^{|y|}$, then let $c = 1$.*

2. *Otherwise, if $Bin'(v_1)1^{|w|-1}$ is not of the form $z \diamond z$ for any z, then let $c \in \{0,1\}$ be such that $Bin'(v_1)1^{|w|-1}c$ is primitive.*

3. *Otherwise, if $Bin'(v_1)1^{|w|-1}$ is of the form $z \diamond z$ for some z, then let $c = \bar{d}$.*

Then $\mathcal{P}'(u') = \mathcal{P}(u') = \mathcal{P}(u) = \mathcal{P}'(u)$ for the binary partial word

$$u' = Bin'(v_1)1^{|w|-1}c\,Bin'(v_1)(H(Bin'(v_1)), d)$$

Example 5.9
The pword $u = abcdab\diamond dabf dabcdabcdabcdabfd$ illustrates Item 2(a)(i)B of Lemma 5.16.

- The partial words found satisfy Lemma 5.7(2) with $k = 1$. Indeed,

$$u = (abcdab\diamond dabfd)(abcd)(abcdab\underline{c}dabfd) = v_1 w v_2$$

and the "b" value is c.

- And $T(v_1) = [\text{Bin}'(v_1), \alpha, \beta]$ is such that $\alpha \neq \square$ and $\beta \neq \square$. Indeed, $T(v_1) = [(01\underline{0}1)(01{\diamond}1)(01\underline{1}1), c, f]$ by Example 5.8.

In this case,

1. Compute $\text{Bin}'(v_1)1^{|w|-1}$ which is equal to $010101{\diamond}10111111$ and is not of the form $z{\diamond}z$ for any z.

2. Find "c" in $\{0, 1\}$ such that $\text{Bin}'(v_1)1^{|w|-1}c$ is primitive. Here $c = 1$ works.

3. Compute h and d. Here $H(v_1) = \{h\} = \{6\}$, and

$$d = \text{Bin}'(v_1)(h - p'(v_1)) = (010101{\diamond}10111)(6 - 4) = 0$$

since $\alpha \neq \square$ and the "b" value is α.

4. Output

$$u' = \text{Bin}'(v_1)1^{|w|-1}c\text{Bin}'(v_1)(H(\text{Bin}'(v_1)), d)$$
$$= (010101{\diamond}10111)(1111)(010101{\diamond}10111)(h, d)$$
$$= (010101{\diamond}10111)(1111)(010101\underline{0}10111)$$

Both u and u' have only the periods 16, 20, 28 and the weak periods 16, 20, 28.

□

LEMMA 5.17 (Item 2(a)(ii)A)
Let u be as in Lemma 5.7(2) with $k = 1$. Assume that $v_1 = xay$ and $v_2 = x{\diamond}y$, and that $T(v_2) = [\text{Bin}'(v_2), \alpha, \beta]$. Also assume that $\alpha = \square$. If $c \in \{0, 1\}$ is such that $\text{Bin}(v_1)1^{|w|-1}c$ is primitive, then $\mathcal{P}'(u') = \mathcal{P}(u') = \mathcal{P}(u) = \mathcal{P}'(u)$ for the binary word

$$u' = \text{Bin}(v_1)1^{|w|-1}c\,\text{Bin}(v_1)$$

PROOF   Again the proof is very similar to that of Lemma 5.10 and is left as an exercise.   □

LEMMA 5.18 (Item 2(a)(ii)B)
Let u be as in Lemma 5.7(2) with $k = 1$. Assume that $v_1 = xay$ and $v_2 = x{\diamond}y$. Assume that $T(v_2) = [\text{Bin}'(v_2), \alpha, \beta]$ with $H(\text{Bin}'(v_2)) \subset H(v_2) = \{h\}$, and that $\alpha \neq \square$. Define d as follows:

$$d = \begin{cases} \overline{Bin'(v_2)(h + p'(v_2))} & \text{if } \beta \neq \square \text{ and } a = \beta \\ Bin'(v_2)(h + p'(v_2)) & \text{if } \beta \neq \square \text{ and } a \neq \beta \\ 0 & \text{otherwise} \end{cases}$$

Let $c \in \{0, 1\}$ be such that $Bin'(v_2)(H(Bin'(v_2)), d)1^{|w|-1}c$ is primitive (let $c = 1$ in the case where $Bin'(v_2) = 0^{|x|}\diamond 1^{|y|}$). Then $\mathcal{P}'(u') = \mathcal{P}(u') = \mathcal{P}(u) = \mathcal{P}'(u)$ for the binary partial word

$$u' = Bin'(v_2)(H(Bin'(v_2)), d)1^{|w|-1}c\,Bin'(v_2)$$

PROOF We prove the lemma when $Bin'(v_2)$ has a hole (when $Bin'(v_2)$ is full, the proof is left as an exercise). Note that $Bin'(v_2)(h, d)$ begins with 0 (otherwise, $h = 0$ and $\alpha = \square$). Note also that in the case where $Bin'(v_2) = 0^{|x|}\diamond 1^{|y|}$, we have that $Bin'(v_2)(h, d)1^{|w|-1}c = 0^{|x|}d1^{|y|+|w|}$ is primitive.

As in the proof of Lemma 5.16, $\mathcal{P}'(u) = \mathcal{P}(u)$ and $\mathcal{P}'(u') = \mathcal{P}(u')$. To see that $\mathcal{P}(u) \subset \mathcal{P}(u')$, first note that all periods q of u satisfy $q \geq p'(u) = |v_1 w| = |Bin'(v_2)(h, d)1^{|w|-1}c|$. Clearly $p'(u) \in \mathcal{P}(u)$ and $p'(u) \in \mathcal{P}(u')$. So put $q = p'(u) + r$ with $r > 0$. We get that r is a weak period of v_2 and hence r is a weak period of $Bin'(v_2)$. If $h + r \geq |v_2|$, then $q \in \mathcal{P}(u')$. If $h + r < |v_2|$, then $\beta \neq \square$ since $h + p'(v_2) \leq h + r < |v_2|$ and $|v_2| \geq p'(v_2) + r$ since $\alpha \neq \square$. By Lemma 5.6, r is a multiple of $p'(v_2)$. We have $v_2(h+r) = v_1(h)$ and so $\beta = v_2(h + p'(v_2)) = v_2(h + r) = v_1(h) = a$. In this case, $d = \overline{Bin'(v_2)(h + p'(v_2))}$ and so $Bin'(v_2)(h + r) = Bin'(v_2)(h, d)(h)$ implying $q \in \mathcal{P}(u')$.

To see that $\mathcal{P}(u') \subset \mathcal{P}(u)$, assume that there exists $q \in \mathcal{P}(u')\backslash\mathcal{P}(u)$ and also that q is minimal with this property. Either $q < |v_1|$ or $|v_1|+|w|-1 \leq q < |u|$, since $Bin'(v_2)(h, d)$ does not begin with 1.

If $q < |v_1|$, then, by the minimality of q, q is the minimal period of u', and Lemma 5.6 implies that $p'(u)$ is a multiple of q, and so we get a contradiction with the choice of c.

If $q = |v_1| + |w| - 1$, then $c = 0$. In this case, if $|w| > 1$ and $d = 1$, then we get $Bin'(v_2)(h, 1)1 = 0Bin'(v_2)$, which is impossible. If $|w| > 1$ and $d = 0$, then we get that $Bin'(v_2)$ looks like $0^{|x|}\diamond 1^{|y|}$ and therefore that $c = 1$, a contradiction. If $|w| = 1$ and $d = 1$, then we get an impossible situation. And if $|w| = 1$ and $d = 0$, then we get that $Bin'(v_2)(h, 0)$ consists of 0's only and therefore that $Bin'(v_2)(h, 0)1^{|w|-1}c = Bin'(v_2)(h, 0)0$ is not primitive.

Therefore $q > |v_1| + |w| - 1$, and $q > p'(u)$ since $p'(u) \notin \mathcal{P}(u') \setminus \mathcal{P}(u)$. Put $q = p'(u) + r$ where $r > 0$. We get that r is a weak period of $Bin'(v_2)$ and hence of v_2. If $h + r \geq |v_2|$, then $q \in \mathcal{P}(u)$. If $h + r < |v_2|$, then $\beta \neq \square$ and $|v_2| \geq p'(v_2)+r$. By Lemma 5.6, r is a multiple of $p'(v_2)$. We get $Bin'(v_2)(h+r) = Bin'(v_2)(h, d)(h) = d$ and so $d = Bin'(v_2)(h+r) = Bin'(v_2)(h + p'(v_2))$. In this case, $a = \beta$ and so $v_1(h) = a = \beta = v_2(h+p'(v_2)) = v_2(h+r)$ implying $q \in \mathcal{P}(u)$. $\square$

LEMMA 5.19 (Item 2(b)(i))

Let u be as in Lemma 5.7(2) with $k > 1$. Assume that $i = 1$. Put $Bin'(v_i w v_{i+1}) =$

$v''w'v'$ where $|v'| = |v''|$ and $|w'| = |w|$. Then $P'(u') = P(u') = P(u) = P'(u)$ for the binary partial word

$$u' = v''(w'v')^k$$

PROOF   First, the equality $P'(u) = P(u)$ is proved as in Lemma 5.11.

Second, the equality $P'(u') = P'(u)$ follows as in Lemma 5.11 once the equality $P(u') = P(u)$ is proved.

Third, let us show the equality $P(u') = P(u)$. The case where $q \in P(u)$ with $q \leq |u| - p'(u)$ is proved as in Lemma 5.11. The case where $q \in P(u')$ with $q \leq |u| - p'(u)$ is proved as follows. We have $|u'| = |u| \geq p'(u) + q$, and thus, by Theorem 3.1, $\gcd(p'(u), q) \in P(u')$. We also have that $\gcd(p'(u), q)$ is a period of $v''w'v'$ and hence of $v_i w v_{i+1}$. So $\gcd(p'(u), q) \in P(u)$ and since q is a multiple of $\gcd(p'(u), q)$, we also get $q \in P(u)$.

The case where $|u| - p'(u) < q < |u|$ is proved as follows. Here $q = (k-1)p'(u) + r$ with $|v_i| < r < p'(u) + |v_i|$. By Lemma 5.9(1), $q \in P(u)$ if and only if $r \in P'(v_i w v_{i+1}) = P'(v''w'v')$ which, by Lemma 5.4 or Lemma 5.9(1), is equivalent with $q \in P(u')$. □

LEMMA 5.20 (Item 2(b)(ii))

Let u be as in Lemma 5.7(2) with $k > 1$. Assume that $i = k + 1$. Put $Bin'(v_{i-1} w v_i) = v'w'v''$ where $|v'| = |v''|$ and $|w'| = |w|$. Then $P'(u') = P(u') = P(u) = P'(u)$ for the binary partial word

$$u' = (v'w')^k v''$$

PROOF   The proof is similar to that of Lemma 5.19 but uses Lemma 5.9(2) instead of Lemma 5.9(1). □

LEMMA 5.21 (Item 2(b)(iii))

Let u be as in Lemma 5.7(2) with $k > 1$. Assume that $1 < i < k+1$ and $a = b$. Also assume that $H(v_i) = \{h\}$. Put $Bin(v_1 w v_1) = v'w'v'$ where $|v'| = |v_1|$ and $|w'| = |w|$. Then $P'(u') = P(u') = P(u) = P'(u)$ for the binary partial word

$$u' = (v'w')^{i-1} v'(h, \diamond)(w'v')^{k-i+1}$$

PROOF   The proof is similar to that of Lemma 5.19 except for the case where $q \in P(u)$ with $|u| - p'(u) < q < |u|$ when we want to prove that $q \in P(u')$. Here $q = (k-1)p'(u) + r$ with $|v_i| < r < p'(u) + |v_i|$. In this case $r \in P(v_1 w v_1)$, and hence $r \in P(v'w'v')$ and $r \in P(v'(h, \diamond)w'v')$. If $h + r \geq |v_1 w v_1|$, then $q \in P(u')$ by Lemma 5.9(1). If $h + r < |v_1 w v_1|$, then $(v_i w v_{i+1})(h + r) = (v_1 w v_1)(h + r) = a$ by Lemma 5.9(1)(b). We get

$(v'(h, \diamond)w'v')(h + r) = (v'w'v')(h + r) = (v'w'v')(h) = v'(h)$, and $q \in \mathcal{P}(u')$ by Lemma 5.9(1). $\quad\square$

The following remarks will be useful for understanding the next lemma.

REMARK 5.3 If u satisfies Lemma 5.9(1) and $T(v_i) = [\text{Bin}'(v_i), \alpha, \beta]$, then the following hold:

- If $\alpha = \square$ and $\beta \neq \square$, then $|x| < |y|$ (otherwise, $h + p'(v_i) > h + h = |x| + |x| \geq |xy|$ which implies $h + p'(v_i) \geq |x \diamond y| = |v_i|$ and thus $\beta = \square$ a contradiction).

- If $\alpha \neq \square$ and $\beta = \square$, then $|x| > |y|$ (otherwise, $h + p'(v_i) \leq h + h = |x| + |x| \leq |xy| < |x \diamond y| = |v_i|$ and so $\beta \neq \square$ a contradiction).

$\quad\square$

REMARK 5.4 If u and $q \in \mathcal{P}(u)$ satisfy the assumptions of Lemma 5.9(2) and $T(v_i) = [\text{Bin}'(v_i), \alpha, \beta]$, then $r \in \mathcal{P}'(v_{i-1}wv_i)$ and the following hold:

- If $r \leq h$, then $p'(v_i) \leq p'(v_{i-1}wv_i) \leq r \leq h$ and so $h - p'(v_i) \geq 0$ and $\alpha \neq \square$.

- If $r \leq h$ and $\beta \neq \square$, then $|v_i| \geq p'(v_i) + r$ (otherwise, $|v_i| < p'(v_i) + r \leq p'(v_i) + h$ implying $\beta = \square$ a contradiction).

$\quad\square$

REMARK 5.5 If u and $q \in \mathcal{P}(u)$ satisfy the assumptions of Lemma 5.9(1) and $T(v_iwv_{i+1}) = [\text{Bin}'(v_iwv_{i+1}), \alpha, \beta]$, then $r \in \mathcal{P}'(v_iwv_{i+1})$ and the following hold:

- If $h + r < |v_iwv_{i+1}|$, then $h + p'(v_iwv_{i+1}) \leq h + r < |v_iwv_{i+1}|$ and so $\beta \neq \square$.

- If $h + r < |v_iwv_{i+1}|$ and $\alpha \neq \square$, then $|v_iwv_{i+1}| \geq p'(v_iwv_{i+1}) + r$ (otherwise, $|v_iwv_{i+1}| < p'(v_iwv_{i+1}) + r \leq h + r < |v_iwv_{i+1}|$ a contradiction).

$\quad\square$

REMARK 5.6 If u and $q \in \mathcal{P}(u)$ satisfy the assumptions of Lemma 5.9(2) and $T(v_{i-1}wv_i) = [\text{Bin}'(v_{i-1}wv_i), \alpha, \beta]$, then $r \in \mathcal{P}'(v_{i-1}wv_i)$ and the following hold:

- If $r \leq h$, then $p'(v_{i-1}wv_i) \leq r \leq h$ and so $h - p'(v_{i-1}wv_i) \geq 0$ and $\alpha \neq \square$.

- If $r \leq h$ and $\beta \neq \square$, then $|v_{i-1}wv_i| \geq p'(v_{i-1}wv_i) + r$ (otherwise, $|v_{i-1}wv_i| < p'(v_{i-1}wv_i) + r \leq p'(v_{i-1}wv_i) + h$ implying $\beta = \square$ which is a contradiction).

□

Being inspired by Lemma 5.21, we describe a comparison routine. Consider the pword

$$cedafstced\diamond fstcedbf$$

which can be factorized as

$$\underset{v_1}{(ceda\underline{f})} \; \underset{w}{(st)} \; \underset{v_2}{(ced\diamond f)} \; \underset{w}{(st)} \; \underset{v_3}{(ced\underline{b}f)}$$

where $k = 2$, $i = 2$, $x = ced$ and $y = f$. Imposing the period 16 results in the following alignment:

$$\begin{array}{l} c\,e\,d\,a\,f\,s\,t\,c\,e\,d\,\diamond\,f\,s\,t\,c\,e \\ d\,b\,f \end{array}$$

which implies the equalities $c = d = f$ and $e = b$. So the pword is

$$\underset{v_1}{(cbc\underline{a}c)} \; \underset{w}{(st)} \; \underset{v_2}{(cbc\diamond c)} \; \underset{w}{(st)} \; \underset{v_3}{(cbc\underline{b}c)}$$

which we call u. Using Algorithm 5.1, we get

$$\mathrm{Bin}(v_1wv_3) = (01011)(11)(11010) = v'w'v''$$

where $|v'| = |v_1|$, $|w'| = |w|$, and $|v''| = |v_3|$. Now, create a word v as follows: First, align v' and v'' where the underlined positions are determined by the imposed period:

$$\begin{array}{l} 0\,\underline{1}\,\underline{0}\,1\,1 \\ 1\,1\,\underline{0}\,\underline{1}\,0 \end{array}$$

Then for each column i of the alignment, $v(i)$ is the element of $\{0, 1\}$ that is underlined if any, otherwise it is 1:

$$0\,1\,0\,1\,0$$

Now, the value d is computed as follows: if the position h of $\diamond$ (which is 3) is underlined in v', then $d = v'(h)$ and if it is underlined in v'', then $\overline{d} = v''(h)$. We claim that the pword

$$v(h, d)w'v(h, \diamond)w'v(h, \overline{d})$$

has the same periods and weak periods as u:

$$(010\underline{00})(11)(010\diamond0)(11)(010\underline{1}0)$$

where $\bar{d} = v''(3) = 1$.

LEMMA 5.22 (Item 2(b)(iv))

Let u be as in Lemma 5.7(2) with $k > 1$. Assume that $1 < i < k+1$ and $a \neq b$. Assume that $T(v_i) = [Bin'(v_i), \alpha, \beta]$ with $H(Bin'(v_i)) \subset H(v_i) = \{h\}$. Then $\mathcal{P}(u') = \mathcal{P}(u)$ and $\mathcal{P}'(u') = \mathcal{P}'(u)$ for the binary partial word u' that gets computed according to one of the following as described in the two tables below:

A. *Put $Bin'(v_{i-1}wv_i) = v'w'v''$ where $|v'| = |v''|$ and $|w'| = |w|$, and put $d = v'(h)$. Then*

$$u' = (v'(h,d)w')^{i-1}v'(h,\diamond)(w'v'(h,\bar{d}))^{k-i+1}$$

B. *Put $Bin'(v_iwv_{i+1}) = v''w'v'$ where $|v'| = |v''|$ and $|w'| = |w|$, and put $d = \overline{v'(h)}$. Then*

$$u' = (v'(h,d)w')^{i-1}v'(h,\diamond)(w'v'(h,\bar{d}))^{k-i+1}$$

C. *Put*

$$u' = rev(Bin'(rev(u)))$$

D. *Put $Bin(v_1wv_{k+1}) = v'w'v''$ where $|v'| = |v''|$ and $|w'| = |w|$.*

- *If $k = 2$, then*

$$u' = v'w'\diamond w'v''$$

- *If $k > 2$, then*
    1. *Find first 0 in v''.*
        (a) *If it exists and is within x, then $d = \overline{v''(h)}$. Otherwise $d = v'(h)$.*
        (b) *If it does not exist, then $d = v'(h)$.*
    2. *Build v as follows: $x\diamond$ from v' and y from v''.*

    Put

$$u' = (v(h,d)w')^{i-1}v(h,\diamond)(w'v(h,\bar{d}))^{k-i+1}$$

E. *Put $Bin(v_1wv_{k+1}) = v'w'v''$ where $|v'| = |v''|$ and $|w'| = |w|$.*

- *If $p'(v_1wv_{k+1}) < |v_1w|$, then do X if E-X is the item.*
- *If $p'(v_1wv_{k+1}) = |v_1wv_{k+1}|$, then $v = 01^{|v|-1}$ and $d = 0$ and put*

$$u' = (v(h,d)w')^{i-1}v(h,\diamond)(w'v(h,\bar{d}))^{k-i+1}$$

- *If $|v_1w| < p'(v_1wv_{k+1}) < |v_1wx\diamond|$, then $d = v'(h)$, and if $|v_1wx\diamond| \leq p'(v_1wv_{k+1}) < |v_1wv_{k+1}|$, then $d = \overline{v''(h)}$. In either case,*

$$u' = (v'(h,d)w')^{i-1}v'(h,\diamond)(w'v'(h,\bar{d}))^{k-i+1}$$

F. Put $Bin(v_1wv_{k+1}) = v'w'v''$ where $|v'| = |v''|$ and $|w'| = |w|$.

- If $p'(v_1wv_{k+1}) \leq \frac{|v_1wv_{k+1}|}{2}$ or $p'(v_1wv_{k+1}) < |v_1|$, then do X if F-X is the item.

- If $p'(v_1wv_{k+1}) > \frac{|v_1wv_{k+1}|}{2}$, then

  1. If v' and v'' differ only in the hole position, then
     $$u' = (v'w')^{i-1}v'(h,\diamond)(w'v'')^{k-i+1}$$

  2. If v' and v'' differ in more than the hole position, then

     (a) If $p'(v_1wv_{k+1}) = |v_1wv_{k+1}|$, then $v = 01^{|v|-1}$ and $d = 0$ and put
        $$u' = (v(h,d)w')^{i-1}v(h,\diamond)(w'v(h,\bar{d}))^{k-i+1}$$

     (b) If $|v_1| \leq p'(v_1wv_{k+1}) < |v_1w|$, then do comparison routine to build v and set $d = v(h)$ (align v' and v'' and 0's win). Put
        $$u' = (v(h,d)w')^{i-1}v(h,\diamond)(w'v(h,\bar{d}))^{k-i+1}$$

     (c) If $|v_1w| \leq p'(v_1wv_{k+1}) < |v_1wx\diamond|$, then $d = v'(h)$, and if $|v_1wx\diamond| \leq p'(v_1wv_{k+1}) < |v_1wv_{k+1}|$, then $d = v''(h)$. In either case,
        $$u' = (v'(h,d)w')^{i-1}v'(h,\diamond)(w'v'(h,\bar{d}))^{k-i+1}$$

$\beta = \square$	$x = \varepsilon$	$y = \varepsilon$	$a = \beta$	Item 2(b)(iv) when $\alpha = \square$
T	T	T		D
T	T	F		A
T	F	T		B
T	F	F		B
F	T	T	T	A
F	T	T	F	A
F	T	F	T	E-A
F	T	F	F	F-A
F	F	T	T	A
F	F	T	F	B
F	F	F	T	A
F	F	F	F	E-B

$\beta = \square$	$x = \varepsilon$	$y = \varepsilon$	$a = \beta$	$b = \alpha$	Item 2(b)(iv) when $\alpha \neq \square$
T	T	T		T	A
T	T	T		F	A
T	T	F		T	A
T	T	F		F	A
T	F	T		T	E-C
T	F	T		F	C
T	F	F		T	B
T	F	F		F	B
F	T	T	T	T	A
F	T	T	T	F	A
F	T	T	F	T	A
F	T	T	F	F	A
F	T	F	T	T	A
F	T	F	T	F	A
F	T	F	F	T	A
F	T	F	F	F	A
F	F	T	T	T	B
F	F	T	T	F	B
F	F	T	F	T	B
F	F	T	F	F	B
F	F	F	T	T	B
F	F	F	T	F	E-A
F	F	F	F	T	E-A
F	F	F	F	F	B

Example 5.10

This example illustrates Item 2(b)(iv)B of Lemma 5.22. Consider

$$u = abcdab f dab f dabcdabcdab \diamond dab f dabcdabcdabcdab f d$$

- The partial words found satisfy Lemma 5.7(2) with $1 < i < k+1$ and $a \neq b$. Indeed,

$$(abcdab\underline{f}dabfd)\ (abcd)\ (abcdab\diamond dabfd)\ (abcd)\ (abcdab\underline{c}dabfd)$$
$$\quad v_1 \qquad\qquad w \qquad\quad v_2 \qquad\qquad w \qquad\quad v_3$$

- $T(v_i) = [\mathrm{Bin}'(v_i), \alpha, \beta]$ is such that $\alpha \neq \square$, $\beta \neq \square$, $x \neq \varepsilon$, $y \neq \varepsilon$, the "a" value is β, and the "b" value is α. Indeed, since $T(v_2) = [010101\diamond10111, c, f]$, we have $\alpha = c \neq \square$, $\beta = f \neq \square$, $x = abcdab \neq \varepsilon$, $y = dabfd \neq \varepsilon$, the "a" value is f which is β, and the "b" value is c which is α.

In this case, u falls into Lemma 5.22(B).

1. Compute $H(v_i) = \{6\}$ and so $h = 6$.

2. Compute $\text{Bin}'(v_i w v_{i+1}) = v''w'v'$ where $|v'| = |v''|$ and $|w'| = |w|$. Here $\text{Bin}'(v_2 w v_3) = v''w'v' = (010101\diamond10111)(1111)(010101010111)$.

3. Compute $d = \overline{v'(h)}$ or $d = \overline{v'(h)} = \overline{(010101010111)(6)} = \overline{0} = 1$.

4. Output

$$u' = (v'(h,d)w')^{i-1}v'(h,\diamond)(w'v'(h,\bar{d}))^{k-i+1}$$
$$= v'(h,1)w'v'(h,\diamond)w'v'(h,\bar{1})$$
$$= (010101\underline{1}10111)(1111)(010101\diamond10111)(1111)(010101\underline{0}10111)$$

Both u and $\text{Bin}'(u)$ have only the periods 36, 44 and the weak periods 16, 36, 44.

$\square$

We now state and prove the existence of a binary equivalent for any given pword with one hole.

THEOREM 5.3
For every partial word u with one hole over an alphabet A, there exists a partial word v of length $|u|$ over the alphabet $\{0,1\}$ such that v does not begin with 1, $H(v) \subset H(u)$, $\mathcal{P}(v) = \mathcal{P}(u)$, and $\mathcal{P}'(v) = \mathcal{P}'(u)$.

PROOF   The proof is by induction on $|u|$. For $|u| \leq 3$, the result is obvious. Assume that the result holds for all partial words with one hole of length less than or equal to $n \geq 3$.

First, assume that u is as in Lemma 5.7(1) with $|u| = n + 1$. For $k = 1$, the word $\text{Bin}(v)$ satisfies $\mathcal{P}(\text{Bin}(v)) = \mathcal{P}(v)$. If $\text{Bin}(v) = \varepsilon$, then $v = \varepsilon$ and $u' = 01^{|u|-1}$ satisfies $\mathcal{P}(u') = \mathcal{P}(u)$ and $\mathcal{P}'(u') = \mathcal{P}'(u)$, since, in this case, $\mathcal{P}(u) = \mathcal{P}'(u) = \{|u|\}$. If $\text{Bin}(v) \neq \varepsilon$, then $\text{Bin}(v)$ begins with 0. By Lemma 5.1, there exists $c \in \{0,1\}$ such that $\text{Bin}(v)1^{|w_1|-1}c$ is primitive. By Lemma 5.10, the word $u' = \text{Bin}(v)1^{|w_1|-1}c\text{Bin}(v)$ satisfies $\mathcal{P}(u') = \mathcal{P}(u)$ and $\mathcal{P}'(u') = \mathcal{P}'(u)$. For $k > 1$, the result follows by Lemmas 5.11, 5.12, 5.13, and 5.14. We have $|vw_iv| \leq n$ and, by the inductive hypothesis, there exists a partial word $\text{Bin}'(vw_iv)$ over the alphabet $\{0,1\}$ such that $\text{Bin}'(vw_iv)$ begins with 0 or $\diamond$, $H(\text{Bin}'(vw_iv)) \subset H(vw_iv) = \{h\}$, $\mathcal{P}(\text{Bin}'(vw_iv)) = \mathcal{P}(vw_iv)$, and $\mathcal{P}'(\text{Bin}'(vw_iv)) = \mathcal{P}'(vw_iv)$. Consider for instance the case where $1 < i < k$ and $a \neq b$ and $\alpha_{vw_iv} = \square$ and $\beta_{vw_iv} \neq \square$ and $x \neq \varepsilon$. By the inductive hypothesis, there exist v' and w' over the alphabet $\{0,1\}$ such that $\text{Bin}'(vw_iv) = v'w'v'$, $|v'| = |v|$ and $|w'| = |w_i|$. The partial word

$$u' = ((v'w')(h,d))^{i-1}(v'w')(h,\diamond)((v'w')(h,\bar{d}))^{k-i}v'$$

where d is defined as in Lemma 5.14(C) satisfies the desired properties. In particular, u' begins with 0. To see this, since $x \neq \varepsilon$ we have $h \geq 1$, and since $v'w'v'$ begins with 0 the result follows. The other cases are handled similarly.

Now, assume that u is as in Lemma 5.7(2) with $|u| = n + 1$. For $k = 1$, first say $v_1 = x \diamond y$ and $v_2 = xby$ (here $u = x \diamond ywxby$). We have $|v_1| \le n$ and, by the inductive hypothesis, there exists a partial word $\text{Bin}'(v_1)$ over the alphabet $\{0, 1\}$ such that $\text{Bin}'(v_1)$ begins with 0 or $\diamond$, $H(\text{Bin}'(v_1)) \subset H(v_1)$, $\mathcal{P}(\text{Bin}'(v_1)) = \mathcal{P}(v_1)$, and $\mathcal{P}'(\text{Bin}'(v_1)) = \mathcal{P}'(v_1)$. If $\beta_{v_1} \ne \square$, then Lemma 5.16 shows the existence of binary numbers c and d such that the partial word $u' = \text{Bin}'(v_1)1^{|w|-1}c\text{Bin}'(v_1)(H(\text{Bin}'(v_1)), d)$ satisfies the desired properties. If $\beta_{v_1} = \square$, then the result follows by Lemma 5.15.

Now say $v_1 = xay$ and $v_2 = x \diamond y$ (here $u = xaywx \diamond y$). We have $|v_2| \le n$ and, by the inductive hypothesis, there exists a partial word $\text{Bin}'(v_2)$ over the alphabet $\{0, 1\}$ such that $\text{Bin}'(v_2)$ does not begin with 1, $H(\text{Bin}'(v_2)) \subset H(v_2)$, $\mathcal{P}(\text{Bin}'(v_2)) = \mathcal{P}(v_2)$, and $\mathcal{P}'(\text{Bin}'(v_2)) = \mathcal{P}'(v_2)$. We first consider the case where $\alpha_{v_2} \ne \square$ (here $x \ne \varepsilon$). In this case, $H(\text{Bin}'(v_2)) \ne \{0\}$. For d defined as in Lemma 5.18, by Lemma 5.1 there exists $c \in \{0, 1\}$ such that $\text{Bin}'(v_2)(H(\text{Bin}'(v_2)), d)1^{|w|-1}c$ is primitive (put $c = 1$ if $\text{Bin}'(v_2) = 0^{|x|} \diamond 1^{|y|}$). By Lemma 5.18, the partial word

$$u' = \text{Bin}'(v_2)(H(\text{Bin}'(v_2)), d)1^{|w|-1}c\text{Bin}'(v_2)$$

satisfies the desired properties. The case of $\alpha_{v_2} = \square$ follows from Lemma 5.17.

For $k > 1$, the result follows by Lemmas 5.19, 5.20, 5.21 and 5.22. For the case where $1 < i < k + 1$ and $a \ne b$ for instance, by Lemma 5.22, we have $|v_i| \le n$ and, by the inductive hypothesis, there exists a partial word $\text{Bin}'(v_i)$ over the alphabet $\{0, 1\}$ such that $\text{Bin}'(v_i)$ begins with 0 or $\diamond$, $H(\text{Bin}'(v_i)) \subset H(v_i) = \{h\}$, $\mathcal{P}(\text{Bin}'(v_i)) = \mathcal{P}(v_i)$, and $\mathcal{P}'(\text{Bin}'(v_i)) = \mathcal{P}'(v_i)$. Consider for instance the case where $\alpha_{v_i} \ne \square$, $\beta_{v_i} \ne \square$, $x \ne \varepsilon$, $y \ne \varepsilon$, $a = \beta_{v_i}$ and $b = \alpha_{v_i}$. Then by Lemma 5.22(B), since $|v_iwv_{i+1}| \le n$, by the inductive hypothesis, there exist v', w', and v'' over the alphabet $\{0, 1\}$ such that $\text{Bin}'(v_iwv_{i+1}) = v''w'v'$, $|v'| = |v''|$ and $|w'| = |w|$, $v''w'v'$ begins with 0, $H(v''w'v') \subset H(v_iwv_{i+1}) = \{h\}$, $\mathcal{P}(v''w'v') = \mathcal{P}(v_iwv_{i+1})$, and $\mathcal{P}'(v''w'v') = \mathcal{P}'(v_iwv_{i+1})$. The partial word

$$u' = (v'(h, d)w')^{i-1}v'(h, \diamond)(w'v'(h, \bar{d}))^{k-i+1}$$

where $d = \overline{v'(h)}$ satisfies the desired properties. In particular, u' begins with 0 since $h \ne 0$ and v' begins with 0. ∎

5.3 The algorithm

As a consequence of Theorem 5.3, we provide a linear time algorithm which, given the partial word u, computes the desired binary partial word v. We first describe the algorithm (note that the output may have to be complemented so that it does not begin with 1).

ALGORITHM 5.2

Let A be an alphabet not containing the special symbol $\square$. Given as input a partial word u with one hole over A, put $H(u) = \{h\}$ where $0 \leq h < |u|$. The following algorithm computes a triple $T(u) = [Bin'(u), \alpha_u, \beta_u]$, where $Bin'(u)$ is a partial word of length $|u|$ over the alphabet $\{0,1\}$ such that $Bin'(u)$ does not begin with 1, where $H(Bin'(u)) \subset \{h\}$, where $\mathcal{P}(Bin'(u)) = \mathcal{P}(u)$ and $\mathcal{P}'(Bin'(u)) = \mathcal{P}'(u)$, and where

$$\alpha_u = \begin{cases} \square & \text{if } h - p'(u) < 0 \\ u(h - p'(u)) & \text{otherwise} \end{cases}$$

and

$$\beta_u = \begin{cases} \square & \text{if } h + p'(u) \geq |u| \\ u(h + p'(u)) & \text{otherwise} \end{cases}$$

Moreover, if $\mathcal{P}(u) \neq \mathcal{P}'(u)$, then $H(Bin'(u)) = \{h\}$ and $\alpha_u = u(h - p'(u)) \neq u(h + p'(u)) = \beta_u$. Also, if $\alpha_u \neq \square$ and $\beta_u \neq \square$, then $H(Bin'(u)) = \{h\}$.

Find the minimal weak period $p'(u)$ of u. If $p'(u) = |u|$, then output $T(u) = [01^{|u|-1}, \square, \square]$. If $p'(u) \neq |u|$, then find partial words satisfying Lemma 5.7(1) or Lemma 5.7(2).

1. *If the partial words found satisfy Lemma 5.7(1), then do one of the following:*

    (a) *If $k = 1$, compute $Bin(v)$, find $c \in \{0,1\}$ such that $Bin(v)1^{|w_1|-1}c$ is primitive, and output*
    $$T(u) = [Bin(v)1^{|w_1|-1}c\,Bin(v), \square, \square]$$

    (b) *If $k > 1$, then compute $T(vw_iv) = [Bin'(vw_iv), \alpha, \beta]$ and compute $h' = h - (i-1)p'(u)$. Then do one of the following:*

    　i. *If $i = 1$, then do one of the following:*

    　　A. *If $\beta = \square$, then compute $Bin(vw_kv) = v'w'v'$ where $|v'| = |v|$ and $|w'| = |w_k|$. Then output*
    $$T(u) = [(v'w')^k v', \square, b]$$

    　　B. *If $\alpha = \square$ and $\beta \neq \square$, then compute $Bin'(vw_iv) = v'w'v'$ where $|v'| = |v|$ and $|w'| = |w_i|$. Put $d = Bin'(vw_iv)(h' + p'(vw_iv))$ and output*
    $$T(u) = [(v'w')(h', \diamond)((v'w')(h', \bar{d}))^{k-1}v', \square, b]$$

    　　C. *Otherwise, compute $Bin'(vw_iv) = v'w'v'$ where $|v'| = |v|$ and $|w'| = |w_i|$. Find $d \in \{0,1\}$ according to the tables of Lemma 5.11(C) and output*
    $$T(u) = [v'w'((v'w')(h', \bar{d}))^{k-1}v', \square, b]$$
    unless the corresponding entry in one of the tables is empty in which case output

$$T(u) = [\,rev(\,Bin'(\,rev(u))),\square,b\,]$$

ii. *If $i = k$, then do one of the following:*

   A. *If $\alpha = \square$, then compute $Bin(vw_1v) = v'w'v'$ where $|v'| = |v|$ and $|w'| = |w_1|$. Then output*

$$T(u) = [(v'w')^k v', a, \square]$$

   B. *If $\alpha \neq \square$ and $\beta = \square$, then compute $Bin'(vw_iv) = v'w'v'$ where $|v'| = |v|$ and $|w'| = |w_i|$. If $v = \varepsilon$ and $y \neq \varepsilon$, then output*

$$T(u) = [\,rev(\,Bin'(\,rev(u))), a, \square]$$

   Otherwise, put $d = \overline{Bin'(vw_iv)(h' - p'(vw_iv))}$ and output

$$T(u) = [((v'w')(h',d))^{k-1}(v'w')(h',\diamond)v', a, \square]$$

   C. *Otherwise, compute $Bin'(vw_iv) = v'w'v'$ where $|v'| = |v|$ and $|w'| = |w_i|$. Find $d \in \{0,1\}$ as follows:*

$$d = \begin{cases} \overline{Bin'(vw_iv)(h' + p'(vw_iv))} & \text{if } a = \beta \\ Bin'(vw_iv)(h' + p'(vw_iv)) & \text{if } a \neq \beta \end{cases}$$

   and output

$$T(u) = [((v'w')(h',d))^{k-1}v'w'v', a, \square]$$

iii. *If $1 < i < k$ and $a = b$, then compute $Bin(vw_1v) = v'w'v'$ where $|v'| = |v|$ and $|w'| = |w_1|$. Then output*

$$T(u) = [(v'w')^{i-1}(v'w')(h',\diamond)(v'w')^{k-i}v', a, b]$$

iv. *If $1 < i < k$ and $a \neq b$, then do one of the following:*

   A. *If $\alpha \neq \square$ and $\beta = \square$, then compute $\overline{Bin(vw_kv)} = v'w'v'$ where $|v'| = |v|$ and $|w'| = |w_k|$, and put $d = \overline{(v'w')(h')}$. Then output $T(u)$ which is equal to*

$$[((v'w')(h',d))^{i-1}(v'w')(h',\diamond)((v'w')(h',\bar{d}))^{k-i}v', a, b]$$

   B. *If $\beta \neq \square$ and $x = \varepsilon$, then compute $Bin(vw_1v) = v'w'v'$ where $|v'| = |v|$ and $|w'| = |w_1|$, and put $d = (v'w')(h')$. Then output $T(u)$ which is equal to*

$$[((v'w')(h',d))^{i-1}(v'w')(h',\diamond)((v'w')(h',\bar{d}))^{k-i}v', a, b]$$

   C. *Otherwise, compute $Bin'(vw_iv) = v'w'v'$ where $|v'| = |v|$ and $|w'| = |w_i|$. If $\beta = \square$ or $x = \varepsilon$, then put $d = 0$. Otherwise find $d \in \{0,1\}$ according to the tables of Lemma 5.14(C). Then output $T(u)$ equal to*

$$[((v'w')(h',d))^{i-1}(v'w')(h',\diamond)((v'w')(h',\bar{d}))^{k-i}v', a, b]$$

   unless the corresponding entry in one of the tables is empty in which case output $T(u)$ equal to

$$[\,rev(\,Bin'(\,rev(u))), a, b]$$

2. *If the partial words found satisfy Lemma 5.7(2), then do one of the following:*

   (a) *If $k = 1$, then do one of the following:*

     i. *If $v_1 = x \diamond y$ and $v_2 = xby$, compute $T(v_1) = [Bin'(v_1), \alpha, \beta]$.*

        A. *If $\beta = \square$, then compute $Bin(v_2)$, find $c \in \{0,1\}$ such that $Bin(v_2)1^{|w|-1}c$ is primitive, and output*

$$T(u) = [Bin(v_2)1^{|w|-1}cBin(v_2), \square, b]$$

        B. *If $\beta \neq \square$, then find $d \in \{0,1\}$ as follows:*

$$d = \begin{cases} \overline{Bin'(v_1)(h - p'(v_1))} & \text{if } \alpha \neq \square \text{ and } b = \alpha \\ Bin'(v_1)(h - p'(v_1)) & \text{if } \alpha \neq \square \text{ and } b \neq \alpha \\ 1 & \text{otherwise} \end{cases}$$

           Find $c \in \{0,1\}$ as follows. If $Bin'(v_1) = 0^{|x|} \diamond 1^{|y|}$, then let $c = 1$. Otherwise, if $Bin'(v_1)1^{|w|-1}$ is not of the form $z \diamond z$, then let c be such that $Bin'(v_1)1^{|w|-1}c$ is primitive. Otherwise, let $c = \bar{d}$. Then output

$$T(u) = [Bin'(v_1)1^{|w|-1}cBin'(v_1)(H(Bin'(v_1)), d), \square, b]$$

     ii. *If $v_1 = xay$ and $v_2 = x \diamond y$, compute $T(v_2) = [Bin'(v_2), \alpha, \beta]$.*

        A. *If $\alpha = \square$, then compute $Bin(v_1)$, find $c \in \{0,1\}$ such that $Bin(v_1)1^{|w|-1}c$ is primitive, and output*

$$T(u) = [Bin(v_1)1^{|w|-1}cBin(v_1), a, \square]$$

        B. *If $\alpha \neq \square$, then compute $h' = h - p'(u)$ and find $d \in \{0,1\}$ as follows:*

$$d = \begin{cases} \overline{Bin'(v_2)(h' + p'(v_2))} & \text{if } \beta \neq \square \text{ and } a = \beta \\ Bin'(v_2)(h' + p'(v_2)) & \text{if } \beta \neq \square \text{ and } a \neq \beta \\ 0 & \text{otherwise} \end{cases}$$

           Find $c \in \{0,1\}$: If $Bin'(v_2) = 0^{|x|} \diamond 1^{|y|}$, let $c = 1$. Otherwise, let c be such that $Bin'(v_2)(H(Bin'(v_2)), d)1^{|w|-1}c$ is primitive. Then output

$$T(u) = [Bin'(v_2)(H(Bin'(v_2)), d)1^{|w|-1}cBin'(v_2), a, \square]$$

(b) *If $k > 1$, then do one of the following:*

     i. *If $i = 1$, compute $Bin'(v_i w v_{i+1}) = v''w'v'$ where $|v'| = |v''|$ and $|w'| = |w|$, and output*

$$T(u) = [v''(w'v')^k, \square, b]$$

     ii. *If $i = k+1$, compute $Bin'(v_{i-1} w v_i) = v'w'v''$ where $|v'| = |v''|$ and $|w'| \Leftrightarrow |w|$, and output*

$$T(u) = [(v'w')^k v'', a, \square]$$

     iii. *If $1 < i < k+1$ and $a = b$, compute $h' = h - (i-1)p'(u)$, compute $Bin(v_1 w v_1) = v'w'v'$ where $|v'| = |v_1|$ and $|w'| = |w|$, and output*

$$T(u) = [(v'w')^{i-1}v'(h', \diamond)(w'v')^{k-i+1}, a, b]$$

     iv. *If $1 < i < k+1$ and $a \neq b$, compute $T(v_i) = [Bin'(v_i), \alpha, \beta]$, and compute $h' = h - (i-1)p'(u)$. Then do one of A to F according to the tables of Lemma 5.22:*

A. Compute $Bin'(v_{i-1}wv_i) = v'w'v''$ where $|v'| = |v''|$ and $|w'| = |w|$, and put $d = v'(h')$. Then output
$$T(u) = (v'(h',d)w')^{i-1}v'(h',\diamond)(w'v'(h',\bar{d}))^{k-i+1}, a, b]$$

B. Compute $Bin'(v_iwv_{i+1}) = v''w'v'$ where $|v'| = |v''|$ and $|w'| = |w|$, and put $d = v'(h')$. Then output
$$T(u) = [(v'(h',d)w')^{i-1}v'(h',\diamond)(w'v'(h',\bar{d}))^{k-i+1}, a, b]$$

$\vdots$

REMARK 5.7   In the above algorithm, note that when a value d gets computed according to the tables of Lemma 5.11(C) say, the h occurring in one of the equalities (5.1), (5.2), (5.3) or (5.4) becomes h'.   ⬜

The correctness of the algorithm follows from the proof of Theorem 5.3. We now consider the complexity of the algorithm.

THEOREM 5.4
Given a partial word u with one hole over A, a partial word $Bin'(u)$ with at most one hole over $\{0,1\}$ with the same periods and weak periods of u can be computed by Algorithm 5.2 optimally in linear time.

PROOF   Let us first compute the complexity of the main functions of Algorithm 5.2.

- *Compute the minimal weak period:* Let us consider finding the minimal weak period of a partial word with one hole. A linear pattern matching algorithm can be easily adapted to compute the minimal period of a given word u. Given words v and w, the algorithm finds the leftmost occurrence, if any, of v as a factor of w. The comparisons done are of the type $a \overset{?}{=} b$, for letters a and b. Such an algorithm can be easily adapted to compute $p'(u)$ for a partial word u with one hole by overloading the comparison operator in $a \overset{?}{=} b$ to return all comparisons of the special symbol $\diamond$ with any letter a or b as true. (For example, both $\diamond \overset{?}{=} b$ and $a \overset{?}{=} \diamond$ returns true for all letters a and b in the alphabet A, while $a \overset{?}{=} b$ only returns true if both a and b are the same symbol.) Overloading the operator does not change the time complexity of the algorithm any more than by a constant factor. Thus, the computing of $p'(u)$ can be performed in linear time.

- *Find partial words satisfying Lemma 5.7:* Finding a positive integer k and partial words $v, w_1, w_2, \ldots, w_k$ satisfying Lemma 5.7(1) (respectively, finding a positive integer k and partial words $w, v_1, v_2, \ldots, v_{k+1}$ satisfying Lemma 5.7(2)) is performed in linear time, since we know that $p'(u) = |vw_1| = |vw_2| = \cdots = |vw_k|$ (respectively, $p'(u) = |v_1w| = |v_2w| = \cdots = |v_kw| = |v_{k+1}w|$) from computing the minimal weak period as described above.

- *Test for primitivity:* Primitivity can be tested in linear time for full words as was shown in Exercise 2.21. Indeed, a word u is primitive if and only if $u^2 = xuy$ implies that either $x = \varepsilon$ or $y = \varepsilon$. This part of the algorithm needs to be altered slightly to handle binary partial words with one hole. By far the easiest approach would be to substitute the hole with a 0 and test the new binary full word for primitivity as above. If the new word is primitive, then substitute the hole for 1 and test this new word for primitivity. If both words are primitive, then the binary partial word with one hole is primitive, otherwise it is not. This change in the algorithm increases the time complexity by at most a constant factor.

Algorithm 5.2 also uses Algorithm 5.1, which is linear, for constructing binary images of given words via Bin.

Algorithm 5.2 is recursive, so let us compute the complexity of a single call of the procedure T, say $f(n)$, where n is the length of the current partial word for this call, say u. Let us consider the call related to Item 2(a)(i)A (the other items are handled similarly). There, u satisfies Lemma 5.7(2) with $k = 1$, $v_1 = x\diamond y$ and $v_2 = xby$. Algorithm 5.2 computes the following functions:

1. Compute $p'(u)$.

2. Find partial words satisfying Lemma 5.7.

3. Compute $\text{Bin}(v_2)$.

4. Test for primitivity.

Since every function used in Algorithm 5.2 requires at most linear time, we have shown so far that a single call of T requires $f(n) \in O(n)$ time.[1] More precisely, there is a constant k such that $f(n) \leq kn$, for any $n \geq 0$.

To calculate the time required for the whole algorithm on an input u of length n, we first determine how fast the length of the current partial word decreases from a call to the next call or the next two calls. Let us examine the worst case of Lemma 5.7(2) following path 2(b)(iv)X with X being any subcase. Consider u_1 and u_2 the current partial words for two consecutive calls of T on u, respectively. For instance, for 2(b)(iv)A, we have that $u = v_1wv_2w\ldots v_kwv_{k+1}$, $u_1 = v_i$, and $u_2 = v_{i-1}wv_i$, and consequently $|u_1| < |u_2| \leq \frac{2}{3}|u|$. Therefore, the time required by Algorithm 5.2 to compute $\text{Bin}'(u)$ is at most

---

[1] For $f, g : \mathbb{Z}^+ \to \mathbb{R}$, we say that f is "big oh of g," and write $f \in O(g)$, if there exist $k \in \mathbb{R}^+$ and a positive integer N such that $|f(n)| \leq k|g(n)|$ for all integers $n \geq N$.

$$2\Sigma_{i \geq 0} f((\tfrac{2}{3})^i n) \leq 2\Sigma_{i \geq 0} k(\tfrac{2}{3})^i n \leq 6kn$$

hence it is linear, as claimed. Finally, it is clear that the algorithm is optimal, as the problem requires at least linear time. □

Exercises

5.1 $\boxed{\text{s}}$ Give a constructive proof of Lemma 5.3 by providing an algorithm that given as input a word u over an alphabet A, outputs the desired factorization of u.

5.2 Run Algorithm 5.1 on input $u = abcabcabca$.

5.3 $\boxed{\text{s}}$ If $u = a \diamond bc$, then 4 is the only period and weak period of u. Show that no partial word v with one hole over $\{0, 1\}$ satisfies the desired properties of Theorem 5.3, but the full word 0111 does.

5.4 For $u = abca \diamond ca$, compute $p'(u)$, α_u and β_u. What is $T(u)$?

5.5 Factor the following partial words according to Lemma 5.7:

- $u_1 = abcdabedabcdab \diamond dabedabcdabedabedabcd$
- $u_2 = adabcd \diamond dabcdadabc$

5.6 Show that a partial word u has the same set of periods and the same set of weak periods as $\mathrm{rev}(u)$.

5.7 What does Algorithm 5.2 output for $u = abb \diamond cbb$?

5.8 Using Algorithm 5.2, compute $T(abbc \diamond caabb)$. Which item does this example illustrate?

5.9 What does Algorithm 5.2 output given the pwords with one hole of length three or less? Which items of the algorithm handle these pwords?

5.10 $\boxed{\text{s}}$ Which item of Algorithm 5.2 does $u = adabcd \diamond dabcdadabc$ illustrate?

5.11 Show that the partial word

$$u = ab \diamond dabedabcd$$

illustrates Item 2(a)(i)B of Lemma 5.16. Show your computations as in Example 5.9.

5.12 $\boxed{\text{s}}$ Show that the partial word

$$u = abcdabedabcdab\diamond dabedabcdabedabedabcd$$

illustrates Item 1(b)(iv)C of Algorithm 5.2. What is the output of the algorithm?

5.13 Let z be a partial word with one hole over $\{0, 1\}$. Say $H(z) = \{h\}$ and let $z(h, a)$ be the partial word obtained from z after replacing h by $a \in \{0, 1\}$. What can be said when

     1. $z(h, 1)1 = 0z$?

     2. $z(h, 0)1 = 0z$?

     3. $\boxed{s}$ $z(h, 1)$ is a prefix of $0z$?

     4. $z(h, 0)$ is a prefix of $0z$?

5.14 What is the output of Algorithm 5.2 on input $cadcastc\diamond dcastcbdca$?

Challenging exercises

5.15 Prove Lemma 5.1.

5.16 Let $u \in A^*$ be factorized as in Lemma 5.3 with $k \geq 1$. Suppose that $v'w'v'$ is a binary equivalent of vwv where $|v'| = |v|$ and $|w'| = |w|$. Then show that $\mathcal{P}(u) = \mathcal{P}(u')$ for the binary word $u' = (v'w')^k v'$.

5.17 $\boxed{s}$ Prove Lemma 5.4.

5.18 Prove Theorem 5.1.

5.19 $\boxed{s}$ Prove Theorem 5.2.

5.20 $\boxed{s}$ Let u be a nonempty partial word over an alphabet A with minimal weak period $p'(u)$. Then there exist a positive integer k, (possibly empty) partial words $v_1, v_2, \ldots, v_{k+1}$, and nonempty partial words $w_1, w_2, \ldots, w_k$ such that

$$u = v_1 w_1 v_2 w_2 \ldots v_k w_k v_{k+1}$$

where $p'(u) = |v_1 w_1| = |v_2 w_2| = \cdots = |v_k w_k|$, where $|v_1| = |v_2| = \cdots = |v_k| = |v_{k+1}|$, and where $v_i \uparrow v_{i+1}$ for all $1 \leq i \leq k$, and $w_i \uparrow w_{i+1}$ for all $1 \leq i < k$.

5.21 Prove Lemma 5.9.

5.22 Give examples for all items of Lemma 5.12.

5.23 Show $\mathcal{P}'(u') = \mathcal{P}'(u)$ for Lemma 5.14.

5.24 $\boxed{\text{s}}$ Prove Lemma 5.15.

5.25 $\boxed{\text{H}}$ Prove Lemma 5.16.

5.26 Prove Lemma 5.17.

5.27 Prove Lemma 5.18 when $\text{Bin}'(v_2)$ is full. Why is $\beta = \square$ in this case?

Programming exercises

5.28 Design an applet that provides an implementation of Algorithm 5.1.

5.29 Write a program that finds pwords satisfying Lemma 5.7. What is the output for running the program on

- *abcdabedabcdab⋄dabedabcdabedabedabcd*
- *adabcd⋄dabcdadabc*

5.30 Write a program that when given as input a partial u with one hole, outputs the item number of Algorithm 5.2 u falls into. That is, Item 1(a), 1(b)(i), 1(b)(ii), 1(b)(iii), 1(b)(iv), 2(a)(i), 2(a)(ii), 2(b)(i), 2(b)(ii), 2(b)(iii) or 2(b)(iv).

5.31 Refine your program of Exercise 5.30 to handle A, B, ... items.

5.32 Write a program to look for a partial word u with two holes that has no binary equivalent v satisfying all the following conditions:

- $|v| = |u|$
- $\mathcal{P}(v) = \mathcal{P}(u)$
- $\mathcal{P}'(v) = \mathcal{P}'(u)$
- $H(v) \subset H(u)$

Website

A World Wide Web server interface at

http://www.uncg.edu/mat/AlgBin

has been established for automated use of Algorithm 5.2. Another one has been established at

`http://www.uncg.edu/mat/bintwo`

that takes as input a partial word u with an arbitrary number of holes and that outputs a binary equivalent v that satisfies the conditions $|v| = |u|$, $\mathcal{P}(v) = \mathcal{P}(u)$, $\mathcal{P}'(v) = \mathcal{P}'(u)$ and $H(v) \subset H(u)$ if such binary equivalent exists.

Bibliographic notes

Theorem 5.1 is from Guibas and Odlyzko [82]. Their proof uses properties of correlations and is somewhat complicated. The algorithmic approach of Section 5.1 that includes Lemmas 5.1, 5.2, 5.3, 5.4, Algorithm 5.1, and Theorem 5.2 is from Halava, Harju and Ilie [87]. Sections 5.2 and 5.3 contain a new version of an algorithm from Blanchet-Sadri and Chriscoe [23].

Part III

PRIMITIVITY

Chapter 6

Primitive Partial Words

In this chapter, we study *primitive partial words*. Recall that a full word over a finite alphabet is primitive if it cannot be written as a power of another word. In the case of a partial word, we have the following definition.

DEFINITION 6.1 *A partial word u is* **primitive** *if there exists no word v such that $u \subset v^i$ with $i \geq 2$.*

Recall that in Exercise 1.18 of Chapter 1, a property was stated that nonempty full words can be written as powers of primitive words. Moreover, if u is a nonempty partial word, then there exists a primitive word v and a positive integer i such that $u \subset v^i$. Uniqueness however holds for full words but not for partial words.

In Section 6.1, we describe a linear time algorithm to test primitivity on partial words. The algorithm is based on the combinatorial result that under some condition, a partial word is primitive if and only if it is not compatible with an inside factor of its square. One of the concepts of speciality discussed in Chapter 2, which relates to commutativity on partial words, is foundational in the design of the algorithm.

The number of primitive words of a fixed length over an alphabet of a fixed size is well known and relates to the Möbius function. In Section 6.2, we discuss a formula for the number of primitive full words of length n over an alphabet of size k, and start counting primitive partial words by considering the case of prime length. Section 6.3 contains several definitions and some important general properties of *exact* periods of partial words that are useful for the counting. In Section 6.4, we present a first counting method which consists in first considering all nonprimitive pwords with h holes obtained by replacing h positions in nonprimitive full words with ◇'s, and then subtracting the pwords that have been doubly counted. There, we express in particular the number of primitive partial words with one or two holes of length n over a k-size alphabet in terms of the number of such full words. Section 6.5 discusses a second method. We count nonprimitive partial words of length n with h holes over a k-size alphabet through a constructive method that refines the counting done in the previous sections.

Finally, Section 6.6 extends several well known basic properties on the existence of primitive words to primitive partial words.

6.1   Testing primitivity on partial words

The property of being primitive is testable on a word of n symbols in $O(n)$ time. A linear time algorithm can be based on the combinatorial property that no primitive word u can be an inside factor of uu (see Exercise 2.21). Indeed, u is primitive if and only if u is not a proper factor of uu, that is, $uu = xuy$ implies $x = \varepsilon$ or $y = \varepsilon$. The following proposition shows that the property also holds for partial words with one hole.

PROPOSITION 6.1

Let u be a partial word with one hole. Then u is primitive if and only if $uu \uparrow xuy$ for some partial words x, y implies $x = \varepsilon$ or $y = \varepsilon$.

PROOF   Assume that u is primitive and that $uu \uparrow xuy$ for some nonempty partial words x, y. Since $|x| < |u|$, by Lemma 1.2, there exist nonempty partial words z, v such that $u = zv$, $z \uparrow x$, and $vu \uparrow uy$. Then $zvzv \uparrow xzvy$ yields $vz \uparrow zv$ by simplification. By Lemma 2.5, v and z are contained in powers of a common word, a contradiction with the fact that u is primitive.

Now, assume that $uu \uparrow xuy$ for some partial words x, y implies $x = \varepsilon$ or $y = \varepsilon$. Suppose to the contrary that u is not primitive. Then there exists a nonempty word v and an integer $i \geq 2$ such that $u \subset v^i$. But then $uu \uparrow v^{i-1}uv$, and using our assumption we get $v^{i-1} = \varepsilon$ or $v = \varepsilon$, a contradiction. $\Box$

In the case of partial words with at least two holes, the following holds.

PROPOSITION 6.2

Let u be a partial word with at least two holes.

1. *If $uu \uparrow xuy$ for some partial words x, y implies $x = \varepsilon$ or $y = \varepsilon$, then u is primitive.*

2. *If $uu \uparrow xuy$ for some nonempty partial words x and y satisfying $|x| \leq |y|$, then the following hold:*

    (a) *If $|x| = |y|$, then u is not primitive.*

    (b) *If u is not $(|x|, |y|)$-special, then u is not primitive (it is contained in a power of a word of length $|x|$).*

    (c) *If u is $(|x|, |y|)$-special, then u is not contained in a power of a word of length $|x|$.*

PROOF   Statement 1 follows as in Proposition 6.1. For Statement 2, assume that $uu \uparrow xuy$ for some nonempty partial words x, y. Let u_1 be the prefix of length $|x|$ of u and u_2 be the suffix of length $|y|$ of u ($u = u_1u_2$). The compatibility relation $u_1u_2u_1u_2 \uparrow xu_1u_2y$ yields $u_1u_2 \uparrow u_2u_1$. For Statement 2(a), since $|x| = |y|$, $u_1u_2 \uparrow u_2u_1$ implies $u_1 \uparrow u_2$. By definition, there exists a partial word w such that $u_1 \subset w$ and $u_2 \subset w$. We get $u = u_1u_2 \subset w^2$, and the statement follows. For Statement 2(b), since $u = u_1u_2$ is not $(|u_1|, |u_2|)$-special, by Theorem 2.6, u_1 and u_2 are contained in powers of a common word, showing that u is not primitive. Here, for $0 \le i < |x|$, $seq_{|x|,|y|}(i)$ is 1-periodic with letter a_i for some $a_i \in A \cup \{\diamond\}$. We conclude that u is contained in a power of $a_0a_1 \ldots a_{|x|-1}$. For Statement 2(c), put $|y| = m|x| + r$ where $0 \le r < |x|$. If $r > 0$, then u is obviously not contained in a power of a word of length $|x|$. And if $r = 0$, then there exists $0 \le i < |x|$ such that $seq_{|x|,|y|}(i) = (i, i+|x|, i+2|x|, \ldots, i+m|x|, i)$ contains two positions that are holes of u while $u(i)u(i+|x|)u(i+2|x|) \ldots u(i+m|x|)u(i)$ is not 1-periodic. □

Example 6.1

This example illustrates Proposition 6.2(2(c)). The primitive partial word $u = ab\diamond bbb\diamond b$ is compatible with an inside factor of its square uu as illustrated in the following diagram:

$$a\ b\ \diamond\ b\ b\ b\ \diamond\ b\ a\ b\ \diamond\ b\ b\ b\ \diamond\ b$$
$$a\ b\ \diamond\ b\ b\ b\ \diamond\ b$$

Here u is $(2, 6)$-special since $seq_{2,6}(0) = (0, 2, 4, 6, 0)$ contains the holes 2 and 6 while $u(0)u(2)u(4)u(6)u(0) = a\diamond b\diamond a$ is not 1-periodic. Here, u is not contained in a power of a word of length 2. □

We now give an algorithm for testing whether a partial word is primitive.

Algorithm Primitivity Testing

```
input:   partial word u
output:  primitive (if u is) and nonprimitive (otherwise)
U ← uu
count ← ‖H(u)‖
if count < 2 then
     check compatiblity of u with a substring of U[1..2|u| − 1]
     if successful then
         return nonprimitive
     else
         return primitive
else
     k ← 1 and l ← |u| − 1
     while k ≤ l do
         check compatibility of u with U[k..k + |u|]
         if successful then
```

```
            if u is (k,l)-special and k < l then
                k ← k + 1 and l ← l − 1
            if u is not (k,l)-special or k = l then
                return nonprimitive
      else
            k ← k + 1 and l ← l − 1
return primitive
```

REMARK 6.1   Note that if u is primitive, then its reversal $\operatorname{rev}(u)$ is also primitive. This fact justifies the while loop being for $k \leq l$.   ▯

The following example illustrates our algorithm.

Example 6.2
Consider the partial word $u = a \diamond\diamond aba \diamond$ where $D(u) = \{0, 3, 4, 5\}$ and $H(u) = \{1, 2, 6\}$. The algorithm proceeds as follows:

$k = 1, l = 6$: Compatibility of u with $U[1..8)$ is nonsuccessful.

$k = 2, l = 5$: Compatibility of u with $U[2..9)$ is successful.

$$a \diamond\diamond a\,b\,a \diamond a \diamond\diamond a\,b\,a \diamond$$
$$a \diamond\diamond a\,b\,a \diamond$$

Here, the partial word u is $(2, 5)$-special.

$k = 3, l = 4$: Compatibility of u with $U[3..10)$ is nonsuccessful.

Thus the partial word u is primitive.
   Now, consider the partial word $u = ab\diamond\diamond bc\diamond bc$ where $D(u) = \{0, 1, 4, 5, 7, 8\}$ and $H(u) = \{2, 3, 6\}$. The algorithm proceeds as follows:

$k = 1, l = 8$: Compatibility of u with $U[1..10)$ is nonsuccessful.

$k = 2, l = 7$: Compatibility of u with $U[2..11)$ is nonsuccessful.

$k = 3, l = 6$: Compatibility of u with $U[3..12)$ is successful.

$$a\,b\diamond\diamond b\,c\diamond b\,c\,a\,b\diamond\diamond b\,c\diamond b\,c$$
$$a\,b\diamond\diamond b\,c\diamond b\,c$$

Here, the partial word u is not $(3, 6)$-special and is thus nonprimitive $(u \subset (abc)^3)$.

In conclusion, the following theorem holds.

THEOREM 6.1
The property of being primitive is testable on a partial word of length n in $O(n)$ time.

PROOF   The correctness of our algorithm follows from Propositions 6.1 and 6.2. To see that primitivity can be tested in linear time in the length of a given partial word u, any linear time pattern matching algorithm can be easily adapted to test whether the string u is compatible with an inside substring of uu. The algorithm finds the leftmost occurrence, if any, of a factor of uu, $U[k..k + |u|)$, compatible with u. For a full word u, the comparisons done are of the type $a \stackrel{?}{=} b$, for letters a and b in the alphabet A. For a partial word u, we can overload the comparison operator in $a \stackrel{?}{=} b$ to return all comparisons of the special symbol $\diamond$ with any letter a or b as true. (For example, both $\diamond \stackrel{?}{=} b$ and $a \stackrel{?}{=} \diamond$ returns true for all letters a and b in A, while $a \stackrel{?}{=} b$ only returns true if both a and b are the same symbol.) Overloading the operator does not change the time complexity of the algorithm any more than by a constant factor. Thus, the discovery of the leftmost occurrence, if any, of a substring $U[k..k + |u|)$ compatible with u can be performed in linear time. This part of the algorithm needs to be altered slightly to handle partial words with at least two holes.

Fixing $k > 0$, the following diagram pictures the alignment of u with $U[k..k + |u|)$:

$$
\begin{array}{ccccccccc}
u(0) & u(1) & \dots & u(|u| - k - 1) & u(|u| - k) & u(|u| - k + 1) & \dots & u(|u| - 1) \\
u(k) & u(k + 1) & \dots & u(|u| - 1) & u(0) & u(1) & \dots & u(k - 1)
\end{array}
$$

Now, let $l = |u| - k$. If $k < l$, then the checking of whether or not u is compatible with $U[k..k + |u|)$ can be done simultaneously with the checking of whether or not u is (k, l)-special. Indeed, for any $0 \leq i < k$, consecutive positions in $seq_{k,l}(i)$ turn out to be aligned positions in the above diagram. The algorithm starts by considering $i = 0$ and repeats the following, increasing i until $i = k$ (whenever $i = k$, both u is compatible with $U[k..k + |u|)$ and u is not (k, l)-special). While considering i, the algorithm computes $seq_{k,l}(i) = (i_0, i_1, i_2, \dots, i_{n+1})$ along with its letter *seqletter* initialized with $u(i)$. Whenever the position i_j is added to the sequence, the algorithm compares $u(i_j)$ with $u(i_{j-1})$. If not compatible, then the compatibility of u with $U[k..k + |u|)$ is nonsuccessful and the algorithm increases k by 1 and decreases l by 1. If compatible, then the algorithm updates *seqletter* depending on the value of $u(i_j)$. There are four cases that can arise while updating *seqletter* (here a, b denote distinct letters in A): (1) *seqletter* $= \diamond$ and $u(i_j) = \diamond$ (no up-

date is needed); (2) *seqletter* $= \diamond$ and $u(i_j) = a$ (*seqletter* is updated with a); (3) *seqletter* $= a$ and $u(i_j) = a$ (no update is needed); and (4) *seqletter* $= a$ and $u(i_j) = b$ (here it is discovered that $u(i_0)u(i_1)u(i_2)\ldots u(i_{n+1})$ is not 1-periodic). If any of Cases (1), (2) or (3) occurs and $j < n + 1$, then the algorithm repeats the process by adding the position i_{j+1} to the sequence. If any of Cases (1), (2) or (3) occurs and $j = n+1$, then the algorithm increases i. If Case (4) occurs, then we claim that the algorithm will increase k by 1 and decrease l by 1. To see this, if the number of holes seen so far in the sequence, or *seqholes*, is not less than 2, then u is (k, l)-special and regardless of whether or not u is compatible with $U[k..k + |u|)$, the algorithm will increase k by 1 and decrease l by 1. If *seqholes* < 2, then u is (k, l)-special or u is not compatible with $U[k..k + |u|)$, and again regardless of which case happens, the algorithm will increase k by 1 and decrease l by 1. These changes in the original algorithm increase the time complexity by at most a constant factor.
□

6.2 Counting primitive partial words

We begin the counting of primitive partial words with some notation. Denote by $P_{h,k}(n)$ (respectively, $N_{h,k}(n)$) the number of primitive (respectively, nonprimitive) partial words with h holes of positive length n over a k-size alphabet A. Also, denote by $\mathcal{P}_{h,k}(n)$ (respectively, $\mathcal{N}_{h,k}(n)$) the set of primitive (respectively, nonprimitive) partial words with h holes of length n over A. Let $T_{h,k}(n)$ denote the total number of partial words of length n with h holes over A, and $\mathcal{T}_{h,k}(n)$ the set of all such partial words. The equality

$$P_{h,k}(n) + N_{h,k}(n) = T_{h,k}(n) \tag{6.1}$$

holds and it is easy to see that

$$T_{h,k}(n) = \binom{n}{h} k^{n-h} = \frac{n!}{h!(n-h)!} k^{n-h} \tag{6.2}$$

The partial word $ab\diamond ca\diamond cc$ belongs to $\mathcal{N}_{2,3}(8)$ while the word $ab\diamond ca\diamond ca$ belongs to $\mathcal{P}_{2,3}(8)$.

REMARK 6.2   Note that $P_{0,k}(1) = k$ while $N_{0,k}(1) = 0$. Note in addition that when $h > n$, we have $P_{h,k}(n) = N_{h,k}(n) = 0$. When $n = h + 1 > 1$, we have $P_{h,k}(n) = 0$ and $N_{h,k}(n) = nk$. And when $h = n$, we have $P_{h,k}(n) = 1$ if $n = 1$, and 0 otherwise, and also $N_{h,k}(n) = 0$ if $n = 1$, and 1 otherwise.   □

We first count primitive full words. We start with a definition.

DEFINITION 6.2 *The **Möbius function**, denoted by μ, is a number theoretic function defined by*

$$\mu(n) = \begin{cases} 1 & \text{if } n = 1 \\ (-1)^i & \text{if } n \text{ is a product of } i \text{ distinct primes} \\ 0 & \text{if } n \text{ is divisible by the square of a prime} \end{cases} \quad (6.3)$$

The first few values of μ are $\mu(1) = 1$, $\mu(2) = (-1)^1 = -1$ since 2 is a prime, $\mu(3) = -1$, $\mu(4) = 0$ since $4 = 2^2$, $\mu(5) = -1$, $\mu(6) = (-1)^2 = 1$ since $6 = 2 \times 3$, $\mu(7) = -1$, $\mu(8) = 0$ since $8 = 2^3$, $\mu(9) = 0$, and $\mu(10) = 1$. Notice that

$$\sum_{d|n} \mu(d) = \begin{cases} 1 & \text{if } n = 1 \\ 0 & \text{if } n \geq 2 \end{cases} \quad (6.4)$$

for $n = 1, \ldots, 10$.

Equality 6.4 is actually always true. One can deduce from it that two functions ϕ, ψ from $\mathbb{P}$ to $\mathbb{Z}$ are related by $\sum_{d|n} \psi(d) = \phi(n)$ if and only if $\sum_{d|n} \mu(d)\phi(\frac{n}{d}) = \psi(n)$ (these are left as exercises for the reader). Since there are exactly k^n words of length n over a k-size alphabet and every nonempty word w has a unique primitive root v for which $w = v^{n/d}$ for some divisor d of n, the following relation holds:

$$\sum_{d|n} P_{0,k}(d) = k^n \quad (6.5)$$

Using Equality 6.5 and setting $\phi(n) = k^n$ and $\psi(d) = P_{0,k}(d)$, we obtain the following expression for $P_{0,k}(n)$:

$$P_{0,k}(n) = \sum_{d|n} \mu(d)k^{n/d} \quad (6.6)$$

Example 6.3
Using Equality 6.6, the number of primitive words of length $n = 10$ over an alphabet of size $k = 2$ is

$$P_{0,2}(10) = \sum_{d|10} \mu(d)2^{10/d}$$
$$= \mu(1)2^{10/1} + \mu(2)2^{10/2} + \mu(5)2^{10/5} + \mu(10)2^{10/10}$$
$$= (1)2^{10} + (-1)2^5 + (-1)2^2 + (1)2^1$$
$$= 1024 - 32 - 4 + 2$$
$$= 990$$

☐

We now count primitive partial words of prime length p. If w is a nonprimitive pword with h holes of length p, then w must consist of a string containing

h ◇'s and $p - h$ a's for some letter $a \in A$. In other words, $w \subset a^p$. There are k choices for the letter a and $\binom{p}{h}$ choices for positioning the ◇'s. Thus

$$N_{h,k}(p) = \binom{p}{h} k \qquad (6.7)$$

$$P_{h,k}(p) = T_{h,k}(p) - N_{h,k}(p) = \binom{p}{h}(k^{p-h} - k) \qquad (6.8)$$

The tables below contain some numerical values for alphabets of sizes $k = 2$ and $k = 3$ where prime numbers n are underlined. These tables were obtained by having a computer generate all possible partial words with zero, one, two or three holes, and count the number of primitive and nonprimitive such words.

TABLE 6.1: Values for alphabet of size $k = 2$ and $h \in \{0, 1\}$.

n	$T_{0,2}(n)$	$P_{0,2}(n)$	$N_{0,2}(n)$	$T_{1,2}(n)$	$P_{1,2}(n)$	$N_{1,2}(n)$
1	2	2	0	1	1	0
2	4	2	2	4	0	4
3	8	6	2	12	6	6
4	16	12	4	32	16	16
5	32	30	2	80	70	10
6	64	54	10	192	132	60
7	128	126	2	448	434	14
8	256	240	16	1024	896	128
9	512	504	8	2304	2232	72
10	1024	990	34	5120	4780	340
11	2048	2046	2	11264	11242	22
12	4096	4020	76	24576	23664	912
13	8192	8190	2	53248	53222	26
14	16384	16254	130	114688	112868	1820
15	32768	32730	38	245760	245190	570
16	65536	65280	256	524288	520192	4096
17	131072	131070	2	1114112	1114078	34
18	262144	261576	568	2359296	2349072	10224
19	524288	524286	2	4980736	4980698	38
20	1048576	1047540	1036	10485760	10465040	20720

6.3 Exact periods

In this section, we discuss the concept of *exact period* which will play a role in our counting of primitive partial words. It is defined as follows.

TABLE 6.2: Values for alphabet of size $k = 2$ and $h \in \{2, 3\}$.

n	$T_{2,2}(n)$	$P_{2,2}(n)$	$N_{2,2}(n)$	$T_{3,2}(n)$	$P_{3,2}(n)$	$N_{3,2}(n)$
1	0	0	0	0	0	0
2	1	0	1	0	0	0
3	6	0	6	1	0	1
4	24	4	20	8	0	8
5	80	60	20	40	20	20
6	240	102	138	160	24	136
7	672	630	42	560	490	70
8	1792	1376	416	1792	1088	704
9	4608	4320	288	5376	4716	660
10	11520	10070	1450	15360	11920	3440
11	28160	28050	110	42240	41910	330
12	67584	62760	4824	112640	97920	14720
13	159744	159588	156	292864	292292	572
14	372736	361354	11382	745472	703528	41944
15	860160	856170	3990	1863680	1846470	17210
16	1966080	1936384	29696	4587520	4458496	129024
17	4456448	4456176	272	11141120	11139760	1360
18	10027008	9942408	84600	26738688	26312400	426288
19	22413312	22412970	342	63504384	63502446	1938
20	49807360	49615640	191720	149422080	148333200	1088880

TABLE 6.3: Values for alphabet of size $k = 3$ and $h = 0$.

n	$T_{0,3}(n)$	$P_{0,3}(n)$	$N_{0,3}(n)$
1	3	3	0
2	9	6	3
3	27	24	3
4	81	72	9
5	243	240	3
6	729	696	33
7	2187	2184	3
8	6561	6480	81
9	19683	19656	27
10	59049	58800	249
11	177147	177144	3
12	531441	530640	801
13	1594323	1594320	3
14	4782969	4780776	2193
15	14348907	14348640	267
16	43046721	43040160	6561
17	129140163	129140160	3
18	387420489	387400104	20385
19	1162261467	1162261464	3
20	3486784401	3486725280	59121

TABLE 6.4: Values for alphabet of size $k = 3$ and $h = 1$.

n	$T_{1,3}(n)$	$P_{1,3}(n)$	$N_{1,3}(n)$
1	1	1	0
2	6	0	6
3	27	18	9
4	108	72	36
5	405	390	15
6	1458	1260	198
7	5103	5082	21
8	17496	16848	648
9	59049	58806	243
10	196830	194340	2490
11	649539	649506	33
12	2125764	2116152	9612
13	6908733	6908694	39
14	22320522	22289820	30702
15	71744535	71740530	4005
16	229582512	229477536	104976
17	731794257	731794206	51
18	2324522934	2324156004	366930
19	7360989291	7360989234	57
20	23245229340	23244046920	1182420

TABLE 6.5: Values for alphabet of size $k = 3$ and $h \in \{2, 3\}$.

n	$T_{2,3}(n)$	$P_{2,3}(n)$	$N_{2,3}(n)$	$T_{3,3}(n)$	$P_{3,3}(n)$	$N_{3,3}(n)$
1	0	0	0	0	0	0
2	1	0	1	0	0	0
3	9	0	9	1	0	1
4	54	12	42	12	0	12
5	270	240	30	90	60	30
6	1215	774	441	540	144	396
7	5103	5040	63	2835	2730	105
8	20412	18360	2052	13608	10368	3240
9	78732	77760	972	61236	59022	2214
10	295245	284850	10395	262440	239040	23400
11	1082565	1082400	165	1082565	1082070	495
12	3897234	3847284	49950	4330260	4183632	146628
13	13817466	13817232	234	16888014	16887156	858
14	48361131	48171774	189357	64481508	63805728	675780

DEFINITION 6.3   *We call a partial word w an $\frac{n}{d}$-**repeat** if w is d-periodic and d is a divisor of n distinct from n. In such case, d is called an* **exact period** *of w.*

Example 6.4
To illustrate the definition, consider the partial word $w_1 = ab\diamond\diamond\diamond b$. Aligning w_1 in rows of length 3 and 2, we can see that w_1 is an $\frac{6}{3}$-repeat as well as an $\frac{6}{2}$-repeat. Both 3 and 2 are exact periods.

$$
\begin{array}{cc}
a\ b\ \diamond & a\ b \\
\diamond\ \diamond\ b & \diamond\ \diamond \\
 & \diamond\ b
\end{array}
$$

$\square$

DEFINITION 6.4   *We call the $\diamond$ in position i of partial word w* **free with respect to exact period** d *if whenever $j \in D(w)$, we have $j \not\equiv i \bmod d$.*

Returning to Example 6.4, none of the $\diamond$'s are free with respect to any of the exact periods. Such is not the case with the following example.

Example 6.5
If we consider the partial word $w_2 = bb\diamond\diamond b\diamond$ and align it with respect to two of its exact periods, namely 3 and 2,

$$
\begin{array}{cc}
b\ b & b\ b\ \diamond \\
\diamond\ \diamond & \diamond\ b\ \diamond \\
b\ \diamond &
\end{array}
$$

we see that the $\diamond$'s in positions 2, 3, and 5 are not free with respect to exact period 2, the $\diamond$ in position 3 is not free with respect to exact period 3, but the $\diamond$'s in positions 2 and 5 are free with respect to exact period 3. $\square$

Continuing with some more terminology, let $w = a_0 \ldots a_{n-1}$ where $a_i \in A \cup \{\diamond\}$. We denote by $\mathcal{D}(n)$ the set of divisors of n distinct from n, by $\mathcal{E}(w)$ the set of *exact* periods of w, that is,

$$\mathcal{E}(w) = \{d \mid d \in \mathcal{P}(w) \text{ and } d \in \mathcal{D}(n)\}$$

and by $\mathcal{R}(w)$ the reduced set of exact periods of w, that is,

$$\mathcal{R}(w) = \{d \mid d \in \mathcal{E}(w) \text{ and there exists no } d' \in \mathcal{E}(w) \cap \mathcal{D}(d)\}$$

If d is an exact period of w, we set

$$B_d(i) = \{a_j \mid 0 \le j < n \text{ and } j \equiv i \bmod d\}$$

Now, assume that w is nonprimitive, and let $i_1 < i_2 < \cdots < i_h$ be the elements in $H(w)$. Suppose that w has exact period d and has no free $\diamond$'s with respect to d. Note that for all j, $B_d(i_j) = \{\diamond, b_{i_j}\}$ for some $b_{i_j} \in A$. We define the function f_d as $f_d(i_1, i_2, \ldots, i_h) = (b_{i_1}, b_{i_2}, \ldots, b_{i_h})$. We also define the function f with domain $\mathcal{E}(w)$ where $d \mapsto f_d(i_1, i_2, \ldots, i_h)$, and set $\nu(w) = \|f(\mathcal{E}(w))\|$.

Returning to Example 6.4, Figure 6.1 depicts the mapping f for $w = ab\diamond\diamond\diamond b$. Here, $\nu(w) = 2$.

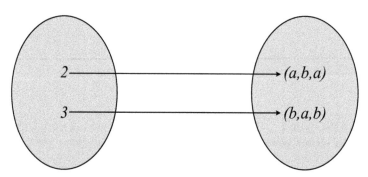

FIGURE 6.1:   Mapping f for $w = ab\diamond\diamond\diamond b$.

LEMMA 6.1

Let w be a nonprimitive partial word that has no free $\diamond$'s with respect to any of its exact periods. Then $\nu(w) = \|\mathcal{R}(w)\|$.

PROOF   Let $i_1 < i_2 < \cdots < i_h$ be the elements in $H(w)$. It is easy to see that $\nu(w) \leq \|\mathcal{R}(w)\|$ because $(i_1, i_2, \ldots, i_h)$ will get mapped to the same h-tuple under both f_d and f_{md} for all integers $m \geq 1$. We now show that f is one-to-one on $\mathcal{R}(w)$. Suppose not, and let $p, q \in \mathcal{R}(w)$ satisfy both $p < q$ and $f_p(i_1, i_2, \ldots, i_h) = f_q(i_1, i_2, \ldots, i_h)$. The bound given by Fine and Wilf's Theorem 3.1 satisfies $p + q - \gcd(p, q) \leq \frac{n}{3} + \frac{n}{2} - \gcd(p, q) = \frac{5n}{6} - \gcd(p, q) < n$ which implies that any full word of length n with exact periods p, q will also have $\gcd(p, q)$ as an exact period. If we now replace in w the hole in position i_j by b_{i_j} for all j, we obtain a full word w' that has exact periods p, q and thus period $\gcd(p, q)$, and so $\gcd(p, q)$ is also a period of w. Thus $q \in \mathcal{E}(w)$ and $\gcd(p, q) \in \mathcal{E}(w) \cap \mathcal{D}(q)$ implying that $q \notin \mathcal{R}(w)$ which leads to a contradiction. Since f is one-to-one on $\mathcal{R}(w)$, it follows that $\nu(w) \geq \|\mathcal{R}(w)\|$ and thus $\nu(w) = \|\mathcal{R}(w)\|$.   ☐

Considering again Figure 6.1 of the mapping f for $w = ab\diamond\diamond\diamond b$, $\|\mathcal{R}(w)\| = \|\{2, 3\}\| = 2$ which coincides with $\nu(w) = 2$ already calculated.

In the sequel, given a pword $w \in \mathcal{N}_{h,k}(n)$ with no free $\diamond$'s with respect to any of its exact periods, the parameter $\nu(w)$ will play an important role. We will obtain words $w' \in \mathcal{N}_{0,k}(n)$ by replacing the $\diamond$'s in w with the corresponding assignments under all possible exact periods of w. Distinct w''s in $\mathcal{N}_{0,k}(n)$ can generate the same $w \in \mathcal{N}_{h,k}(n)$ whenever $\|\mathcal{R}(w)\| > 1$. Indeed, looking again at Figure 6.1 where $w = ab\diamond\diamond\diamond b$ has no free $\diamond$'s but satisfies $\|\mathcal{R}(w)\| > 1$, we see that w is a nonprimitive partial word with 3 holes obtained by replacing 3 positions with $\diamond$'s in *2* (which equals $\nu(w)$) nonprimitive full words: $w'_1 = (ab)^3 = ab\underline{ab}ab$ comes from $2 \mapsto (a, b, a)$ and $w'_2 = (abb)^2 = abb\underline{ab}b$ comes from $3 \mapsto (b, a, b)$.

The following lemma relates to the computation of $\nu(w)$ for nonprimitive pwords w with one hole.

LEMMA 6.2
If $w \in \mathcal{N}_{1,k}(n)$, then $\nu(w) = 1$.

PROOF   Since w is a nonprimitive partial word with one hole, it has no free $\diamond$'s with respect to any of its exact periods. By Lemma 6.1, we have $\nu(w) = \|\mathcal{R}(w)\|$. Now, let $p, q \in \mathcal{R}(w)$ satisfy $p < q$. Since p, q are exact periods, we have $p + q \leq \frac{n}{3} + \frac{n}{2} = \frac{5n}{6} < n$ and Theorem 3.1(2) implies that w has $\gcd(p, q)$ as period. But since $q \in \mathcal{E}(w)$, we have that $\gcd(p, q) \in \mathcal{E}(w)$. Since $\gcd(p, q) \neq q$ and $\gcd(p, q)$ divides q, we get a contradiction with the fact that $q \in \mathcal{R}(w)$.   ⬜

For nonprimitive pwords w with two holes, we have the following.

LEMMA 6.3
If $w \in \mathcal{N}_{2,k}(n)$, then $\|\mathcal{R}(w)\| = 1$. As a consequence, if n is odd, then $\nu(w) = 1$.

PROOF   Theorem 3.5 for two holes gives the optimal bound for the length of w given $p, q \in \mathcal{R}(w)$, that is, $L_{(2,p,q)} = 2p + q - \gcd(p, q)$ with $p < q$. Since p, q are exact periods, we have $p, q \leq \frac{n}{2}$.

- If $p \leq \frac{n}{4}$ and $q \leq \frac{n}{2}$, then $L_{(2,p,q)} \leq \frac{2n}{4} + \frac{n}{2} - \gcd(p, q) = n - \gcd(p, q) < n$.

- If $p = \frac{n}{3}$ and $q = \frac{n}{2}$, then $L_{(2,\frac{n}{3},\frac{n}{2})} = \frac{2n}{3} + \frac{n}{2} - \gcd(\frac{2n}{6}, \frac{3n}{6}) = \frac{7n}{6} - \frac{n}{6} = n$.

Thus $L_{(2,p,q)} \leq n$ and Theorem 3.5 implies that w has $\gcd(p, q)$ as period. Again we get a contradiction as in Lemma 6.2.

Now, if w is a nonprimitive partial word with two holes of odd length n, then w has no free $\diamond$'s with respect to any of its exact periods. By Lemma 6.1, we have $\nu(w) = \|\mathcal{R}(w)\|$.   ⬜

If $w \in \mathcal{N}_{2,k}(n)$ and n is even and w has two free $\diamond$'s with respect to $d = \frac{n}{2}$, then w has no free $\diamond$'s with respect to any of its exact periods distinct from $\frac{n}{2}$. The smallest exact period being a divisor of $\frac{n}{2}$ by Lemma 6.3, in such case, we also define $\nu(w) = 1$.

6.4   First counting method

In this section, we first consider all nonprimitive pwords with h holes obtained by replacing h positions in nonprimitive full words with $\diamond$'s, and then subtract the pwords that have been doubly counted. In particular, we express $N_{1,k}(n)$ and $N_{2,k}(n)$ in terms of $N_{0,k}(n)$.

Let $w = a_0 a_1 \ldots a_{n-1}$ be a full word of length n over an alphabet A of size k. Let $0 \leq i_1 < i_2 < \cdots < i_h < n$ and denote by $w_{i_1,\ldots,i_h}$ the partial word built from w by replacing positions $i_1, \ldots, i_h$ with $\diamond$'s. Setting

$$\mathcal{S}_h(w) = \{w_{i_1,\ldots,i_h} \mid 0 \leq i_1 < i_2 < \cdots < i_h < n\}$$

we say that w generates each element in $\mathcal{S}_h(w)$. For any set X of partial words, we denote by $\mathcal{N}(X)$ the set of nonprimitive pwords in X, that is,

$$\mathcal{N}(X) = \{w \mid w \text{ is nonprimitive and } w \in X\}$$

LEMMA 6.4
If $w \in \mathcal{N}_{0,k}(n)$, then $\mathcal{S}_h(w) \subset \mathcal{N}_{h,k}(n)$.

PROOF   Since $w \in \mathcal{N}_{0,k}(n)$, there exists a word v such that $w = v^i$ for some $i \geq 2$. If $0 \leq i_1 < \cdots < i_h < n$, then $w_{i_1,\ldots,i_h} \subset w = v^i$. It follows that $\mathcal{S}_h(w) \subset \mathcal{N}_{h,k}(n)$.   □

Denote by $\mathcal{W}_{h,k}(n)$ the set of all nonprimitive partial words with h holes of length n over A obtained by replacing any h positions with $\diamond$'s in nonprimitive full words of length n over A. The following holds:

$$\mathcal{W}_{h,k}(n) = \bigcup_{w \in \mathcal{N}_{0,k}(n)} \mathcal{N}(\mathcal{S}_h(w)) = \bigcup_{w \in \mathcal{N}_{0,k}(n)} \mathcal{S}_h(w)$$

Obviously,

$$\|\mathcal{W}_{h,k}(n)\| \leq \binom{n}{h} N_{0,k}(n)$$

The following lemma states that, given w a full primitive word, the nonprimitive partial word obtained by replacing h positions in w with $\diamond$'s must be in $\mathcal{S}_h(v)$ for some nonprimitive full word v.

LEMMA 6.5
If $w \in \mathcal{P}_{0,k}(n)$, then $\mathcal{S}_h(w) \subset \mathcal{S}_h(v) \cup \mathcal{P}_{h,k}(n)$ for some $v \in \mathcal{N}_{0,k}(n)$.

PROOF   Let $w \in \mathcal{P}_{0,k}(n)$. If $w_{i_1,\ldots,i_h} \in \mathcal{S}_h(w)$ is nonprimitive, then there exists a full word u such that $w_{i_1,\ldots,i_h} \subset u^i$ for some $i \geq 2$. The word $v = u^i$ is such that $w_{i_1,\ldots,i_h} \in \mathcal{S}_h(v)$. ▯

We will now concentrate on the one- and two-hole cases. We will prove the case of two holes and leave the case of one hole to the reader.

THEOREM 6.2
The equality $N_{1,k}(n) = nN_{0,k}(n)$ holds.

We can deduce the following corollary.

COROLLARY 6.1
The equality $P_{1,k}(n) = n(P_{0,k}(n) + k^{n-1} - k^n)$ holds.

The two-hole case is stated in the next two theorems.

THEOREM 6.3
For an odd positive integer n, the following equality holds:

$$N_{2,k}(n) = \binom{n}{2} N_{0,k}(n)$$

PROOF   If u, v are distinct nonprimitive full words of length n, then $S_2(u) \cap S_2(v) = \emptyset$. Indeed, suppose there exists a pword $w \in S_2(u) \cap S_2(v)$ such that $u_{i_1,i_2} = w$ and $v_{i_1,i_2} = w$. Thus $u(i) = v(i)$ for all $0 \leq i < n, i \neq i_1, i_2$. Since $w \in \mathcal{N}_{2,k}(n)$ and n is odd, w is not an 2-repeat. It is easy to see that in this case, there are no free ◇'s in w, that is, there exist $j_1, j_2 \in D(w)$ such that $j_1 \equiv i_1 \bmod d$ and $j_2 \equiv i_2 \bmod d$ for each exact period d. This means that in the words u and v, there exists only one pair of assignments for $u(i_1), u(i_2)$ and $v(i_1), v(i_2)$ respectively, since we have already shown in Lemma 6.3 that $\nu(w) = 1$. It follows that $u(i_1) = v(i_1)$ and $u(i_2) = v(i_2)$ implying that $u = v$ which contradicts our assumption. Since the sets in the union $\bigcup_{w \in \mathcal{N}_{0,k}(n)} S_h(w)$ are pairwise disjoint, we may conclude that

$$N_{2,k}(n) = \|\mathcal{W}_{2,k}(n)\| = \binom{n}{2} N_{0,k}(n)$$

▯

THEOREM 6.4

For an even positive integer n, the following equality holds:

$$N_{2,k}(n) = \binom{n}{2} N_{0,k}(n) - (k-1)T_{1,k}\left(\frac{n}{2}\right)$$

PROOF   It suffices to show that a number of $T_{1,k}(\frac{n}{2})$ words are counted k times each. Let w be a nonprimitive word of even length that generates the partial word $w_{i,j}$. Assume $n \geq 4$. If $w_{i,j}$ is not an $\frac{n}{2}$-repeat, then there are at least three occurrences of the base of length $\leq \frac{n}{3}$. It follows that there are no free $\diamond$'s, which means that the generator w is unique. Assume now that $w_{i,j}$ has an exact period $d = \frac{n}{2}$. If i and j do not belong to the same class modulo d, then again $w_{i,j}$ is uniquely generated since there are no free $\diamond$'s. Now, suppose $i \equiv j \bmod d$ and consider again the pword $w_{i,j}$:

$$
\begin{aligned}
w_{i,j} = a_0 \quad a_1 \quad &\ldots \quad a_{i-1} \diamond a_{i+1} \ldots a_{d-1} \\
a_d \, a_{d+1} &\ldots a_{j-1} \diamond a_{j+1} \ldots a_{n-1}
\end{aligned}
$$

Note that in this case we have a pair of free $\diamond$'s, which means that in the initial word w, the letter at positions i and j can be any letter in the alphabet, thus a total of k possibilities. The number of partial words u of length $\frac{n}{2}$ with one hole is $T_{1,k}(\frac{n}{2})$. Note that all possible pwords of the form uu can each be generated by k different words in $N_{0,k}(n)$. Removing $k-1$ copies of such words leaves us with a total of $\frac{1}{2}n(n-1)N_{0,k}(n) - (k-1)T_{1,k}(\frac{n}{2})$, which is what we wanted. $\qquad\square$

COROLLARY 6.2

The following holds:

$$
P_{2,k}(n) = \begin{cases}
\binom{n}{2}(P_{0,k}(n) + k^{n-2} - k^n) & \text{if } n \text{ is odd} \\
\binom{n}{2}(P_{0,k}(n) + k^{n-2} - k^n) + (k-1)T_{1,k}(\frac{n}{2}) & \text{if } n \text{ is even}
\end{cases}
$$

PROOF   If n is odd, then using Theorem 6.3 we have the following list of equalities:

$$
\begin{aligned}
P_{2,k}(n) &= T_{2,k}(n) - N_{2,k}(n) \\
&= T_{2,k}(n) - \binom{n}{2}N_{0,k}(n) \\
&= \binom{n}{2}k^{n-2} - \binom{n}{2}(T_{0,k}(n) - P_{0,k}(n)) \\
&= \binom{n}{2}k^{n-2} - \binom{n}{2}(k^n - P_{0,k}(n)) \\
&= \binom{n}{2}(P_{0,k}(n) + k^{n-2} - k^n)
\end{aligned}
$$

The case when n is even follows from Theorem 6.4. ▯

We end this section with three propositions: the first holding for any number of holes and the other two for three holes.

PROPOSITION 6.3

If $w \in \mathcal{N}_{h,k}(n)$ and w has no free $\diamond$'s with respect to any of its exact periods, then there exist $\nu(w)$ words in $\mathcal{N}_{0,k}(n)$ that generate w.

PROOF   Let $i_1 < i_2 < \cdots < i_h$ be the elements in $H(w)$. For $p \in \mathcal{R}(w)$, let the h-tuple $(b_{i_1}, \ldots, b_{i_h})$ be the image of $(i_1, \ldots, i_h)$ under f_p (here $b_{i_j} \in A$ for all j). Obviously, replacing for all j the $\diamond$ in position i_j with the corresponding letter b_{i_j} yields a full nonprimitive word that generates w. Since we showed in Lemma 6.1 that f is bijective on $\mathcal{R}(w)$, it follows that there are $\nu(w) = \|\mathcal{R}(w)\|$ full words that generate w. ▯

PROPOSITION 6.4

If $w \in \mathcal{N}_{3,k}(n)$ and w has three free $\diamond$'s, then there exist $kT_{1,k}(\frac{n}{3})$ words in $\mathcal{N}_{0,k}(n)$ that generate w.

PROOF   The pword w has three free $\diamond$'s only if it is an 3-repeat, that is, $w \subset v^3$ for some pword $v \in \mathcal{T}_{1,k}(\frac{n}{3})$. There also must exist some $0 \le i < \frac{n}{3}$ such that $B_{\frac{n}{3}}(i) = \{\diamond\}$. Let v' denote the full word obtained by replacing the $\diamond$ at position i in v with any letter in A. The resulting full word $(v')^3$ is a generator for w. Since there are k choices to replace the $\diamond$ in v with a letter, it means that k possible full words generate w. Since the total number of words in $\mathcal{N}_{3,k}(n)$ that are 3-repeats is given by $T_{1,k}(\frac{n}{3})$, it follows that $kT_{1,k}(\frac{n}{3})$ words in $\mathcal{N}_{0,k}(n)$ generate w. ▯

PROPOSITION 6.5

If $w \in \mathcal{N}_{3,k}(n)$ and w has two free $\diamond$'s, then there exist $k(n-2)T_{1,k}(\frac{n}{2})$ words in $\mathcal{N}_{0,k}(n)$ that generate w.

PROOF   If w has two free $\diamond$'s, then it must be an 2-repeat, that is, $w \subset v^2$ for some pword $v \in \mathcal{T}_{1,k}(\frac{n}{2})$. There must exist some i, j, $0 \le i, j < \frac{n}{2}$ such that $B_{\frac{n}{2}}(i) = \{\diamond\}$ and $B_{\frac{n}{2}}(j) = \{\diamond, a\}$ for some $a \in A$. Let v' denote a full word obtained by replacing the $\diamond$'s at positions $i, i + \frac{n}{2}$ within w with any letter in A and the $\diamond$ at position j with the letter a. There are k choices to replace the $\diamond$'s at positions $i, i + \frac{n}{2}$ with a letter. Also, note that there are $(n-2)T_{1,k}(\frac{n}{2})$ words in $\mathcal{N}_{3,k}(n)$ that have a pair of free $\diamond$'s since there are $n - 2$ positions where we can place the third $\diamond$. Overall, there are $k(n-2)T_{1,k}(\frac{n}{2})$ words in

$\mathcal{N}_{0,k}(n)$ that generate w. $\qquad\Box$

6.5 Second counting method

We now count nonprimitive partial words of length n with h holes over a k-size alphabet A through a constructive method that refines the counting done in the previous sections.

DEFINITION 6.5 *If w is a nonprimitive pword of length n with h holes and d is the smallest integer such that there exists a pword v satisfying $w \subset v^{n/d}$, then the **proot** of w is the pword $w[0..d)$.*

Example 6.6
Consider the nonprimitive partial word $w = a\diamond ab\diamond bab$ with two holes of length 8 over the binary alphabet $\{a, b\}$. The containments $w \subset (ab)^{8/2}$ and $w \subset (abab)^{8/4}$ hold, and $d = 2$ is the smallest integer satisfying $w \subset v^{8/d}$ for some v. Thus the proot of w is $w[0..d) = w[0..2) = a\diamond$. $\qquad\Box$

Let $\mathcal{RP}_{h,k}(n, d, h')$ denote the set of nonprimitive pwords of length n with h holes over an alphabet of size k with a primitive proot having length d and containing h' holes, and let $\mathcal{RN}_{h,k}(n, d, h'))$ denote a similar set except that the proot is nonprimitive. Denote by $\mathcal{R}_{h,k}(n, d)$ the set of nonprimitive pwords with h holes of length n over an alphabet of size k with a proot of length d. Using the convention adopted earlier, $RP_{h,k}(n, d, h')$ will denote the cardinality of $\mathcal{RP}_{h,k}(n, d, h')$. We define $RN_{h,k}(n, d, h')$ similarly. In addition, $R_{h,k}(n, d)$ will denote the cardinality of $\mathcal{R}_{h,k}(n, d)$.

We obtain the equality

$$R_{h,k}(n, d) = \sum_{h'=0}^{h} (RP_{h,k}(n, d, h') + RN_{h,k}(n, d, h')) \qquad (6.9)$$

The set $\mathcal{N}_{h,k}(n)$ will be generated by considering all possible proots of length $d \in \mathcal{D}(n)$. Different cases occur: The proot belongs to $\mathcal{P}_{h',k}(d)$ for some $h' = 0, \ldots, h$, or the proot belongs to $\mathcal{N}_{h',k}(d)$ for some $h' = 1, \ldots, h$.

REMARK 6.3 Note that, in order to avoid double counting, we will never generate nonprimitive pwords starting with a nonprimitive full proot. Therefore, we may always assume that $RN_{h,k}(n, d, 0) = 0$, hence the missing $h' = 0$ for $\mathcal{N}_{h',k}(d)$. $\qquad\Box$

Given a proot $w[0..d)$ with h' holes, we build the corresponding *temporary* pword $t = (w[0..d))^{n/d}$. We transform t to generate nonprimitive pwords by replacing letters with $\diamond$'s, or vice versa, while the proot remains unchanged. There result pwords containing h holes and having proot $w[0..d)$.

TABLE 6.6: Partial words in $\mathcal{N}_{1,2}(8)$: "1" refers to length of proot is 1; "2p" to length of primitive proot is 2; "2n" to length of nonprimitive proot is 2; "4p0" (respectively, "4p1") to length of primitive full (respectively, one-hole) proot is 4; and "4n" to length of nonprimitive proot is 4.

1	$aaaaaaa\diamond$	4n	$aba\diamond abaa$	4p1	$baa\diamond baaa$	1	$bbbbb\diamond bb$
1	$aaaaaa\diamond a$	2p	$aba\diamond abab$	4p1	$baa\diamond baab$	1	$bbbb\diamond bbb$
1	$aaaaa\diamond aa$	4p0	$abbaabb\diamond$	2p	$bababab\diamond$	4n	$bbb\diamond bbba$
1	$aaaa\diamond aaa$	4p0	$abbaab\diamond a$	2p	$bababa\diamond a$	1	$bbb\diamond bbbb$
4p0	$aaabaaa\diamond$	4p0	$abbaa\diamond ba$	2p	$babab\diamond ba$	4p1	$bb\diamond abbaa$
4p0	$aaabaa\diamond b$	4p0	$abba\diamond bba$	2p	$baba\diamond aba$	4p1	$bb\diamond abbba$
4p0	$aaaba\diamond ab$	4p0	$abbbabb\diamond$	4p0	$babbbab\diamond$	4n	$bb\diamond bbbab$
4p0	$aaab\diamond aab$	4p0	$abbbab\diamond b$	4p0	$babbba\diamond b$	1	$bb\diamond bbbbb$
1	$aaa\diamond aaaa$	4p0	$abbba\diamond bb$	4p0	$babbb\diamond bb$	4p1	$b\diamond aabaaa$
4n	$aaa\diamond aaab$	4p0	$abbb\diamond bbb$	4p0	$babb\diamond abb$	4p1	$b\diamond aabbaa$
4p0	$aabaaab\diamond$	4p1	$abb\diamond abba$	2p	$bab\diamond baba$	4p1	$b\diamond abbaab$
4p0	$aabaaa\diamond a$	4p1	$abb\diamond abbb$	4n	$bab\diamond babb$	4p1	$b\diamond abbbab$
4p0	$aabaa\diamond ba$	4p1	$ab\diamond aabaa$	4n	$ba\diamond abaaa$	2n	$b\diamond bababa$
4p0	$aaba\diamond aba$	2p	$ab\diamond aabba$	4n	$ba\diamond ababa$	4n	$b\diamond babbba$
4p0	$aabbaab\diamond$	4p1	$ab\diamond babab$	4p1	$ba\diamond bbaab$	4n	$b\diamond bbbabb$
4p0	$aabbaa\diamond b$	4n	$ab\diamond babbb$	4p1	$ba\diamond bbabb$	1	$b\diamond bbbbbb$
4p0	$aabba\diamond bb$	1	$a\diamond aaaaaa$	4p0	$bbaabba\diamond$	1	$\diamond aaaaaaa$
4p0	$aabb\diamond abb$	4n	$a\diamond aaabaa$	4p0	$bbaabb\diamond a$	4n	$\diamond aaabaaa$
4p1	$aab\diamond aaba$	4n	$a\diamond abaaab$	4p1	$bbaab\diamond aa$	4p1	$\diamond aabaaab$
4p1	$aab\diamond aabb$	2n	$a\diamond ababab$	4p1	$bbaa\diamond baa$	4p1	$\diamond aabbaab$
1	$aa\diamond aaaaa$	4p1	$a\diamond baaaba$	4n	$bbabbba\diamond$	4n	$\diamond abaaaba$
4n	$aa\diamond aaaba$	4p1	$a\diamond baabba$	2n	$bbabbb\diamond b$	2n	$\diamond abababa$
4p1	$aa\diamond baaab$	4p1	$a\diamond bbaabb$	4p1	$bbabb\diamond ab$	4p1	$\diamond abbaabb$
4p1	$aa\diamond baabb$	4p1	$a\diamond bbabbb$	4p1	$bbab\diamond bab$	4p1	$\diamond abbbabb$
4p0	$abaaaba\diamond$	4p0	$baaabaa\diamond$	4p1	$bba\diamond bbaa$	4p1	$\diamond baaabaa$
4p0	$abaaab\diamond a$	4p0	$baaaba\diamond a$	4p1	$bba\diamond bbab$	4p1	$\diamond baabbaa$
4p0	$abaaa\diamond aa$	4p0	$baaab\diamond aa$	4p0	$bbbabbb\diamond$	2n	$\diamond bababab$
4p0	$abaa\diamond baa$	4p0	$baaa\diamond aaa$	4p0	$bbbabb\diamond a$	4n	$\diamond babbbab$
2p	$abababa\diamond$	4p0	$baabbaa\diamond$	4p0	$bbbab\diamond ba$	4p1	$\diamond bbaabba$
2p	$ababab\diamond b$	4p0	$baabba\diamond b$	4p0	$bbba\diamond bba$	4p1	$\diamond bbabbba$
2p	$ababa\diamond ab$	4p0	$baabb\diamond ab$	1	$bbbbbbb\diamond$	4n	$\diamond bbbabbb$
2p	$abab\diamond bab$	4p0	$baab\diamond aab$	1	$bbbbbb\diamond b$	1	$\diamond bbbbbbb$

Example 6.7

We illustrate the abovementioned ideas by computing $N_{1,2}(8)$ where the set of lengths of proots is $\{1, 2, 4\}$. We set $A = \{a, b\}$ as our alphabet.

If the length of the proot is 1, then $w \subset a^8$ or $w \subset b^8$. There are $2 \times 8 = 16$ ways to build such a pword of length 8 with one hole over A. The 2 representing the two distinct letters a and b and the 8 representing the length of the strings we are counting. Thus $R_{1,2}(8,1) = 16$. Examples of such strings include $aaaaaaa\diamond$ and $bb\diamond bbbbb$. They are the "1"'s in the table.

Now, if the length of the proot is 2, then $w \subset v^4$ for some v. Note that since $P_{1,2}(2) = 0$, the proot cannot be a primitive partial word with one hole and consequently $RP_{1,2}(8, 2, 1) = 0$. Also recall that the proot cannot be a nonprimitive full word. Therefore, we split the nonprimitive pwords with a proot of length 2 into two sets: the ones with a proot that is a primitive full word and the ones with a nonprimitive proot with one hole.

- If the proot is a primitive full word, then it belongs to the set $\{ab, ba\}$. To obtain nonprimitive partial words with one hole from the temporary words $t_1 = abababab$ and $t_2 = babababa$, we replace the letter in any of the last six positions of t_1 or t_2 with $\diamond$. Note that replacing any letter in any of the first two positions with $\diamond$, thus in the proot, would bring us back to the previous case when the proot has length 1 and we would be doubly counting. Six new nonprimitive pwords can be derived from t_1 and the same is true for t_2, thus $RP_{1,2}(8, 2, 0) = 12$. They are the "2p"'s in the table.

- If the proot is a nonprimitive partial word with one hole, then it belongs to the 4-element set $\{a\diamond, b\diamond, \diamond a, \diamond b\}$. There is only one way to build nonprimitive partial words with such proots. They are the "2n"'s in the table. For example, if the proot is $\diamond b$, then the only possibility is $\diamond bababab$. Note that $\diamond bbbbbbb$ is not a possibility since it has already been taken into account. Thus $RN_{1,2}(8, 2, 1) = 4$.

We obtain the equality

$$R_{1,2}(8,2) = RP_{1,2}(8,2,0) + RP_{1,2}(8,2,1) + RN_{1,2}(8,2,1) = 16$$

Last, if the length of the proot is 4, then $w \subset v^2$ and again, we split all possible nonprimitive partial words with a proot of length 4 into three sets.

- If the proot is a primitive full word, then it belongs to a set of cardinality $P_{0,2}(4) = 12$. To obtain nonprimitive partial words with one hole, we may replace any of the last four positions with $\diamond$ and $RP_{1,2}(8, 4, 0) = 48$. They are the "4p0"'s in the table.

- If the proot is a primitive partial word with one hole, then it belongs to a set of cardinality $P_{1,2}(4) = 16$, the "4p1"'s in the table. For example, if the proot is $\diamond abb$, then the temporary pword is $t = \diamond abb\diamond abb$. In place of the second $\diamond$, we can put either an a or a b thus obtaining $\diamond abbaabb$ and $\diamond abbbabb$, both nonprimitive partial words with one hole. Thus, $RP_{1,2}(8,4,1) = 32$.

- If the proot is a nonprimitive partial word with one hole, then it belongs to the set

$$\{aaa\diamond, aa\diamond a, a\diamond aa, \diamond aaa, aba\diamond, ab\diamond b, a\diamond ab, \diamond bab\}$$

unioned with the set containing the pwords obtained by switching a with b, for a total of $N_{1,2}(4) = 16$ proots. There is only one way to build a nonprimitive pword with one hole from such proots. If the proot is $aaa\diamond$, then the only possibility is $aaa\diamond aaab$. Similarly, for the proot $aba\diamond$, the only nonprimitive pword with one hole that can be built with this proot is $aba\diamond abaa$. Note that the temporary pword in this case is $t = aba\diamond aba\diamond$, but the second $\diamond$ can be replaced by any letter, except the *one* letter which will make the pword an 4-repeat with a proot of length 2 (this case has already been taken into account). Thus, $RN_{1,2}(8,4,1) = 16$, they are the "4n"'s in the table.

We obtain the equality

$$R_{1,2}(8,4) = RP_{1,2}(8,4,0) + RP_{1,2}(8,4,1) + RN_{1,2}(8,4,1) = 48 + 32 + 16 = 96$$

The above computations lead to

$$N_{1,2}(8) = R_{1,2}(8,1) + R_{1,2}(8,2) + R_{1,2}(8,4) = 16 + 16 + 96 = 128$$

agreeing with the corresponding value in the table.

Note that $\mathcal{R}_{1,2}(8,d_1) \cap \mathcal{R}_{1,2}(8,d_2)$ is empty for any two distinct d_1, d_2. ☐

The following lemma proves that we are not doubly counting any nonprimitive pwords.

LEMMA 6.6

Given a proot $w[0..d)$ of length $d \in \mathcal{D}(n)$, the nonprimitive partial words (with one or two holes) generated from $w[0..d)$ have their smallest exact period equal to d.

PROOF   The analysis we are about to perform is similar for the case when we count nonprimitive pwords with two holes. Suppose that during the process of transforming a temporary pword t the resulting pword $w \in \mathcal{N}_{1,k}(n)$ has an exact period d' with $d' < d$. There are four cases we need to consider, depending on whether the proot is a primitive or nonprimitive pword or whether d' is a divisor of d or not. If d' is not a divisor of d, then let $l = \gcd(d, d')$ with $l < d'$.

Case 1. $w[0..d)$ is nonprimitive and $d'|d$

This case cannot occur since, we subtract the value of $\nu(w[0..d)) = 1$ from the total number of options available to replace the extra ⋄'s in t.

Case 2. $w[0..d)$ is nonprimitive and $d' \nmid d$

Since $d, d' \in \mathcal{E}(w)$ it follows from Theorem 3.1 that $l \in \mathcal{E}(w)$. We are now back in the previous case and no double counting occurs. The reason is that, the pwords w which we are trying to avoid when transforming t have already been avoided in the previous case.

Case 3. $w[0..d)$ is primitive and $d'|d$

This case is easy to deal with simply because of the primitivity of $w[0..d)$. Since w is d'-periodic and $d'|d$, then $w[0..d)$ must also be d'-periodic and thus nonprimitive, which is a contradiction.

Case 4. $w[0..d)$ is primitive and $d' \nmid d$

Since $d, d' \in \mathcal{E}(w)$ it follows from Theorem 3.1 that $l \in \mathcal{E}(w)$. Since $w = w[0..d)v$ for some pword v and w is d-periodic and l-periodic and $l|d$ it follows that $w[0..d)$ has l as exact period and is thus nonprimitive. This again involves a contradiction with $w[0..d)$ being a primitive pword.

We have now proved that given a proot of length d, the nonprimitive pwords derived from it will always have their smallest exact period equal to d.   ☐

6.5.1   The one-hole case

The following theorem gives the main result on counting nonprimitive partial words with one hole of length n over A.

THEOREM 6.5
The following equality holds:

$$N_{1,k}(n) = kn + \sum_{d|n,\, d \neq 1,n} ((n-d)P_{0,k}(d) + kP_{1,k}(d) + (k-1)N_{1,k}(d)) \quad (6.10)$$

PROOF   Let w be a nonprimitive pword of length n with one hole over A. Let d be the smallest integer such that there exists a pword v satisfying $w \subset v^{n/d}$. Note that $d \in \mathcal{D}(n)$. The case when $d = 1$ can be easily dealt with. There are kn ways we can build a nonprimitive pword of length n with one hole over A, and thus $R_{1,k}(n,1) = kn$. Consider now the case when $d \in \mathcal{D}(n) \setminus \{1\}$. We split the proof into three cases based on the nature of the proot $w[0..d)$, and we set $t = (w[0..d))^{n/d}$.

Case 1. First, if the proot is a primitive full word, then it belongs to a set of $P_{0,k}(d)$ elements. Transforming t into a nonprimitive pword with one hole requires that we place a ⋄ anywhere in t, except in the positions

$0, \ldots, d-1$. Since there is a total of $n-d$ such positions, we get

$$RP_{1,k}(n,d,0) = (n-d)P_{0,k}(d)$$

Case 2. Now, if the proot is a primitive partial word with one hole, then it belongs to a set of $P_{1,k}(d)$ elements. To obtain a nonprimitive pword of length n with one hole, we need to replace in t all the holes, except the first one, with letters in A. Note that once a hole has been replaced with a letter, all remaining holes must be replaced by the same letter. There are k ways we can replace a hole with a letter, thus

$$RP_{1,k}(n,d,1) = kP_{1,k}(d)$$

Case 3. Finally, if the proot is a nonprimitive partial word with one hole, then it belongs to a set of $N_{1,k}(d)$ elements. Transforming t into a nonprimitive partial word with one hole requires that all holes, except the first one (in the proot), be replaced by a letter in A. Note that once the second hole is replaced by a letter, all remaining holes need to be replaced by the same letter. When replacing the holes, we have all k letters available, except that set of letters that would lead to a nonprimitive pword with one hole and a proot shorter than d, a case that we have already taken into account. Since $\nu(w[0..d)) = 1$, there are $k - \nu(w[0..d)) = k-1$ nonprimitive partial words with one hole that can be obtained from the temporary pword t above and which have not been counted in previous cases. Since there are $N_{1,k}(d)$ such temporary pwords, it follows that

$$RN_{1,k}(n,d,1) = (k-1)N_{1,k}(d)$$

Therefore, the total number of nonprimitive partial words with one hole of length n over A with a proot of length d is

$$\begin{aligned} R_{1,k}(n,d) &= RP_{1,k}(n,d,0) + RP_{1,k}(n,d,1) + RN_{1,k}(n,d,1) \\ &= (n-d)P_{0,k}(d) + kP_{1,k}(d) + (k-1)N_{1,k}(d) \end{aligned}$$

Denoting by N the right hand side of Equality 6.10, we want to prove that $N_{1,k}(n) = N$. Note that for a given d, the three cases above cover all possible proots of length d. We do not consider the case of nonprimitive full roots because this falls into the case of full primitive proots with length d' satisfying $d' < d$. Also, once a proot is fixed, we always consider all possible ways the temporary pword t can be transformed into a nonprimitive partial word with one hole, provided we keep the proot unchanged. Modifying the proot by substituting a letter for a $\diamond$ or vice versa, would lead to a nonprimitive word with a different proot (shorter or longer), something that has already been accounted in a different case. We are thus covering all possible nonprimitive partial words with one hole, which implies that $N \geq N_{1,k}(n)$.

We must now prove that $N \leq N_{1,k}(n)$. For a given proot of length d, it holds that the sets $\mathcal{RP}_{1,k}(n,d,0), \mathcal{RP}_{1,k}(n,d,1)$ and $\mathcal{RN}_{1,k}(n,d,1)$ are pairwise disjoint. The reason is that the generating proot for each of the sets are different, as they belong to three different pairwise disjoint sets: $\mathcal{P}_{0,k}(d), \mathcal{P}_{1,k}(d)$ and $\mathcal{N}_{1,k}(d)$. In each of the three cases, the proot is different to start with, and recall that proots remain unchanged throughout the process of transforming a temporary pword into a nonprimitive partial word. Thus, for all three cases, the resulting nonprimitive partial words will be different. Let r_1, r_2 be proots of length $d_1, d_2 \in \mathcal{D}(n)$, with $1 < d_1 < d_2 < n$, and u, v be any pwords such that $u \in \mathcal{R}_{1,k}(n,d_1)$ and $v \in \mathcal{R}_{1,k}(n,d_2)$. Using Lemma 6.6, u and v have their smallest exact period equal to d_1, respectively d_2. From $d_1 \neq d_2$ it follows that $u \neq v$. Since u, v were any words in $\mathcal{R}_{1,k}(n,d_1)$, respectively $\mathcal{R}_{1,k}(n,d_2)$, it follows that $\mathcal{R}_{1,k}(n,d_1) \cap \mathcal{R}_{1,k}(n,d_2) = \emptyset$. Since d_1, d_2 are any proper divisors of n, it holds that

$$\bigcap_{d_i \in \mathcal{D}(n)} \mathcal{R}_{1,k}(n,d_i) = \emptyset$$

This proves that no double counting occurs and thus $N \leq N_{1,k}(n)$.   ☐

Example 6.8

Theorem 6.5 implies that $N_{1,2}(8) = 128$ which matches the computations of the previous section. Indeed,

$$N_{1,2}(8) = 16 + \sum_{d \in \{2,4\}} ((8-d)P_{0,2}(d) + 2P_{1,2}(d) + (2-1)N_{1,2}(d))$$
$$= 16 + 12 + 0 + 4 + 48 + 32 + 16$$
$$= 128$$

☐

The formula of Theorem 6.5 can be further reduced.

COROLLARY 6.3

The equality $N_{1,k}(n) = nN_{0,k}(n)$ holds.

PROOF   We prove the equality $N_{1,k}(n) = nN_{0,k}(n)$ by induction on n using Theorem 6.5. For $n = 1$, the result trivially holds since $N_{1,k}(1) = N_{0,k}(1) = 0$. Assuming the equality holds for all positive integers smaller than n, we get the following sequence of equalities for $N_{1,k}(n)$:

$$kn + \sum_{d|n,\ d\neq 1,n} ((n-d)P_{0,k}(d) + kP_{1,k}(d) + (k-1)N_{1,k}(d))$$

$$= kn + k \sum_{d|n,\ d\neq 1,n} (P_{1,k}(d) + N_{1,k}(d)) + \sum_{d|n,d\neq 1,n} ((n-d)P_{0,k}(d) - N_{1,k}(d))$$

$$= kn + k \sum_{d|n,\ d\neq 1,n} T_{1,k}(d) + \sum_{d|n,d\neq 1,n} ((n-d)P_{0,k}(d) - dN_{0,k}(d))$$

$$= kn + k \sum_{d|n,\ d\neq 1,n} T_{1,k}(d) + n \sum_{d|n,d\neq 1,n} P_{0,k}(d) - \sum_{d|n,d\neq 1,n} (dP_{0,k}(d) + dN_{0,k}(d))$$

$$= kn + k \sum_{d|n,\ d\neq 1,n} dk^{d-1} + n \sum_{d|n,d\neq 1,n} P_{0,k}(d) - \sum_{d|n,d\neq 1,n} dT_{0,k}(d)$$

$$= kn + \sum_{d|n,\ d\neq 1,n} dk^{d} + n \sum_{d|n,d\neq 1,n} P_{0,k}(d) - \sum_{d|n,d\neq 1,n} dk^{d}$$

$$= kn + n \sum_{d|n,\ d\neq 1,n} P_{0,k}(d)$$

$$= n(\sum_{d|n,\ d\neq 1,n} P_{0,k}(d) + k)$$

$$= n(\sum_{d|n,\ d\neq 1,n} P_{0,k}(d) + P_{0,k}(1) + P_{0,k}(n) - P_{0,k}(n))$$

$$= n(\sum_{d|n} P_{0,k}(d) - P_{0,k}(n))$$

$$= n(k^n - P_{0,k}(n))$$

$$= n(T_{0,k}(n) - P_{0,k}(n))$$

$$= nN_{0,k}(n)$$

▯

6.5.2 The two-hole case

The following theorem holds.

THEOREM 6.6

The number of nonprimitive partial words with two holes of length n over a k-size alphabet, $N_{2,k}(n)$, is equal to

$$\binom{n}{2}k + \sum_{d|n,d\neq 1,n}(RP_{2,k}(n,d,0) + RP_{2,k}(n,d,1) + RP_{2,k}(n,d,2) + RN_{2,k}(n,d,1) + RN_{2,k}(n,d,2))$$

where

$$RP_{2,k}(n, d, 0) = \binom{n-d}{2} P_{0,k}(d) \qquad (6.11)$$

$$RP_{2,k}(n, d, 1) = \begin{cases} k(n-d)P_{1,k}(d) & \text{if } d \neq \frac{n}{2} \\ (k(n-d) - (k-1))P_{1,k}(d) & \text{if } d = \frac{n}{2} \end{cases} \qquad (6.12)$$

$$RN_{2,k}(n, d, 1) = \begin{cases} (k-1)(n-d)N_{1,k}(d) & \text{if } d \neq \frac{n}{2} \\ (k-1)(d-1)N_{1,k}(d) & \text{if } d = \frac{n}{2} \end{cases} \qquad (6.13)$$

$$RP_{2,k}(n, d, 2) = k^2 P_{2,k}(d) \qquad (6.14)$$

$$RN_{2,k}(n, d, 2) = \begin{cases} (k^2-1)N_{2,k}(d) - (k-1)T_{1,k}(\frac{d}{2}) & \text{if } d \text{ is even} \\ (k^2-1)N_{2,k}(d) & \text{if } d \text{ is odd} \end{cases} \qquad (6.15)$$

PROOF   We give a constructive algorithm for nonprimitive pwords and prove that, along the process of building them, no pwords are missed or double counted.

Let w be a nonprimitive pword of length n with two holes over A, and let d be the smallest integer such that there exists a pword v satisfying $w \subset v^{n/d}$. The case when $d = 1$ can be easily dealt with since there are $\binom{n}{2}k$ ways of building such a nonprimitive pword. Consider now the case when $d \in \mathcal{D}(n) \setminus \{1\}$. We split the proof into five cases based on the nature of the proot $w[0..d) = a_0 a_1 \ldots a_{d-1}$, and we set $t = (w[0..d))^{n/d}$. If $w[0..d)$ has h' holes, then let $0 \leq i_1 < i_2 < \cdots < i_{h'} < d$ be such that $a_{i_j} = \diamond$. Define

$$C_d = \{l \mid d \leq l < n \text{ and } l \not\equiv i_1 \bmod d, \ldots, l \not\equiv i_{h'} \bmod d\}$$
$$D_d(i_j) = \{l \mid d \leq l < n \text{ and } l \equiv i_j \bmod d\} \text{ for all } 1 \leq j \leq h'$$

Note that $\|C_d\| = (\frac{n}{d}-1)(d-h')$ and $\|D_d(i_j)\| = \frac{n}{d}-1$. Recall that Lemma 6.6 guarantees that no double counting will occur.

Case 1. $w[0..d) \in \mathcal{P}_{0,k}(d)$

We need to replace two positions by $\diamond$'s anywhere in t, except in the proot. There is a total of $n - d$ such positions and thus Equality (6.11) holds.

Case 2. $w[0..d) \in \mathcal{P}_{1,k}(d)$

At this point, all symbols at positions from set $D_d(i_1)$ are $\diamond$'s and all those in C_d are letters. After transforming t into a pword in $\mathcal{N}_{2,h}(n)$, there must remain only one $\diamond$ in the last $n - d$ positions. This can be achieved in two ways, by placing a $\diamond$ in position j with either $j \in C_d$ or $j \in D_d(i_1)$.

Let us first consider the case where $j \in C_d$. There are $\|C_d\|$ options where to place the second $\diamond$ and k choices to pick a letter to replace the positions in $D_d(i_1)$. Note that once a position from set $D_d(i_1)$ has been replaced, all others must be replaced by the same letter. This case yields a total of $k(\frac{n}{d}-1)(d-1)$ choices.

Let us now consider the case where $j \in D_d(i_1)$. Note that this can be done in $\|D_d(i_1)\|$ ways and that all positions in C_d remain unchanged. We are now

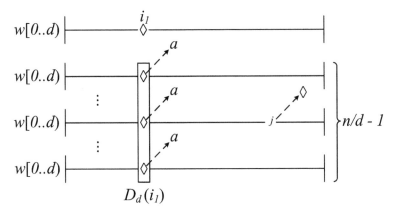

FIGURE 6.2: Representation of Case 2 when $j \in C_d$.

left with $\|D_d(i_1)\| - 1$ ◇'s to be replaced with the same letter. This can be done in k ways provided that $\|D_d(i_1)\| - 1 > 0$, thus a total of $k\|D_d(i_1)\|$ options. If $\|D_d(i_1)\| - 1 = 0$, which implies that $d = n/2$, then this case yields only $\|D_d(i_1)\| = 1$ option, that is $w = w[0..d)w[0..d)$.

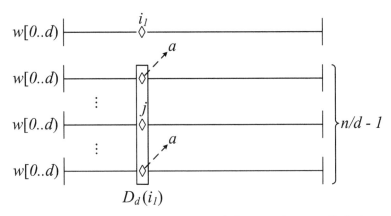

FIGURE 6.3: Representation of Case 2 when $j \in D_d(i_1)$.

Thus, for $d \neq \frac{n}{2}$ we have

$$RP_{2,k}(n, d, 1) = (k(\tfrac{n}{d} - 1)(d - 1) + k(\tfrac{n}{d} - 1))P_{1,k}(d) = k(n - d)P_{1,k}(d)$$

and if $d = \frac{n}{2}$ then

$$RP_{2,k}(n, d, 1) = (k(d - 1) + 1)P_{1,k}(d) = (k(n - d) - (k - 1))P_{1,k}(d)$$

Putting the two cases together we have that Equality (6.12) holds.

Case 3. $w[0..d) \in \mathcal{N}_{1,k}(d)$

The approach for this case is similar to the one for Case 2. We replace position j in t with $\diamond$.

Let us first consider the case where $j \in C_d$. There are $\|C_d\|$ options where to place the second $\diamond$, but this time only $k - \nu(w[0..d))$ letters available to replace the $\diamond$'s at positions from $D_d(i_1)$. This last restraint guarantees that the generated pword w will not have an exact period less than d, in other words the proot of w remains unchanged.

Let us now consider the case where $j \in D_d(i_1)$. If $d \neq \frac{n}{2}$, then there are $\|D_d(i_1)\|$ options to place the second $\diamond$ and $k - \nu(w[0..d))$ letters available to replace the remaining $\diamond$'s from positions within set $D_d(i_1)$, thus a total of $(k - \nu(w[0..d)))(\frac{n}{d} - 1)$ options. If $d = \frac{n}{2}$, then there is no solution since $w = w[0..d)w[0..d)$ would have a shorter proot.

Thus for $d \neq \frac{n}{2}$, $RN_{2,k}(n, d, 1) = ((k - \nu(w[0..d)))(\frac{n}{d} - 1)(d - 1) + (k - \nu(w[0..d)))(\frac{n}{d} - 1))N_{1,k}(d) = (k - \nu(w[0..d)))(n - d)N_{1,k}(d)$. If $d = \frac{n}{2}$, then $RN_{2,k}(n, d, 1) = (k - \nu(w[0..d)))(d - 1)N_{1,k}(d)$. Keeping in mind that $\nu(w[0..d)) = 1$, we have that Equality (6.13) holds.

Case 4. $w[0..d) \in \mathcal{P}_{2,k}(d)$

We must replace all positions from the set $D_d(i_1)$ with the same letter, and similarly for $D_d(i_2)$. There are k^2 options to choose these two letters and thus Equality (6.14) holds.

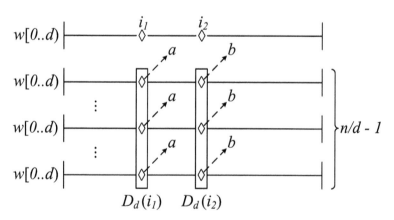

FIGURE 6.4:   Representation of Case 4.

Case 5. $w[0..d) \in \mathcal{N}_{2,k}(d)$

First, let us consider the case where $w[0..d)$ has a pair of free $\diamond$'s. Of course, this can happen only when $w[0..d)$ is $\frac{d}{2}$-periodic. It is easy to see that the number of pwords $w[0..d)$ of the form $w[0..d) = uu$, where u is a pword with one hole, is equal to $T_{1,k}(d/2)$. We now have k options to replace the $\diamond$'s from

6.6   Existence of primitive partial words

In this section, we discuss some fundamental properties of primitive partial
words.

First, Theorem 3.1(1) implies the following result.

PROPOSITION 6.6

*Let u, v be nonempty words and let m, n be integers. If u^m and v^n have a
common prefix (respectively, suffix) of length at least $|u| + |v| - \gcd(|u|, |v|)$,
then there exists a word x of length not greater than $\gcd(|u|, |v|)$ such that
$u = x^k$ and $v = x^l$ for some integers k, l.*

PROOF   The proof is left as an exercise.   □

The following two corollaries hold.

COROLLARY 6.5

*Let u and v be words. If $u^k = v^l$ for some positive integers k, l, then there
exists a word w such that $u = w^m$ and $v = w^n$ for some integers m, n.*

Let Q be the set of all primitive words over A. Let $Q_1 = Q \cup \{\epsilon\}$, and for
any $i \geq 2$ let $Q_i = \{u^i \mid u \in Q\}$.

COROLLARY 6.6

Let m, n be positive integers. If $m \neq n$, then $Q_m \cap Q_n = \emptyset$.

REMARK 6.4   Corollary 6.5 will be extended to partial words in Chapter 10.   □

We now give three propositions that are extensions of Proposition 6.6 to
partial words.

PROPOSITION 6.7

*Let u, v be nonempty words, let y, z be partial words, and let w be a word
satisfying $|w| \geq |u| + |v| - \gcd(|u|, |v|)$. If $wy \subset u^m$ and $wz \subset v^n$ (respectively,
$yw \subset u^m$ and $zw \subset v^n$) for some integers m, n, then there exists a word x
of length not greater than $\gcd(|u|, |v|)$ such that $u = x^k$ and $v = x^l$ for some
integers k, l.*

PROOF   See Exercise 3.15.   ⬜

PROPOSITION 6.8
Let u, v be nonempty words, let y, z be partial words, and let w be a partial word with one hole satisfying $|w| \geq |u| + |v|$. If $wy \subset u^m$ and $wz \subset v^n$ (respectively, $yw \subset u^m$ and $zw \subset v^n$) for some integers m, n, then there exists a word x of length not greater than $\gcd(|u|, |v|)$ such that $u = x^k$ and $v = x^l$ for some integers k, l.

PROOF   The proof is left as an exercise.   ⬜

PROPOSITION 6.9
Let u, v be words satisfying $0 < |u| < |v|$, let y, z be partial words, and let w be a non $(\|H(w)\|, |u|, |v|)$-special partial word satisfying $\|H(w)\| \geq 2$ and $|w| \geq l_{(\|H(w)\|, |u|, |v|)}$. If $wy \subset u^m$ and $wz \subset v^n$ (respectively, $yw \subset u^m$ and $zw \subset v^n$) for some integers m, n, then there exists a word x of length not greater than $\gcd(|u|, |v|)$ such that $u = x^k$ and $v = x^l$ for some integers k, l.

PROOF   Let w' be the prefix of length $l_{(\|H(w)\|, |u|, |v|)}$ of w. Both $|u|$ and $|v|$ are periods of w'. By Theorem 3.1 or Theorem 3.4, $\gcd(|u|, |v|)$ is also a period of w', and hence there exists a word x of length $\gcd(|u|, |v|)$ such that w' is contained in a power of x. If $H(w') = \emptyset$, then the result clearly follows. Otherwise, let $i \in H(w')$. Let r, $0 \leq r < |x|$, be the remainder of the division of i by $|x|$. There exists an integer i' such that $i + i'|x| \notin H(w')$ and $w'(i + i'|x|) = x(r)$. Hence for all $0 \leq j < |x|$, we have $j \notin H(w')$ and $x(j) = w'(j)$, or $j \in H(w')$ and there exists an integer j' satisfying $j + j'|x| \notin H(w')$ and $x(j) = w'(j + j'|x|)$. Since $|x|$ divides both $|u|$ and $|v|$, we conclude that $u = x^k$ and $v = x^l$ for some integers k, l.   ⬜

Second, it turns out that for two words u and v, the primitiveness of uv implies the primitiveness of vu as stated in the following result.

PROPOSITION 6.10
Let u and v be words. If there exists a primitive word x such that $uv = x^n$ for some positive integer n, then there exists a primitive word y such that $vu = y^n$. In particular, if uv is primitive, then vu is primitive.

PROOF   The proof is left as an exercise.   ⬜

A similar result holds for partial words.

PROPOSITION 6.11

Let u and v be partial words. If there exists a primitive word x such that $uv \subset x^n$ for some positive integer n, then there exists a primitive word y such that $vu \subset y^n$. Moreover, if uv is primitive, then vu is primitive.

PROOF First, assume that $n = 1$. Let x be a primitive word such that $uv \subset x$. Put $x = u'v'$ where $|u'| = |u|$ and $|v'| = |v|$. By Proposition 6.10, since $u'v'$ is primitive, $v'u'$ is also primitive. The result follows with $y = v'u'$.

Now, assume that $n > 1$. Since $uv \subset x^n$, there exist words x_1, x_2 such that $x = x_1 x_2$, $u \subset (x_1 x_2)^k x_1$ and $v \subset x_2 (x_1 x_2)^l$ with $k + l = n - 1$. Since $x = x_1 x_2$ is primitive, $x_2 x_1$ is also primitive by Proposition 6.10. The result follows since $vu \subset (x_2 x_1)^n$.

Now, suppose that uv is a primitive partial word. If vu is not primitive, then there exists a word y such that $vu \subset y^m$ for some $m \geq 2$. So there exist words y_1, y_2 such that $y = y_1 y_2$, $v \subset (y_1 y_2)^k y_1$ and $u \subset y_2 (y_1 y_2)^l$ with $k + l = m - 1$. Hence $uv \subset (y_2 y_1)^m$ and uv is not primitive, a contradiction. Therefore, if uv is primitive, then vu is primitive. ▯

Example 6.9

The partial words $u = a \diamond c \diamond$ and $v = bc \diamond \diamond c$ illustrate Proposition 6.11. Indeed, we have $uv \subset (abc)^3$ and $vu \subset (bca)^3$. ▯

Third, Proposition 6.6 implies the following result.

PROPOSITION 6.12

Let u be a word such that $\|\alpha(u)\| \geq 2$. If a is any letter, then u or ua is primitive.

Proposition 6.10 and Proposition 6.12 imply the following result.

COROLLARY 6.7

Let u_1, u_2 be nonempty words such that $\|\alpha(u_1 u_2)\| \geq 2$. Then for any letter a, $u_1 u_2$ or $u_1 a u_2$ is primitive.

The following results hold for partial words with one hole.

PROPOSITION 6.13

Let u be a partial word with one hole such that $\|\alpha(u)\| \geq 2$. If a is any letter, then u or ua is primitive.

PROOF   Suppose $ua \subset v^m$ and $u \subset w^n$ with v, w full words and $m \geq 2, n \geq 2$. Then $|v| = (|u| + 1)/m$ and $|w| = |u|/n$. Hence $|v| + |w| = |u|(1/m + 1/n) + 1/m < |u| + 1$. Therefore $|u| \geq |v| + |w|$. By Proposition 6.8, there exists a word x such that $v = x^k$ and $w = x^l$ for some integers k, l. It follows that $ua \subset x^{km}$ and $u \subset x^{nl}$, which implies that $\alpha(u) \subseteq \{a\}$, a contradiction. ☐

COROLLARY 6.8

Let u_1, u_2 be nonempty partial words such that $u_1 u_2$ has one hole and $\|\alpha(u_1 u_2)\| \geq 2$. Then for any letter a, $u_1 u_2$ or $u_1 a u_2$ is primitive.

PROOF   By Proposition 6.13, $u_2 u_1$ or $u_2 u_1 a$ is primitive. By Proposition 6.11, if $u_2 u_1$ is primitive, then $u_1 u_2$ is primitive, and if $(u_2)(u_1 a)$ is primitive, then $(u_1 a)(u_2)$ is primitive. The result follows. ☐

The following result holds for partial words with at least two holes.

PROPOSITION 6.14

Let u be a partial word with at least two holes such that $\|\alpha(u)\| \geq 2$. Let a be any letter and assume that $ua \subset v^m$ and $u \subset w^n$ with v, w full words and integers $m \geq 2, n \geq 2$. For all integers H satisfying $0 \leq H \leq \|H(u)\|$, let u_H be the longest prefix of u that contains exactly H holes. Then the following hold:

1. *$|u_0| < |v| + |w| - \gcd(|v|, |w|)$.*

2. *$|u_1| < |v| + |w|$.*

3. *If $|v| < |w|$, then for all integers H satisfying $2 \leq H \leq \|H(u)\|$, u_H is $(H, |v|, |w|)$-special or $|u_H| < l_{(H,|v|,|w|)}$.*

4. *If $|w| < |v|$, then for all integers H satisfying $2 \leq H \leq \|H(u)\|$, u_H is $(H, |w|, |v|)$-special or $|u_H| < l_{(H,|w|,|v|)}$.*

PROOF   Both $|v|$ and $|w|$ are periods of u. Since v ends with a, put $v = xa$. We get $u \subset (xa)^{m-1}x$ and $u \subset w^n$. We consider the following cases:

    Case 1. $m = n$

If $m = n$, then $m|x| + m - 1 = m|w|$. The latter implies $|w| = |x| + (m-1)/m$, which is impossible.

    Case 2. $m < n$

Since $n|w| + 1 = m|v|$ and $m < n$, we have $|w| < |v|$. If $|u_0| \geq |v| + |w| - \gcd(|v|, |w|)$, then by Proposition 6.6 there exists a word y such that $v = y^k$ and $w = y^l$ for some integers k, l. Therefore $ua \subset y^{km}$ and $u \subset y^{nl}$ which is contradictory since $\|\alpha(u)\| \geq 2$, and Statement 1 follows. If $|u_1| \geq$

$|v|+|w|$, then Statement 2 similarly follows using Proposition 6.8. If u_H is non $(H, |w|, |v|)$-special and $|u_H| \geq l_{(H,|w|,|v|)}$, then Statement 4 similarly follows using Proposition 6.9.

Case 3. $m > n$

Since $n|w| + 1 = m|v|$ and $m > n$, we have $|w| \geq |v|$. If $|w| > |v|$, then this case is similar to Case 2. If $|w| = |v|$, then $m = n + 1$ and $|v| = 1$. This implies that $v = a$ and $\|\alpha(u)\| \leq 1$, a contradiction. ◻

Example 6.10

If $u = b \diamond abba \diamond b$, then $ua \subset v^3$ and $u \subset w^2$ where $v = bba$ and $w = baab$. Here $|u_0| = |b| < |v| + |w| - \gcd(|v|, |w|)$, $|u_1| = |b \diamond abba| < |v| + |w|$, and $|u_2| = |b \diamond abba \diamond b| < l(2, 3, 4)$. ◻

Fourth, the following result has several interesting consequences, proving in some sense that there exist very many primitive words.

PROPOSITION 6.15

Let u be a word. If a and b are distinct letters, then ua or ub is primitive.

COROLLARY 6.9

1. *Let u be a word. Then at most one of the words ua with $a \in A$ is not primitive.*

2. *Let u_1 and u_2 be words. Then at most one of the words $u_1 a u_2$ with $a \in A$ is not primitive.*

COROLLARY 6.10

If the set $X \subset A^*$ is infinite, then there exists $a \in A$ such that $X\{a\}$ contains infinitely many primitive words.

Recall from Lemma 5.5 of Chapter 5 that if u is a partial word with one hole which is not of the form $x \diamond x$ for any word x and a, b are distinct letters, then ua or ub is primitive. The exclusion of pwords of the form $x \diamond x$ is needed since neither $x \diamond xa$ nor $x \diamond xb$ is primitive since $x \diamond xa \subset (xa)^2$ and $x \diamond xb \subset (xb)^2$.

We now describe a result that holds for any partial word u with at least two holes. Let H denote $\|H(u)\|$. Put $u = u_1 \diamond u_2 \diamond \ldots u_H \diamond u_{H+1}$ where the u_j's do not contain any holes. We define a set S_H as follows:

Do this for all $2 \leq m \leq H + 1$. If there exist a word x and integers $0 = i_0 < i_1 < i_2 < \cdots < i_{m-1} \leq H$ such that

$$u_{i_0+1} \diamond \ldots \diamond u_{i_1} \subset x$$

$$u_{i_1+1}\diamond\ldots\diamond u_{i_2} \subset x$$

$$\vdots$$

$$u_{i_{m-2}+1}\diamond\ldots\diamond u_{i_{m-1}} \subset x$$

$$u_{i_{m-1}+1}\diamond\ldots\diamond u_{H+1} \subset x$$

then put u in the set S_H. Otherwise, do not put u in S_H.

Example 6.11

Let us describe the set S_2. Here $u = u_1\diamond u_2\diamond u_3$ where the u_j's do not contain any holes.

$m = 2$: There exist a word x and integers $0 = i_0 < i_1 \le 2$ such that

$$u_{i_0+1}\diamond\ldots\diamond u_{i_1} \subset x$$
$$u_{i_1+1}\diamond\ldots\diamond u_3 \subset x$$

Here there are two possibilities: (1) $i_0 = 0$ and $i_1 = 1$; and (2) $i_0 = 0$ and $i_1 = 2$. Possibility (1) leads to $u_1 \subset x$ and $u_2\diamond u_3 \subset x$, while Possibility (2) to $u_1\diamond u_2 \subset x$ and $u_3 \subset x$. Consequently, the set S_2 contains partial words of the form $x_1 a x_2 \diamond x_1 \diamond x_2$ or $x_1 \diamond x_2 \diamond x_1 a x_2$ for words x_1, x_2 and letter a.

$m = 3$: There exist a word x and integers $0 = i_0 < i_1 < i_2 \le 2$ such that

$$u_{i_0+1}\diamond\ldots\diamond u_{i_1} \subset x$$
$$u_{i_1+1}\diamond\ldots\diamond u_{i_2} \subset x$$
$$u_{i_2+1}\diamond\ldots\diamond u_3 \subset x$$

There is only one possibility here, that is, $i_0 = 0, i_1 = 1$ and $i_2 = 2$. We get $u_1 \subset x$, $u_2 \subset x$ and $u_3 \subset x$ resulting in partial words belonging to S_2 of the form $x\diamond x\diamond x$ for some word x.

$\square$

THEOREM 6.7

Let u be a partial word with at least two holes which is not in $S_{\|H(u)\|}$. If a and b are distinct letters, then ua or ub is primitive.

PROOF   Set $\|H(u)\| = H$. Assume that $ua \subset v^k$, $ub \subset w^l$ for some primitive full words v, w and integers $k, l \ge 2$. Both $|v|$ and $|w|$ are periods of u, and, since $k, l \ge 2$, $|u| = k|v| - 1 = l|w| - 1 \ge 2\max\{|v|, |w|\} - 1 \ge |v| + |w| - 1$. Without loss of generality, we can assume that $k \ge l$ or $|v| \le |w|$. Set $u = u_1\diamond u_2\diamond\ldots u_H\diamond u_{H+1}$ where the u_j's do not contain any holes. Since v

ends with a and w with b, write $v = xa$ and $w = yb$. We have $u \subset (xa)^{k-1}x$ and $u \subset (yb)^{l-1}y$.

Case 1. $k = l$

Here $|v| = |w|$ and $|x| = |y|$. Note that $2 \leq k = l \leq H + 1$. First, assume that $k = l = H + 1$. In this case, it is clear that $u_1 = u_2 = \cdots = u_{H+1} = x$, a contradiction since $u \notin S_H$. Now, assume that $k = l \leq H$. There exist integers $0 = i_0 < i_1 < i_2 < \cdots < i_{k-1} \leq H$ such that

$$u_{i_0+1}\diamond\ldots\diamond u_{i_1}\diamond \subset xa \text{ and } u_{i_0+1}\diamond\ldots\diamond u_{i_1}\diamond \subset yb,$$

$$u_{i_1+1}\diamond\ldots\diamond u_{i_2}\diamond \subset xa \text{ and } u_{i_1+1}\diamond\ldots\diamond u_{i_2}\diamond \subset yb,$$

$$\vdots$$

$$u_{i_{k-2}+1}\diamond\ldots\diamond u_{i_{k-1}}\diamond \subset xa \text{ and } u_{i_{k-2}+1}\diamond\ldots\diamond u_{i_{k-1}}\diamond \subset yb,$$

$$u_{i_{k-1}+1}\diamond\ldots\diamond u_{H+1} \subset x \text{ and } u_{i_{k-1}+1}\diamond\ldots\diamond u_{H+1} \subset y.$$

We get

$$u_{i_0+1}\diamond\ldots\diamond u_{i_1} \subset x,$$

$$u_{i_1+1}\diamond\ldots\diamond u_{i_2} \subset x,$$

$$\vdots$$

$$u_{i_{k-2}+1}\diamond\ldots\diamond u_{i_{k-1}} \subset x,$$

$$u_{i_{k-1}+1}\diamond\ldots\diamond u_{H+1} \subset x,$$

a contradiction with the fact that $u \notin S_H$.

Case 2. $k > l$

Here $|v| < |w|$ and $|u| \geq |v| + |w|$ (otherwise, $|u| = |v| + |w| - 1$ and $k = l = 2$).

First, assume that $|u| \geq L_{(H,|v|,|w|)}$. Referring to Chapter 3, u is also $\gcd(|v|, |w|)$-periodic. However, $\gcd(|v|, |w|)$ divides $|v|$ and $|w|$, and so $u \subset z^m$ with $|z| = \gcd(|v|, |w|)$. Since v ends with a and w with b, we get that z ends with a and b, a contradiction.

Now, assume that $|u| < L_{(H,|v|,|w|)}$. Set $k = lp + r$ where $0 \leq r < l$. We consider the case where $r = 0$ (the case where $r > 0$ is left to the reader). We have that $k = lp$. The latter and the fact that $k > l$ imply that $p > 1$. Since $ua \subset (xa)^{lp}$ and $ub \subset (yb)^l$, we can write $y = x_1b_1x_2b_2\ldots x_{p-1}b_{p-1}x_p$ where $|x_1| = \cdots = |x_p| = |x|$ and $b_1,\ldots,b_{p-1} \in A$. The containments $u \subset (xa)^{lp-1}x$ and $u \subset (x_1b_1x_2b_2\ldots x_{p-1}b_{p-1}x_pb)^{l-1}x_1b_1x_2b_2\ldots x_{p-1}b_{p-1}x_p$ allow us to write $u = v_1\diamond v_2\diamond\ldots v_{l-1}\diamond v_l$ where

$$v_j \subset x \ a \ x \ a \ \ldots \ x \quad a \quad x$$
$$v_j \subset x_1 \ b_1 \ x_2 \ b_2 \ \ldots \ x_{p-1} \ b_{p-1} \ x_p$$

for all $1 \le j \le l$. If $l - 1 = H$, then $v_j = u_j = (xa)^{p-1}x$ for all j, and we obtain a contradiction with the fact that $u \notin S_H$. If $l - 1 < H$, then there exist integers $0 = i_0 < i_1 < i_2 < \cdots < i_{l-1} \le H$ such that

$$u_{i_0+1} \diamond \ldots \diamond u_{i_1} \diamond = v_1,$$

$$u_{i_1+1} \diamond \ldots \diamond u_{i_2} \diamond = v_2,$$

$$\vdots$$

$$u_{i_{l-2}+1} \diamond \ldots \diamond u_{i_{l-1}} \diamond = v_{l-1},$$

$$u_{i_{l-1}+1} \diamond \ldots \diamond u_{H+1} = v_l.$$

We get

$$u_{i_0+1} \diamond \ldots \diamond u_{i_1} \subset (xa)^{p-1}x,$$

$$u_{i_1+1} \diamond \ldots \diamond u_{i_2} \subset (xa)^{p-1}x,$$

$$\vdots$$

$$u_{i_{l-2}+1} \diamond \ldots \diamond u_{i_{l-1}} \subset (xa)^{p-1}x,$$

$$u_{i_{l-1}+1} \diamond \ldots \diamond u_{H+1} \subset (xa)^{p-1}x,$$

a contradiction with the fact that $u \notin S_H$. ▯

We end this chapter with the following two corollaries.

COROLLARY 6.11

1. Let u be a partial word which is not in $S_{\|H(u)\|}$. Then at most one of the pwords ua with $a \in A$ is not primitive.

2. Let u_1, u_2 be partial words such that u_2u_1 is not in $S_{\|H(u)\|}$. Then at most one of the pwords u_1au_2 with $a \in A$ is not primitive.

PROOF Let us first prove Statement 1. For a given partial word u which is not in $S_{\|H(u)\|}$, apply Theorem 6.7 for two symbols a and b in A. If ua is primitive, mark the symbol a, and if ub is primitive, mark the symbol b. At least a symbol is marked in this way. Continue by considering any two unmarked symbols. Eventually, at most one symbol remains unmarked, and this completes the proof. For Statement 2, at most one of the partial words u_2u_1a with $a \in A$ is not primitive. The result then follows from Proposition 6.11 since $(u_2)(u_1a)$ not primitive yields $(u_1a)(u_2)$ not primitive. ▯

COROLLARY 6.12
Let $X \subset W(A)$ not containing any partial word u in $S_{\|H(u)\|}$. If X is infinite, then there exists $a \in A$ such that $X\{a\}$ contains infinitely many primitive partial words.

PROOF Let a and b be in A. If both $X\{a\}$ and $X\{b\}$ contain only a finite number of primitive partial words, then for some integer n all the partial words of the form ua, ub with $|u| \geq n$ will be nonprimitive. However, by Theorem 6.7, $\{u\}A$ contains at most one nonprimitive partial word, a contradiction. ▯

Exercises

6.1 ⬚ Run Algorithm Primitivity Testing on the partial word $u = abca\diamond\diamond\diamond bc$. Is u primitive?

6.2 Repeat Exercise 6.1 on $u = \diamond ba\diamond aaabb$.

6.3 Describe the behaviour of Algorithm Primitivity Testing on input partial word $u = abca\diamond\diamond a$ as is done in Example 6.5.

6.4 Compute $P_{0,3}(10)$ using Equality 6.6.

6.5 Using the formulas in this chapter, compute $T_{2,3}(n)$, $P_{2,3}(n)$ and $N_{2,3}(n)$ for $n = 15$ and $n = 16$.

6.6 ⬚ Compute $N_{2,3}(17)$ and $N_{2,3}(19)$.

6.7 What are the periods, weak periods, and exact periods of the partial word $ab\diamond\diamond bba\diamond babb$?

6.8 Show that $\nu(w)$ is not necessarily equal to $\|\mathcal{E}(w)\|$.

6.9 ⬚ Show that the equality $P_{1,k}(n) = n(P_{0,k}(n) + k^{n-1} - k^n)$ holds.

6.10 Prove Proposition 6.6.

6.11 Proposition 6.12 holds for partial words with at least two holes. True or False?

6.12 ⬚ Give an example to show that Proposition 6.15 does not hold for partial words with at least two holes.

Challenging exercises

6.13 Prove that Equality 6.4 is always true.

6.14 Deduce from Equality 6.4 that two functions ϕ, ψ from $\mathbb{P}$ to $\mathbb{Z}$ are related by $\sum_{d|n} \psi(d) = \phi(n)$ if and only if $\sum_{d|n} \mu(d)\phi(\frac{n}{d}) = \psi(n)$.

6.15 Prove Theorem 6.2.

6.16 Do a case analysis for $N_{1,2}(6)$ as is done for $N_{1,2}(8)$ in Example 6.7.

6.17 Prove Lemma 6.6 for nonprimitive pwords with two holes.

6.18 [s] Prove Proposition 6.8.

6.19 [s] Prove Proposition 6.10.

6.20 [s] Describe the set S_3.

6.21 Let u be a partial word with at least two holes which is not in $S_{\|H(u)\|}$. Let a, b be distinct letters and assume that $ua \subset v^m$ and $ub \subset w^n$ with v, w full words and integers $m \geq 2, n \geq 2$. For all integers H satisfying $0 \leq H \leq \|H(u)\|$, let v_H be the longest prefix of u that contains exactly H holes. Then show that the following hold:

1. $|v_0| < |v| + |w| - \gcd(|v|, |w|)$.
2. $|v_1| < |v| + |w|$.
3. If $|v| < |w|$, then for all integers H satisfying $2 \leq H \leq \|H(u)\|$, v_H is $(H, |v|, |w|)$-special or $|v_H| < l_{(H,|v|,|w|)}$.
4. If $|w| < |v|$, then for all integers H satisfying $2 \leq H \leq \|H(u)\|$, v_H is $(H, |w|, |v|)$-special or $|v_H| < l_{(H,|w|,|v|)}$.

6.22 Prove the case where $r > 0$ of Theorem 6.7.

Programming exercises

6.23 Give pseudo programming language code for the upgrade of *Algorithm Primitivity Testing* when the checking of whether or not u is compatible with $U[k..k + |u|)$ is done simultaneously with the checking of whether or not u is (k, l)-special as described in the proof of Theorem 6.1.

6.24 Obtain numerical values for $P_{h,k}(n)$ and $N_{h,k}(n)$ by having a computer generate and count all possible primitive and nonprimitive partial words over an alphabet of size $k = 4$ with number of holes $h \in \{0, 1, 2, 3\}$ and length n ranging from 1 to 10.

6.25 Write a program that lists the partial words in the sets $\mathcal{T}_{h,k}(n), \mathcal{P}_{h,k}(n)$ and $\mathcal{N}_{h,k}(n)$. Test your program on $h = 1, k = 3$ and $n = 4$.

6.26 Fill the entries in the table when $h = 2$ and $k = 3$ for the lengths $n = 18$ and $n = 20$.

6.27 Write a program that computes the $R_{h,k}(n, d)$'s of Equality 6.9. Run your program on $h = 2$, $k = 3$ and $n = 12$.

Website

A World Wide Web server interface at

$$\texttt{http://www.uncg.edu/mat/primitive}$$

has been established for automated use of Algorithm Primitivity Testing. Another at

$$\texttt{http://www.uncg.edu/mat/research/primitive2}$$

implements the formulas of Sections 6.2, 6.3, 6.4 and 6.5.

Bibliographic notes

Primitive words play an important role in numerous research areas including formal language theory [66, 67], coding theory [16, 133], and combinatorics on words [51, 106, 107, 108, 109].

The zero-hole case of primitivity testing described in Section 6.1 is discussed in Choffrut and Karhumäki [51] and the one-hole case in Blanchet-Sadri and Chriscoe [23]. As described in Chapter 5, Blanchet-Sadri and Chriscoe extended to partial words with one hole the well known result of Guibas and Odlyzko that states that the sets of periods of words are independent of the alphabet size. They obtained, as a consequence of their constructive proof, a linear time algorithm which, given a partial word with one hole, computes a binary one with the same sets of periods and the same sets of weak periods.

The algorithm requires primitivity testing of partial words with one hole. The arbitrary case of primitivity testing is from Blanchet-Sadri and Anavekar [18].

Proposition 6.6 is from Lothaire [106] and Corollary 6.5 from Lyndon and Schützenberger [109]. Proposition 6.10 is from Shyr and Thierrin [133] and Proposition 6.12 from Shyr [132]. Proposition 6.15, Corollary 6.9 and Corollary 6.10 are from Păun et al. [118]. Theorem 6.7 is from Blanchet-Sadri, Corcoran and Nyberg [25]. The other results of Section 6.6 are from Blanchet-Sadri [17] as well as Exercise 6.21.

The counting of primitive full words is discussed in [119] and several sequences based on that counting appear in Sloane's database. Exercises 6.13 and 6.14 are from Lothaire's book [106]. The counting results of the rest of the chapter are from Blanchet-Sadri and Cucuringu [26].

Chapter 7

Unbordered Partial Words

In this chapter, we study *unbordered partial words* which turn out to be a particularly interesting class of primitive partial words (see Exercise 1.11). Recall that a full word u is *unbordered* if none of its proper prefixes is one of its suffixes. In the case of a partial word u, we have the following definition.

DEFINITION 7.1 *We call a partial word u **unbordered** if no nonempty words x, v, w exist such that $u \subset xv$ and $u \subset wx$. If such nonempty words x, v, w exist, then we call u **bordered** and call x a **border** of u. A border x of u is called **minimal** if $|x| > |y|$ implies that y is not a border of u.*

Note that there are two types of borders. Writing u as $x_1 v = w x_2$ where $x_1 \subset x$ and $x_2 \subset x$, we say that x is an *overlapping* border if $|x| > |v|$, and a *nonoverlapping* border otherwise. Figures 7.1 and 7.2 highlight these definitions.

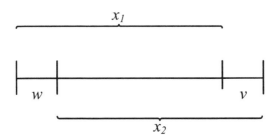

FIGURE 7.1: An overlapping border.

The following example illustrates the above concepts.

Example 7.1

The partial word $u = ab\diamond c\diamond ba$ is bordered with borders a and aba, the first one being minimal, while the partial word $ab\diamond c$ is unbordered. The pword $a\diamond b$ is bordered with overlapping border ab

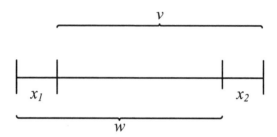

FIGURE 7.2:   A nonoverlapping border.

$$
\begin{array}{c}
a \diamond b \\
\underline{\quad a \diamond b} \\
a\ b
\end{array}
$$

Here $x_1 = a\diamond$ and $x_2 = \diamond b$.   ⬚

7.1   Concatenations of prefixes

We start with a definition.

DEFINITION 7.2   *For partial words u, v, we write $\boldsymbol{u} \ll \boldsymbol{v}$ if there exists a sequence $v_0, \ldots, v_{n-1}$ of prefixes of v such that $u = v_0 \ldots v_{n-1}$.*

Obviously, $\varepsilon \ll u$ and $u \ll u$. The reader can check that if $u \ll v$ and $v \ll w$, then $u \ll w$.

THEOREM 7.1
Let u, v be full words such that $u \neq \varepsilon$ and $u \ll v$. Then there exists a unique sequence $v_0, \ldots, v_{n-1}$ of nonempty unbordered prefixes of v such that $u = v_0 \ldots v_{n-1}$.

PROOF   The proof is left as an exercise.   ⬚

In this section, we extend Theorem 7.1 to partial words. In order to do this, we introduce two types of bordered partial words: the *well bordered* and the *badly bordered* partial words.

DEFINITION 7.3   *Let u be a nonempty bordered partial word. Let x be a minimal border of u, and set $u = x_1 v = w x_2$ where $x_1 \subset x$ and $x_2 \subset x$.*

We call u **well bordered** if x_1 is unbordered. Otherwise, we call u **badly bordered**.

Example 7.2
First, consider the partial word $u = ab\diamond$. We can factorize u as $x_1v = (a)(b\diamond) = (ab)(\diamond) = wx_2$ where $x_1 \subset a$ and $x_2 \subset a$. Since $x_1 = a$ is unbordered, u is well bordered. However, the partial word $a\diamond b$ is badly bordered. Indeed, $a\diamond b = x_1v = (a\diamond)(b) = (a)(\diamond b) = wx_2$ with $x_1 \subset ab$ and $x_2 \subset ab$. In this case, x_1 is bordered. □

For convenience, we will at times refer to a minimal border of a well bordered partial word as a *good border* and of a badly bordered partial word as a *bad border*.

As a result of x being a bad border, we have the following Lemma.

LEMMA 7.1
Let u be a nonempty badly bordered partial word. Let x be a minimal border of u, and set $u = x_1v = wx_2$ where $x_1 \subset x$ and $x_2 \subset x$. Then there exists i such that $i \in H(x_1)$ and $i \in D(x_2)$.

PROOF   Since x_1 is bordered, $x_1 = r_1s_1 = s_2r_2$ for nonempty partial words r_1, r_2, s_1, s_2 where $s_1 \subset s$ and $s_2 \subset s$ for some s. If no i exists such that $i \in H(x_1)$ and $i \in D(x_2)$, then x_2 must also be bordered. So $x_2 = r_1's_1' = s_2'r_2'$ where $r_1' \subset r_1$, $r_2' \subset r_2$, $s_1' \subset s$ and $s_2' \subset s$, thus $s_2 \uparrow s_1'$. This means that there exists a border of u of length shorter that $|x|$ which contradicts the fact that x is a minimal border of u. □

Our goal is to extend Theorem 7.1 to partial words or to construct, given any partial words u and v satisfying $u \ll v$, a sequence of nonempty unbordered prefixes of v, $v_0, \ldots, v_{n-1}$, such that $u \uparrow v_0 \ldots v_{n-1}$. We will see that if during the construction of the sequence a badly bordered prefix is encountered, then the desired sequence may not exist. We first prove two propositions.

PROPOSITION 7.1
If v is a partial word, then there do not exist two distinct compatible sequences of nonempty unbordered prefixes of v.

PROOF   Suppose that $v_0 \ldots v_{n-1} \uparrow v_0' \ldots v_{m-1}'$ where each v_i and each v_i' is a nonempty unbordered prefix of v. If there exists $i \geq 0$ such that $|v_0| = |v_0'|, \ldots, |v_{i-1}| = |v_{i-1}'|$ and $|v_i| < |v_i'|$, then $v_0 = v_0', \ldots, v_{i-1} = v_{i-1}'$ and v_i is a prefix of v_i'. By simplification, $v_i \ldots v_jx \uparrow v_i'$ where $i \leq j < n-1$ and x is a nonempty prefix of v_{j+1}. The fact that x, v_i' are prefixes of v satisfying

$|v_i'| > |x|$ implies that x is a prefix of v_i'. In addition, x is compatible with the suffix of length $|x|$ of v_i', and consequently v_i' is bordered. Similarly, there exists no $i \geq 0$ such that $|v_0| = |v_0'|, \ldots, |v_{i-1}| = |v_{i-1}'|$ and $|v_i| > |v_i'|$. Clearly, $n = m$ and uniqueness follows. □

PROPOSITION 7.2

Let u be a nonempty bordered partial word. Let x be a minimal border of u, and set $u = x_1 v = w x_2$ where $x_1 \subset x$ and $x_2 \subset x$. Then the following hold:

1. *The partial word x is unbordered.*

2. *If u is well bordered, then $u = x_1 u' x_2 \subset x u' x$ for some u'.*

PROOF   For Statement 1, assume that r is a border of x, that is, $x \subset rs$ and $x \subset s'r$ for some nonempty partial words r, s, s'. Since $u \subset xv$ and $x \subset rs$, we have $u \subset rsv$, and similarly, since $u \subset wx$ and $x \subset s'r$, we have $u \subset ws'r$. Then r is a border of u. Since x is a minimal border of u, we have $|x| \leq |r|$ contradicting the fact that $|r| < |x|$. This proves (1).

For Statement 2, if $|v| < |x|$, then $u = wtv$ for some t. Here $x_1 = wt = t'w'$ for some t', w' satisfying $|t| = |t'|$ and $|w| = |w'|$. Since $x_1 \uparrow x_2$, we have $t'w' \uparrow tv$ and by simplification, $t' \uparrow t$. The latter implies the existence of a partial word t'' such that $t' \subset t''$ and $t \subset t''$. So $x_1 = t'w' \subset t''w'$ and $x_1 = wt \subset wt''$. Then t'' is a border of x_1 and x_1 is bordered. According to the definition of u being well bordered, x_1 is an unbordered partial word and this leads to a contradiction. Hence, we have $|v| \geq |x|$ and, for some u', we have $v = u'x_2$ and $w = x_1 u'$, and $u = wx_2 = x_1 u' x_2 \subset xu'x$. This proves (2). □

Note that Proposition 7.2 implies that if u is a nonempty bordered full word, then u is well bordered. In this case, $u = xu'x$ where x is the minimal border of u.

LEMMA 7.2

If u, v are nonempty partial words such that $u = v_0 \ldots v_{n-1}$ where $v_0, \ldots, v_{n-1}$ is a sequence of nonempty unbordered prefixes of v, then there exists a unique sequence $v_0', \ldots, v_{m-1}'$ of nonempty unbordered prefixes of v such that $u \uparrow v_0' \ldots v_{m-1}'$ (the desired sequence is just $v_0, \ldots, v_{n-1}$).

PROOF   If each prefix v_i is unbordered, then the sequence $v_0, \ldots, v_{n-1}$ of nonempty unbordered prefixes of v is such that $u = v_0 \ldots v_{n-1} \uparrow v_0 \ldots v_{n-1}$ and so the existence follows. To show uniqueness, assume $v_0', \ldots, v_{m-1}'$ is another such sequence. We get $u = v_0 \ldots v_{n-1} \uparrow v_0' \ldots v_{m-1}'$ contradicting Proposition 7.1. □

The badly bordered partial words are now split into the *specially bordered* and the *nonspecially bordered* partial words according to the following definition.

DEFINITION 7.4   *Let u be a nonempty partial word that is badly bordered. Let x be a minimal border of u, and set $u = x_1 v = w x_2$ where $x_1 \subset x$ and $x_2 \subset x$. If there exists a proper factor x' of u such that $x_1 \not\uparrow x'$ and $x' \uparrow x_2$, then we call u **specially bordered**. Otherwise, we call u **nonspecially bordered**.*

LEMMA 7.3
Let u, v be nonempty partial words such that $u = v_0 \ldots v_{n-1}$ where $v_0, \ldots, v_{n-1}$ is a sequence of nonempty prefixes of v with some v_i badly bordered. Let y be a minimal border of v_i, and set $v_i = xw' = wx'$ where $x \subset y$ and $x' \subset y$ (and thus $x \uparrow x'$). If there exists a sequence $v'_0, \ldots, v'_{m-1}$ of nonempty unbordered prefixes of v such that $v_i \uparrow v'_0 \ldots v'_{m-1}$, then $|x| < |v'_{m-1}|$ and v_i is specially bordered.

PROOF   By Definition 7.3, x is bordered. If $|x| = |v'_{m-1}|$, then both x and v'_{m-1} are prefixes of v, and thus $x = v'_{m-1}$. We get that x is unbordered, a contradiction. If $|x| > |v'_{m-1}|$, then set $x' = zv'$ where $|v'| = |v'_{m-1}|$. Since both x and v'_{m-1} are prefixes of v, we get that v'_{m-1} is a prefix of x. So $x = v'_{m-1} z'$ for some z', and $v_i = v'_{m-1} z'w' = wzv'$ with $v'_{m-1} \uparrow v'$. Thus v_i has a border of length $|v'_{m-1}| < |x| = |y|$ contradicting the fact that y is a minimal border. And so $|x| < |v'_{m-1}|$.

Since $v_i \uparrow v'_0 \ldots v'_{m-1}$, we have $|v'_{m-1}| \leq |v_i|$. Both v_i and v'_{m-1} being prefixes of v, it results that v'_{m-1} is a prefix of v_i. Hence, since $v_i = xw$ and $|v'_{m-1}| > |x|$ there exists z such that $xz = v'_{m-1}$. Since $v_i = wx'$ and v'_{m-1} is compatible with a suffix of v_i, we have $v'_{m-1} \uparrow z'x'$ for some z'. Thus, we get that $v'_{m-1} = xz \uparrow z'x'$. Since $v'_{m-1} \uparrow z'x'$, set $v'_{m-1} = z''x''$ where $z'' \uparrow z'$ and $x'' \uparrow x'$. So $v'_{m-1} = z''x'' = xz$. If $x'' \uparrow x$, then v'_{m-1} is bordered, a contradiction with the fact that v'_{m-1} is unbordered. Thus $x'' \not\uparrow x$, and since v'_{m-1} is a prefix of v_i, we have that v_i is specially bordered. $\square$

The following example illustrates Lemma 7.3.

Example 7.3
Consider the partial words

$$u = aaaa \diamond aabbaaaaa \diamond baa \text{ and } v = aa \diamond aabbaaaaa \diamond b$$

The factorization $u = (a)(a)(aa \diamond aabbaaaaa \diamond b)(a)(a)$ shows that u can be written as a sequence of nonempty prefixes of v. Here, the third factor is specially bordered and is compatible with a sequence of unbordered prefixes of v. Indeed, the compatibility

$$aa \diamond aabbaaaaa \diamond b \uparrow (aa \diamond aabb)(aa \diamond aabb)$$

holds. The shortest border of that factor is aab which has length shorter than $aa \diamond aabb$. ⬜

LEMMA 7.4

Let u, v be nonempty partial words such that $u = v_0 \ldots v_{n-1}$ where $v_0, \ldots, v_{n-1}$ is a sequence of nonempty prefixes of v with some v_i well bordered. Then there exists a longest sequence $v'_0, v'_1, \ldots, v'_{m-1}$ of nonempty prefixes of v such that $v_i \uparrow v'_0 v'_1 \ldots v'_{m-1}$, v'_j is unbordered for every $1 \le j < m$, and v'_0 is unbordered or badly bordered. Moreover, the following hold:

1. *If v'_0 is unbordered, then a sequence of nonempty unbordered prefixes of v exists that is compatible with v_i.*

2. *If v'_0 is badly bordered, then no sequence of nonempty unbordered prefixes of v exists that is compatible with v_i.*

PROOF　Let $y_{i,0}$ be a minimal border of $w_{i,0} = v_i$, and set $w_{i,0} = x_{i,0} w'_{i,0} = w_{i,1} x'_{i,0}$ where $x_{i,0} \subset y_{i,0}$ and $x'_{i,0} \subset y_{i,0}$ (and thus $x_{i,0} \uparrow x'_{i,0}$). By Definition 7.3, $x_{i,0}$ is unbordered, and

$$v_i = w_{i,1} x'_{i,0} \uparrow w_{i,1} x_{i,0} \tag{7.1}$$

where both $w_{i,1}$ and $x_{i,0}$ are prefixes of $w_{i,0}$ (and hence of v). If $w_{i,1}$ is unbordered, then v_i is compatible with a sequence of nonempty unbordered prefixes of v.

If $w_{i,1}$ is badly bordered, then no sequence $v''_0, \ldots, v''_{m'-1}$ of nonempty unbordered prefixes of v exists that is compatible with $w_{i,1}$ unless $w_{i,1}$ is specially bordered and $|y_{i,1}| < |v''_{m'-1}|$ by Lemma 7.3 (here $y_{i,1}$ is a minimal border of $w_{i,1}$). If this is the case, then $w_{i,1}$ may be compatible with such a sequence of nonempty unbordered prefixes of v, and if so replace $w_{i,1}$ on the right hand side of the compatibility in (1) by $v''_0 \ldots v''_{m'-1}$. If this is not the case, then no sequence of nonempty unbordered prefixes of v exists that is compatible with v_i.

If $w_{i,1}$ is well bordered, then repeat the process. Let $w_{i,0}, w_{i,1}, \ldots, w_{i,j-1}$ be the longest sequence of nonempty well bordered prefixes defined in this manner. For all $0 \le k < j$, let $y_{i,k}$ be a minimal border of $w_{i,k}$, and set $w_{i,k} = x_{i,k} w'_{i,k+1} = w_{i,k+1} x'_{i,k}$ where $x_{i,k} \subset y_{i,k}$ and $x'_{i,k} \subset y_{i,k}$ (and thus $x_{i,k} \uparrow x'_{i,k}$). By Definition 7.3, $x_{i,0}, \ldots, x_{i,j-1}$ are unbordered. We have $w_{i,j-1} = w_{i,j} x'_{i,j-1} \uparrow w_{i,j} x_{i,j-1}$ and thus by induction,

$$v_i = w_{i,j} x'_{i,j-1} \ldots x'_{i,0} \uparrow w_{i,j} x_{i,j-1} \ldots x_{i,0} \tag{7.2}$$

where $w_{i,j}, x_{i,j-1}, \ldots, x_{i,0}$ are prefixes of $w_{i,0}$ (and hence of v). Now, if $w_{i,j}$ is unbordered, then v_i is compatible with a sequence of nonempty unbordered prefixes of v. If $w_{i,j}$ is badly bordered, then proceed as in the case above when $w_{i,1}$ is badly bordered.

We can thus equate v_i with sequences of shorter and shorter factors that are prefixes of v or compatible with prefixes of v and the existence of the required sequence $v'_0, \ldots, v'_{m-1}$ is established. $\square$

THEOREM 7.2

If u, v be nonempty partial words such that $u \ll v$, then let $v_0, \ldots, v_{n-1}$ be a sequence of nonempty prefixes of v such that $u = v_0 \ldots v_{n-1}$. Then one of the following holds:

1. *There exists a sequence $v'_0, \ldots, v'_{m-1}$ of nonempty unbordered prefixes of v such that $u \uparrow v'_0 \ldots v'_{m-1}$.*

2. *There exists a longest sequence $v'_0, \ldots, v'_{m-1}$ of nonempty unbordered or badly bordered prefixes of v such that $u \uparrow v'_0 \ldots v'_{m-1}$ with some v'_j badly bordered, and no sequence of nonempty unbordered prefixes of v exists that is compatible with u.*

PROOF Given a sequence $v_0, \ldots, v_{n-1}$ of nonempty prefixes of v such that $u = v_0 \ldots v_{n-1}$, we wish to construct a sequence $v'_0, \ldots, v'_{m-1}$ of nonempty unbordered prefixes of v such that $u \uparrow v'_0 \ldots v'_{m-1}$. If each prefix v_i is unbordered, then proceed as in Lemma 7.2. If some v_i is badly bordered (respectively, well bordered), then proceed as in Lemma 7.3 (respectively, Lemma 7.4). $\square$

Example 7.4
Consider the partial words

$$u = aaaa \diamond babbaaaaa \diamond baa \text{ and } v = aa \diamond babbaaaaa \diamond b$$

We have a factorization of u in terms of nonempty prefixes of v. Here, the compatibility

$$u \uparrow (a)(a)(aa \diamond babbaaaaa \diamond b)(a)(a)$$

consists of unbordered and badly bordered prefixes of v and is a longest such sequence ($aa \diamond babbaaaaa \diamond b$ is specially bordered and is not compatible with any sequence of nonempty unbordered prefixes of v). We can check that no sequence of nonempty unbordered prefixes of v exists that is compatible with u. $\square$

We now describe an algorithm based on Theorem 7.2 that is given as input a partial word v and a sequence of prefixes of v and that outputs (if it exists) a sequence of unbordered prefixes of v compatible with the given sequence.

ALGORITHM 7.1

The algorithm consists of five steps.

Step 1: *Compute the length of the input partial word v denoted by n.*

Step 2: *Create the set S of all nonempty prefixes of v.*

Step 3: *Create the set S' containing all nonempty unbordered prefixes of v. For each prefix s in S, first compute its length which is denoted by m. If the length of the object is one, then put this object in the set S'. Otherwise, check the object to see if it is bordered. If none of the object's proper prefixes are suffixes, the enter the object into the set S'.*

Step 4: *Create the ordered multiset T' containing the input sequence in the order inputted.*

Step 5: *For each object t in T', first denote the object's length by l. If the length of the object is $l = 1$, then the object is put into an ordered multiset T. If $l > 1$, then check to see whether the object is bordered. If there is no proper prefix of the object that is compatible with a suffix, then enter the object into T. Otherwise, let $t[0..i]$ be the shortest prefix for which this happens. If $t[0..i]$ is unbordered, then replace the object t with two shorter objects, $t[0..l-i-2]$ and $t[0..i]$ (the order in which the new objects are entered in T is important). If $t[0..i]$ is bordered, then the algorithm checks to see if there exists a sequence of nonempty unbordered prefixes of v that is compatible with t. If yes, then T is updated with the sequence. If no, then the algorithm returns "No sequence exists" and exits. When all objects in T' have been examined, update T' with T. If T' is a subset of S', then T is a set of unbordered prefixes of v and the algorithm returns T'. Otherwise, repeat Step 5.*

When all computations are done, the algorithm either returns a sequence in T' or returns "No sequence exists."

We illustrate the algorithm with the following two examples.

Example 7.5

First consider $v = ab\diamond c\diamond ba$ and the sequence

$$ab\diamond c, a, ab\diamond, ab\diamond c\diamond ba, a$$

of prefixes of v. The following table depicts the information submitted:

partial word v	$ab\diamond c\diamond ba$
prefix sequence	$(ab\diamond c, a, ab\diamond, ab\diamond c\diamond ba, a)$

The set S contains all nonempty prefixes of v, while the set S' contains all nonempty unbordered prefixes of v.

$$S = \{a, ab, ab\diamond, ab\diamond c, ab\diamond c\diamond, ab\diamond c\diamond b, ab\diamond c\diamond ba\}$$
$$S' = \{a, ab, ab\diamond c\}$$

In the first iteration, the multiset T' contains the input sequence. During subsequent iterations, it is determined whether each object in T' is well ordered, badly bordered, or unbordered. If the object is well bordered, it is split into two smaller objects, and T' is updated. Otherwise T' is updated and the algorithm continues until either a badly bordered object is found or $T' \subset S'$.

Iteration	T'
1	$\{ab\diamond c, a, ab\diamond, ab\diamond c\diamond ba, a\}$
2	$\{ab\diamond c, a, ab, a, ab\diamond c\diamond b, a, a\}$
3	$\{ab\diamond c, a, ab, a, ab\diamond c, ab, a, a\}$

Since $T' \subset S'$, a sequence of unbordered prefixes of v does exist that is compatible with the original sequence:

$$ab\diamond c, a, ab, a, ab\diamond c, ab, a, a$$

⬜

Example 7.6

Now consider $w = a\diamond\diamond bacb$ and the sequence $a\diamond\diamond, a\diamond, a\diamond\diamond ba$ of prefixes of w. Here

$$S = \{a, a\diamond, a\diamond\diamond, a\diamond\diamond b, a\diamond\diamond ba, a\diamond\diamond bac, a\diamond\diamond bacb\}$$
$$S' = \{a\}$$

and we get

Iteration	T'
1	$\{a\diamond\diamond, a\diamond, a\diamond\diamond ba\}$
2	$\{a\diamond, a, a, a, a\diamond\diamond b, a\}$

The partial word $a\diamond\diamond b \in T'$ is badly bordered (nonspecially bordered), therefore no sequence exists.

⬜

7.2 More results on concatenations of prefixes

We start with a definition.

DEFINITION 7.5 *If u is a nonempty partial word, then $\mathrm{unb}(u)$ denotes the longest unbordered prefix of u.*

Example 7.7
The partial word $u = ab\diamond b$ has $\varepsilon, a, ab, ab\diamond$, and $ab\diamond b$ has prefixes. The latter two are bordered, while ab is unbordered. Therefore, the longest unbordered prefix of u is $\text{unb}(u) = ab$. ▯

It is left as an exercise to show that if u, v are full words such that $u = \text{unb}(u)v$, then $v \ll \text{unb}(u)$ (see Exercise 7.18). This does not extend to partial words as $u = (ab)(\diamond b) = \text{unb}(u)v$ provides a counterexample. However, the following lemma does hold.

LEMMA 7.5
Let u, v be partial words such that $u \neq \varepsilon$ and $u = \text{unb}(u)v$. Then $u \ll \text{unb}(u)$ if and only if $v \ll \text{unb}(u)$.

PROOF If $v \ll \text{unb}(u)$, then obviously $u \ll \text{unb}(u)$. For the other direction, since $u \ll \text{unb}(u)$, we can write $u = u_0 u_1 \ldots u_{n-1}$ where each u_i is a nonempty prefix of $\text{unb}(u)$. We can suppose that $v \neq \varepsilon$. Then $\text{unb}(u) = u_0 \ldots u_k u'$ for some $k < n - 1$ and some prefix u' of u_{k+1}. Since $\text{unb}(u)$ is unbordered, we have that $u' = \varepsilon$, that $k = 0$, and hence that $\text{unb}(u) = u_0$. It follows that $v = u_1 \ldots u_{n-1}$ and $v \ll \text{unb}(u)$. ▯

We get the following corollary.

COROLLARY 7.1
Let u, v be partial words with v nonempty. Then the following hold:

1. *If $u \ll \text{unb}(v)$, then $u \ll v$.*

2. *If w is a partial word such that $v = \text{unb}(v)w$ and $w \ll \text{unb}(v)$, then $u \ll v$ if and only if $u \ll \text{unb}(v)$.*

PROOF Statement 1 holds trivially. For Statement 2, by Lemma 7.5, $w \ll \text{unb}(v)$ if and only if $v \ll \text{unb}(v)$. Now, if $u \ll v$, then since $v \ll \text{unb}(v)$, by transitivity we get $u \ll \text{unb}(v)$. ▯

REMARK 7.1 Statement 2 of Corollary 7.1 is not true in general. Indeed, $u = ababac\diamond aab$ and $v = abac\diamond aba$ provide a counterexample. To see this, $v = (abac)(\diamond aba) = \text{unb}(v)w$ and we have $u \ll v$ since $u = (ab)(abac\diamond a)(ab)$ where ab and $abac\diamond a$ are prefixes of v. However $u \not\ll \text{unb}(v)$ (here $w \not\ll \text{unb}(v)$). For full words u, v, $u \ll v$ if and only if $u \ll \text{unb}(v)$ (see Exercise 7.19). ▯

DEFINITION 7.6 *For partial words u and v, when both $u \ll v$ and $v \ll u$ we write $\boldsymbol{u \approx v}$.*

Note that the relation $\approx$ is an equivalence relation. For full words u and v, $u \approx v$ if and only if $\mathrm{unb}(u) = \mathrm{unb}(v)$ (see Exercise 7.20). For partial words, the following holds.

PROPOSITION 7.3
For partial words u and v, if $u \approx v$, then $\mathrm{unb}(u) = \mathrm{unb}(v)$.

PROOF   Suppose that $u \approx v$. Set $v = \mathrm{unb}(v)w$ for some partial word w. Since $u \ll v$, we can write $u = v_0 \ldots v_{n-1}$ where each v_i is a nonempty prefix of v. Since $v \ll u$, there exists a sequence of nonempty prefixes of u, say $u_0, \ldots, u_{m-1}$, such that $v = u_0 u_1 \ldots u_{m-1}$. Since $\mathrm{unb}(v)$ is a prefix of v, we have $\mathrm{unb}(v) = u_0 \ldots u_k u'$ where u' is a prefix of u_{k+1} and $k < m-1$. Since $\mathrm{unb}(v)$ is unbordered, we have $u' = \varepsilon$, $k = 0$, and $\mathrm{unb}(v) = u_0$. Therefore, both $u = \mathrm{unb}(v)x$ and $\mathrm{unb}(u) = \mathrm{unb}(v)y$ hold for some x, y. It follows that $\mathrm{unb}(v)$ is a prefix of $\mathrm{unb}(u)$. Similarly, $\mathrm{unb}(u)$ is a prefix of $\mathrm{unb}(v)$.   ◻

REMARK 7.2   The converse of Proposition 7.3 does not necessarily hold for partial words as is seen by considering $u = aba\diamond$ and $v = ab\diamond b$. We have $\mathrm{unb}(u) = ab = \mathrm{unb}(v)$ but $u \not\approx v$.   ◻

It is left as an exercise to show that if v is an unbordered word and w is a proper prefix of v for which $u \ll w$, then uv and wv are unbordered. For partial words, we can prove the following.

LEMMA 7.6
Let u be an unbordered partial word. Then the following hold:

  1. *If $v \in P(u)$ and $v \neq u$, then vu is unbordered.*

  2. *If $v \in S(u)$ and $v \neq u$, then uv is unbordered.*

PROOF   Let us prove Statement 1 (the proof of Statement 2 is similar). Set $u = vx$ for some x. If $vu = vvx$ is bordered, then there exist nonempty partial words r, s, s' such that $vvx \subset rs$ and $vvx \subset s'r$. If $|r| \leq |v|$, then $u = vx$ is bordered by r. And if $|r| > |v|$, then $r = v'y$ where $|v'| = |v|$ and this implies that $u = vx$ is bordered by y. In either case, we get a contradiction with the assumption that u is unbordered.   ◻

LEMMA 7.7
If v is an unbordered partial word and $u \ll v$ and $u \neq v$, then uv is unbordered.

PROOF   Since $u \ll v$, we can write $u = v_0 v_1 \ldots v_{n-1}$ where each v_i is a prefix of v. Therefore, any prefix of u is a concatenation of prefixes of v. Assume that uv is bordered by y. If $|y| > |u|$, then set $y = u'y'$ with $u \subset u'$. We get y' a border of v contradicting the fact that v is unbordered. If $|y| \leq |u|$, then we have the following two cases:

Case 1. y contains a prefix of v_0

Here y contains a prefix of v and also a suffix of v and therefore, y is a border of the unbordered word v.

Case 2. $v_0 \dots v_k v' \subset y$ where v' is a prefix of v_{k+1}

If $v' = \varepsilon$, then $v_0 \dots v_k \subset y$ where v_k is a prefix of v. This results in a suffix of y containing both a prefix and a suffix of v. Similarly, if $v' \neq \varepsilon$, then factor y as $y = y_1 y_2$ where $v' \subset y_2$. Because v' is a prefix of v, we can write $v = v'z \subset y_2 z$. But because $|y_2| < |v|$ and we have assumed that uv is bordered by $y = y_1 y_2$, we must have that $v = z'v''$ with $v'' \subset y_2$. Therefore y_2 is a border for v. In either case, we get a contradiction with the fact that v is unbordered. ☐

It is left as an exercise to show that if $u - xv$ is a nonempty unbordered word where x is the longest unbordered proper prefix of u, then v is unbordered. The partial word $u = ab \diamond ac$ where $x = ab$ and $v = \diamond ac$ and the partial word $u = abaca \diamond c$ where $x = abac$ and $v = a \diamond c$ provide counterexamples for partial words. However, when v is full, the following theorem does hold.

THEOREM 7.3
Let u be a nonempty unbordered partial word. Then the following hold:

1. *Let x be the longest proper unbordered prefix of u and let v be such that $u = xv$. If v is a full word, then v is unbordered.*

2. *Let y be the longest proper unbordered suffix of u and let w be such that $u = wy$. If w is a full word, then w is unbordered.*

PROOF We prove Statement 1 (Statement 2 can be proved similarly). Assume that v is bordered. Since v is full, there exist nonempty words z, v' such that $v = zv'z$ where z is the minimal border of v. Then $u = xzv'z$, so that xz is a proper prefix of u such that $|xz| > |x|$. It follows that xz is bordered, and there exist nonempty partial words r, r_1, r_2, s_1, s_2 such that $xz = r_1 s_1 = s_2 r_2$, $r_1 \subset r$ and $r_2 \subset r$ (here r is a minimal border). Let us consider the following two cases.

Case 1. $|r| > |z|$

In this case, $r_2 = x'z$ where x' is a nonempty suffix of x. Since $r_1 \uparrow r_2$, there exist partial words x'', z' such that $r_1 = x''z'$ where $x'' \uparrow x'$ and $z' \uparrow z$. But then, $x''z's_1 = r_1 s_1 = xz = s_2 r_2 = s_2 x'z$. It follows that x'' is a prefix of x and x' is a suffix of x that are compatible. As a result, x is bordered.

Case 2. $|r| \leq |z|$

In this case, r_2 is a suffix of z and set $z = sr_2$ for some s. We get $u = xzv'z = r_1s_1v'sr_2 \subset rs_1v'sr$, whence r is a border of the unbordered partial word u. ☐

A closer look at the proof of Theorem 7.3 allows us to show the following.

THEOREM 7.4

Let u be a nonempty partial word. Then the following hold:

1. *Let x be the longest proper unbordered prefix of u and let v be such that $u = xv$. If v is bordered, then set $v = z_1v_1 = v_2z_2$ where $z_1 \subset z, z_2 \subset z$ and where z is a minimal border of v. Then xz_1 has a minimal border r such that $|r| \leq |z|$. Moreover, if v is well bordered, then $|x| \geq |r|$.*

2. *Let y be the longest proper unbordered suffix of u and let w be such that $u = wy$. If w is bordered, then set $w = z_1v_1 = v_2z_2$ where $z_1 \subset z, z_2 \subset z$ and z is a minimal border of w. Then z_2y has a minimal border r such that $|r| \leq |z|$. Moreover, if w is well bordered, then $|y| \geq |r|$.*

PROOF We prove Statement 1 (Statement 2 can be proved similarly). Then $u = xz_1v_1$, so that xz_1 is a proper prefix of u longer than x. It follows that xz_1 is bordered, and there exist nonempty partial words r, r_1, r_2, s_1, s_2 such that $xz_1 = r_1s_1 = s_2r_2$, $r_1 \subset r$ and $r_2 \subset r$ with r a minimal border. If $|r| > |z|$, then $r_2 = x'z_1$ where x' is a nonempty suffix of x. Since $r_1 \uparrow r_2$, there exist partial words x'', z' such that $r_1 = x''z'$ where $x'' \uparrow x'$ and $z' \uparrow z_1$. But then, $x''z's_1 = r_1s_1 = xz_1 = s_2r_2 = s_2x'z_1$. It follows that x'' is a prefix of x and x' is a suffix of x that are compatible. As a result, x is bordered, which contradicts that x is the longest unbordered proper prefix of u. And so $|r| \leq |z|$ and r_2 is a suffix of z_1. Set $z_1 = sr_2$ for some suffix s of s_2 ($s_2 = xs$). If we further assume that v is well bordered, then we claim that $|x| \geq |r|$. To see this, if $|x| < |r|$, then set $r_1 = xt$ and $z_1 = ts_1$ for some t. Since $r_1 \uparrow r_2$, there exist x', t' such that $r_2 = x't'$ and $x \uparrow x'$ and $t \uparrow t'$. Since r_2 is a suffix of z_1, we have that t' is a suffix of z_1. Consequently, t is a prefix of z_1 and t' is a suffix of z_1 that are compatible. So z_1 is bordered and we get a contradiction with v's well borderedness, establishing our claim. ☐

We now investigate the relationship between the minimal weak period of a given partial word u and the maximum length of its unbordered factors.

DEFINITION 7.7 *The maximum length of the unbordered factors of a partial word u is denoted by $\mu(u)$.*

PROPOSITION 7.4

For all partial words u, $\mu(u) \le p'(u)$.

PROOF   Let w be a subword of u such that $|w| > p'(u)$. Factor w as $w = xw_1 = w_2 y$ where $|w_1| = |w_2| = p'(u)$. We have $x(i) = w(i)$ and $y(i) = w(i + p'(u))$ whenever $i, i + p'(u) \in D(w)$. This means that whenever $x(i) \ne y(i)$, $i \in H(x)$ or $i \in H(y)$. So we can construct a word that contains both x and y. Therefore $x \uparrow y$ and w is bordered. So we must have that $\mu(u) \le p'(u)$.   ⧮

REMARK 7.3   For any partial word u, Proposition 7.4 gives an upper bound for the maximum length of the unbordered factors of u: $\mu(u) \le p'(u)$. This relationship cannot be replaced by $\mu(u) < p'(u)$ as is seen by considering $u = aba\diamond$ with $\mu(u) = p'(u) = 2$.   ⧮

For any partial words v, w, if there exists a partial word u such that $u \ll w$ and $u \subset v$, then we say that v contains a concatenation of prefixes of w. Otherwise, we say that v contains no concatenation of prefixes of w. Similarly, if $u \in P(w)$ and $u \subset v$, then we say that v contains a prefix of w.

PROPOSITION 7.5

Let u, v be partial words such that $u = hvh$ where h abbreviates unb(u). If h is not compatible with any factor of v, then vh is unbordered if one of the following holds:

1. *v is full,*

2. *v contains a prefix of h or a concatenation of prefixes of h.*

PROOF   For Statement 1, suppose that v is full and there exist nonempty x, w_1, w_2 such that $vh \subset xw_1$ and $vh \subset w_2 x$. We must have that $|x| \le |v|$ or else h, which is unbordered, would be bordered by a factor of x. If $|h| < |x|$, then there exists $x' \in S(x)$ such that $h \subset x'$ and because $|x| \le |v|$, there exists v' a factor of v with $v' \subset x'$ and this says that $v' \uparrow h$, contradicting our assumption. Now, if $|h| \ge |x|$, then set $v = rv'$ and $h = h's$ where $|r| = |s| = |x|$. In this case, $r \subset x$ and $s \subset x$, and there exist nonempty $r \in P(v)$ and $s \in S(h)$ such that $r \uparrow s$. But r is full and so $r \uparrow s$ implies that $s \subset r$. But then, by Lemma 7.6, we have that hs is unbordered, and so hr is an unbordered prefix of u with length greater than $|h|$. This contradicts the assumption that $h = $ unb(u), hence vh must be unbordered.

For Statement 2, first assume that v contains a prefix of h. Let $v' \in P(h)$ be such that $v' \subset v$. By Lemma 7.6, since h is unbordered, we have that $v'h$ is unbordered. Now, assume that v contains a concatenation of prefixes of h.

Let v' be such that $v' \ll h$ and $v' \subset v$. By Lemma 7.7, since h is unbordered and $v' \ll h$, we have that $v'h$ is unbordered. In either case, since $v' \subset v$, vh is unbordered as well. ▯

7.3 Critical factorizations

In this section, we investigate some of the properties of an unbordered partial word of length at least two and how they relate to its critical factorizations (if any).

DEFINITION 7.8 *Let u, v be nonempty partial words. We say that u and v **overlap** if there exist partial words r, s satisfying one of the following conditions:*

1. $r \uparrow s$ with $u = ru'$ and $v = v's$,

2. $r \uparrow s$ with $u = u'r$ and $v = sv'$,

3. $u = ru's$ with $u' \uparrow v$,

4. $v = rv's$ with $v' \uparrow u$.

Otherwise we say that u and v do not overlap.

Figures 7.3, 7.4, 7.5 and 7.6 depict the different overlaps of Definition 7.8.

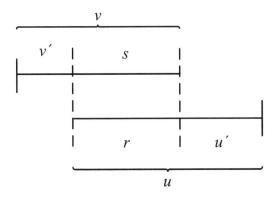

FIGURE 7.3: Overlap of Type 1.

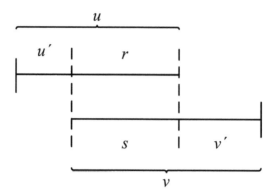

FIGURE 7.4:   Overlap of Type 2.

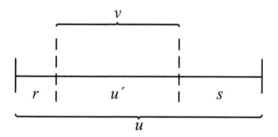

FIGURE 7.5:   Overlap of Type 3.

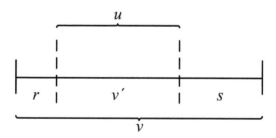

FIGURE 7.6:   Overlap of Type 4.

Example 7.8
The partial words $u = a \diamond bc \diamond$ and $v = ba \diamond c$ overlap since $u = ru' = (a \diamond)(bc \diamond)$ and $v = v's = (ba)(\diamond c)$ with $r \uparrow s$. ⬜

The following proposition helps us produce examples of partial words that do not overlap.

PROPOSITION 7.6
Let u, v be nonempty partial words. If $w = uv$ is unbordered, then $|u| - 1$ is a critical point of w if and only if u and v do not overlap.

PROOF Let us first suppose that u and v overlap. If we have Type 1 overlap, then $w = ru'v's$ and $r \uparrow s$ for some partial words r, s, u', v'. This contradicts the fact that w is unbordered. If we have Type 2 overlap, then $w = u'rsv'$ and there is an internal square at position $|u| - 1$ of length $k = |r| = |s|$, so $p(w, |u| - 1) \leq k$. But because w is unbordered, $p'(w) = |w|$. Of course we have that $k < |w|$ (otherwise we have Type 1 overlap), so this contradicts that $|u| - 1$ is a critical point of w. If we have Type 3 overlap, then $w = ru'sv$ and there is a right-external square of length $|u's|$ at position $|u| - 1$. Because $v \neq \varepsilon$, $|u's| < |w| = p'(w)$ and we have that $|u| - 1$ cannot be a critical point of w, a contradiction. The case for Type 4 overlap is very similar to Type 3.

For the other direction we have that u and v do not overlap and let us suppose that $|u| - 1$ is not a critical point of w. Since $|u| - 1$ is not a critical point, there exist x and y defined as in Definition 4.2, with the length of x strictly smaller than the minimal weak period of w. Let us look at all the four cases of the definition. If we have an internal square, then according to Definition 7.8 we have a Type 2 overlap of u and v, which contradicts our assumption. For a left-external, respectively right-external, square we get that either u is compatible with a factor of v, or v is compatible with a factor of u. Both cases contradict with the fact that u and v do not overlap, giving us a Type 4, respectively Type 3, overlap.

In the case we have a left- and right-external square we get that $x = ru$ and $y = vs$, where $x \uparrow y$ and $r, s \neq \varepsilon$. If $|r| < |v|$, then there exists v' such that $|v'| > 0$ and $v = rv'$. Hence, since $ru \uparrow rv's$ we get a Type 2 overlap, $u \uparrow v's$, which is a contradiction. If $|r| \geq |v|$, then there exists r' such that $r = vr'$. This implies that

$$|w| = |uv| \leq |vr'u| = |ru| = |x| < p'(w) \leq |w|$$

a contradiction. ⬜

Example 7.9
The partial word $w = ab \diamond bcac$ is unbordered with minimal weak period

$p'(w) = 7$. Here $(u, v) = (ab \diamond bc, ac)$ is a critical factorization of w. We can check that u and v do not overlap. $\Box$

We obtain the following two corollaries.

COROLLARY 7.2

Let u, v be nonempty partial words. If $w = uv$ is unbordered and $|u| - 1$ is a critical point of w, then $w' = vu$ is unbordered as well.

PROOF   This is immediately implied by Proposition 7.6 and the fact that if $w' = vu$ is bordered, then u and v must overlap. $\Box$

Example 7.10

Returning to Example 7.9, we see that $w' = vu = acab \diamond bc$ is unbordered. $\Box$

COROLLARY 7.3

Let u, v be nonempty partial words. If $w = uv$ is unbordered and $|u| - 1$ is a critical point of w, then $|v| - 1$ is a critical point of $w' = vu$.

PROOF   By Corollary 7.2, we have that w' is unbordered, and so $p'(w') = |w'|$. Suppose that $p(w', |v| - 1) = p < |w'|$ and let us show that u and v overlap. We consider the case where $x = rv$ and $y = us$ with r, s, x, y nonempty partial words satisfying $x \uparrow y$ and $|x| = |y| = p$. Here we have that $|x| = p < |w'|$ and $rv \uparrow us$. We must have that $|r| < |u|$ and so it is possible to write $u = r'u'$ with $|r'| = |r|$. Simplifying $rv \uparrow us = r'u's$ gives that $v \uparrow u's$. We can then factor v as $v = v's'$ with $|v'| = |u'|$. Simplifying again gives us that $u' \uparrow v'$ and we have that u and v overlap. This contradicts Proposition 7.6, so we must have that $|v| - 1$ is a critical point of w'. $\Box$

Example 7.11

Returning one last time to Example 7.9, position $|v| - 1 = 1$ is a critical point of the factorization $(v, u) = (ac, ab \diamond bc)$ of $w' = vu$. $\Box$

7.4   Conjugates

Referring to Exercise 1.13, we call a word u a *conjugate* of v, and we write $u \sim v$, if $u = xy$ while $v = yx$ for some x and y. Equivalently, u and v are conjugate if and only if there exists a word z such that $uz = zv$. Indeed, if

u and v are conjugate, then $z = x$ satisfies the equation $uz = zv$. For the converse, we can use Lemma 2.1. In Exercise 1.13, the reader was asked to show that $\sim$ is an equivalence relation. The reader now can check that for two words u and v, $(\sqrt{u})^m \sim (\sqrt{v})^n$ if and only if both $m = n$ and $\sqrt{u} \sim \sqrt{v}$ (see Exercise 1.1 for the definition of $\sqrt{}$). Thus, every conjugate of a nonprimitive nonempty word is bordered. This however does not hold for primitive words as the following shows.

THEOREM 7.5

If u is a word such that $u = \sqrt{u}$ and $a \in \alpha(u)$, then there exists an unbordered conjugate av of u. In other words, if u is such that $u = \sqrt{u}$ and $a \in \alpha(u)$, then there exist x, y such that $u = xay$ and $v = ayx$ is unbordered.

PROOF The proof is left as an exercise. ▯

Example 7.12

If $u = aba$, then $x = ab$ and $y = \varepsilon$ work for the letter $a \in \alpha(u)$. ▯

We now give a version of Theorem 7.5 for partial words. Referring again to Exercise 1.13, u and v are *conjugate* if there exist partial words x and y such that $u \subset xy$ and $v \subset yx$. Again, we denote u is a conjugate of v by $u \sim v$. Here, the relation $\sim$ is not an equivalence relation: it is both reflexive and symmetric, but not transitive. Note that the conjugates $a \diamond b$, $\diamond ba$ and $ba \diamond$ of $u = a \diamond b$ are bordered. However, the following result holds.

THEOREM 7.6

Let u be a primitive partial word. Let a be any letter in A appearing in the spelling of u. Then there is an unbordered full conjugate $v = ax$ of u.

PROOF Let u be a primitive partial word and let $a \in A$ be a letter that appears in the spelling of u. Let u' be a full word such that $u \subset u'$. Since u is primitive, u' is primitive as well. The latter and the fact that $a \in \alpha(u')$ imply the existence of words y, z such that $u' = yaz$ and $v = azy$ is unbordered. But v is also a conjugate of u since $u \subset (y)(az)$ and $v \subset (az)(y)$. ▯

Exercises

7.1 Prove that if u is unbordered and $u \subset u'$, then u' is unbordered as well.

7.2 Give all borders of $u = a◊◊◊acb$.

7.3 Give an example of a partial word having at least two minimal borders.

7.4 Classify the following partial words as well bordered or badly bordered:

- $a◊$
- $a◊◊ba$
- $a◊◊b$
- $a◊ab$

7.5 $\boxed{\text{s}}$ We call a bordered pword u *simply bordered* if a minimal border x exists satisfying $|u| \geq 2|x|$. Show that a bordered full word is always simply bordered.

7.6 Prove that every bordered full word of length n has a unique minimal border x. Moreover, x is unbordered and $|x| \leq \lfloor \frac{n}{2} \rfloor$.

7.7 Show that for partial words u, v and w, if $u \ll v$ and $v \ll w$, then $u \ll w$.

7.8 $\boxed{\text{s}}$ Consider the words $u = abaaabaabaaaca$ and $v = abaaacc$. Does $u \ll v$ hold?

7.9 $\boxed{\text{s}}$ Run Algorithm 7.1 on input $v = ab◊◊ba◊◊abba$ and sequence

$$ab◊◊, ab, ab◊◊ba◊, a, ab◊◊ba◊◊a$$

of prefixes of v. Display your output as in Example 7.5 or Example 7.6.

7.10 Repeat Exercise 7.9 for input $v = a◊babca$ and sequence

$$a, a◊ba, a◊babc$$

7.11 Do $u = ab◊◊aba$ and $v = bba◊◊◊◊bab$ overlap? Why or why not?

7.12 $\boxed{\text{s}}$ Consider the unbordered partial word $w = aabc◊bc$. Produce a critical factorization (u, v) of w such that $w' = vu$ is unbordered. What can be said about position $|v| - 1$ of w'?

7.13 Handle the case of Type 4 overlap in the proof of Proposition 7.6.

7.14 Let u, v be nonempty partial words such that $w = uv$ is unbordered. Show that $|u| - 1$ is a critical point of w if and only if the minimal square at position $|u| - 1$ is a left- and right-external square of length $p'(w)$. Check this with Example 7.9.

Challenging exercises

7.15 [s] Show that the problem of enumerating all unbordered full words of length n over a k-letter alphabet yields to a conceptually simple and elegant recursive formula $U_k(n)$: $U_k(0) = 1$, $U_k(1) = k$, and for $n > 0$,

$$U_k(2n) = kU_k(2n-1) - U_k(n)$$
$$U_k(2n+1) = kU_k(2n)$$

7.16 [s] Using the formulas of Exercise 7.15 and Proposition 7.2, obtain a formula for counting bordered full words.

7.17 Prove Theorem 7.1.

7.18 Prove that if u, v are full words such that $u = \text{unb}(u)v$, then $v \ll \text{unb}(u)$.

7.19 Prove that for full words u and v, $u \ll v$ if and only if $u \ll \text{unb}(v)$.

7.20 Prove that for full words u and v, $u \approx v$ if and only if $\text{unb}(u) = \text{unb}(v)$.

7.21 Show that if v is an unbordered word and w is a proper prefix of v for which $u \ll w$, then uv and wv are unbordered.

7.22 Show that if $u = xv$ is a nonempty unbordered word where x is the longest unbordered proper prefix of u, then v is unbordered.

7.23 [s] If the following assumptions hold:

1. w is well bordered,
2. a, b are letters, with $a \neq b$,
3. $au' = \text{unb}(au)$,
4. au is a prefix of w and bu' is a suffix of w,

then show that au is contained in a proper prefix of any minimal border of w.

7.24 [s] Let $u = hvh$ be a partial word with $|h| = \mu(u)$. If h is unbordered and v is full, then u has the form $u = h(h')^{k-2}h$ for some $k \geq 2$ and $h' \supset h$ and $p'(u) = \mu(u)$. Show that the equality $u = h(h')^{k-2}h$ cannot be replaced by $u = h^k$.

7.25 Prove that for two words u and v, $(\sqrt{u})^m \sim (\sqrt{v})^n$ if and only if both $m = n$ and $\sqrt{u} \sim \sqrt{v}$. Thus, every conjugate of a nonprimitive nonempty word is bordered.

7.26 Prove Theorem 7.5.

Programming exercises

7.27 Design an applet that takes as input a nonempty partial word u, and that outputs unb(u), the longest unbordered prefix of u.

7.28 Give pseudo code for an algorithm that computes the maximum length $\mu(u)$ of the unbordered factors of a given partial word u.

7.29 Write a program to find out if two partial words u and v satisfy the following relationships:

- $u \sim v$
- $u \ll v$
- $u \approx v$

7.30 Write a program that when given two nonempty partial words u and v determines whether or not u and v overlap. Run your program on the following pair of partial words:

$$u = ab\diamond\diamond aba \text{ and } v = bba\diamond\diamond\diamond\diamond bab$$

7.31 Referring to Exercises 7.5 and 7.15, let $S_{h,k}(n)$ be the number of pwords with h holes, of length n, over a k-letter alphabet that are not simply bordered. If $h = 0$, then $S_{0,k}(n) = U_k(n)$ since a non simply bordered full word is an unbordered full word. It is easy to see that $S_{1,k}(0) = 0$, $S_{1,k}(1) = 1$, $S_{1,k}(2) = 0$, and for $h > 1$ that $S_{h,k}(1) = 0$ and $S_{h,k}(2) = 0$. Now, for $h > 0$, the following formula holds for odd integers $n = 2m+1$:

$$S_{h,k}(2m + 1) = kS_{h,k}(2m) + S_{h-1,k}(2m)$$

Run computer experiments to approximate $S_{h,k}(2m)$ for even integers $n = 2m$.

Website

A World Wide Web server interface at

http://www.uncg.edu/mat/border

has been established for automated use of Algorithm 7.1. Another website related to unbordered partial words is

http://www.uncg.edu/cmp/research/bordercorrelation

Bibliographic notes

Periodicity and borderedness are two fundamental properties of words that play a role in several research areas including string searching algorithms [44, 57, 58, 60, 79, 98], data compression [56, 137, 142], theory of codes [12], sequence assembly [112] and superstrings [45] in computational biology, and serial data communication systems [46]. It turns out that these two word properties do not exist independently from each other.

Ehrenfeucht and Silberger initiated a line of research to explore the relationship between the minimal period of a full word u, $p(u)$, and $\mu(u)$ [74]. Exercises 7.18, 7.19, 7.20, 7.21 and 7.22, Theorems 7.1 and 7.5 are from Ehrenfeucht and Silberger.

Unbordered partial words were introduced by Blanchet-Sadri [17] and were further studied by Blanchet-Sadri, Davis, Dodge, Merchas and Moorefield [28]. The results on partial words in this chapter are from there. Proposition 7.5 extends a result of Duval [69].

Part IV

CODING

Chapter 8

Pcodes of Partial Words

Codes play an important role in the study of the combinatorics on words. In this chapter, we discuss *pcodes* that play a role in the study of the combinatorics on partial words. While a code of words X does not allow two distinct decipherings of some word in X^+, a pcode of partial words Y does not allow two distinct compatible decipherings in Y^+. In Sections 8.1 to 8.6, the definitions and some important general properties of pcodes and the monoids they generate are presented. There, we describe various ways of defining and analyzing pcodes. In particular, many pcodes can be obtained as antichains with respect to certain partial orderings. We investigate in particular the *Defect Theorem* for partial words. In Section 8.7, we introduce the circular pcodes which take into account, in a natural way, the conjugacy operation that was discussed in Chapter 2. The main feature of these pcodes is that they define a unique factorization of partial words written on a circle. Throughout the chapter proofs will be sometimes omitted. They are left as exercises for the reader.

8.1 Binary relations

We assume that the cardinality of the finite alphabet A is at least two (unless it is stated otherwise).

A binary relation ρ defined on an arbitrary set $S \subset W(A)$ is a subset of $S \times S$. Instead of denoting $(u, v) \in \rho$, we often write $u\rho v$. The relation ρ is called *reflexive* if $u\rho u$ for all $u \in S$; *symmetric* if $u\rho v$ implies $v\rho u$ for all $u, v \in S$; *antisymmetric* if $u\rho v$ and $v\rho u$ imply $u = v$ for all $u, v \in S$; *transitive* if $u\rho v$ and $v\rho w$ imply $u\rho w$ for all $u, v, w \in S$, and *positive* if $\varepsilon\rho u$ for all $u \in S$. It is called *strict* if it satisfies the following conditions for all $u, v \in S$:

$u\rho u$,

$u\rho v$ implies $|u| \leq |v|$,

$u\rho v$ and $|u| = |v|$ imply $v \subset u$.

A strict binary relation is reflexive and antisymmetric, but not necessarily transitive. A reflexive, antisymmetric, and transitive relation ρ defined on S

is called a *partial ordering*, and (S, ρ) is called a *partially ordered set* or *poset*. A partial ordering ρ on S is called *right* (respectively, *left*) *compatible* if $u\rho v$ implies $uw\rho vw$ (respectively, $u\rho v$ implies $wu\rho wv$) for all $u, v, w \in S$. It is called *compatible* if it is both right and left compatible. For any two binary relations ρ_1 and ρ_2 on S, we denote by $(\rho_1) \subset (\rho_2)$ if $u\rho_1 v$ implies $u\rho_2 v$ for all $u, v \in S$ (or the subset inclusion), and by $(\rho_1) \sqsubset (\rho_2)$ if $(\rho_1) \subset (\rho_2)$ but $(\rho_1) \neq (\rho_2)$.

An important notion on binary relations is that of an *antichain*. A nonempty subset X of S is called an *antichain* with respect to a particular binary relation ρ on S (or an ρ-antichain) if for all distinct $u, v \in X$, $(u, v) \notin \rho$ and $(v, u) \notin \rho$. The class of all ρ-antichains of S is denoted by $\mathcal{A}(\rho)$. For every partial word u of S, $\{u\}$ is in $\mathcal{A}(\rho)$.

PROPOSITION 8.1

Let ρ_1, ρ_2 be two binary relations defined on $W(A)$. Then

1. *If $\rho_1 \subset \rho_2$, then $\mathcal{A}(\rho_2) \subset \mathcal{A}(\rho_1)$.*

2. *If ρ_1, ρ_2 are strict and $\mathcal{A}(\rho_2) \subset \mathcal{A}(\rho_1)$, then $\rho_1 \subset \rho_2$.*

PROOF For Statement 1, let $X \in \mathcal{A}(\rho_2)$. If X is a singleton set, then $X \in \mathcal{A}(\rho_1)$. Now suppose that X is not a singleton set and let $u, v \in X$ be such that $u \neq v$ and $u\rho_1 v$. Then $u\rho_2 v$ by assumption. Since X is an antichain with respect to ρ_2, we have $u = v$, a contradiction. Thus $X \in \mathcal{A}(\rho_1)$ and $\mathcal{A}(\rho_2) \subset \mathcal{A}(\rho_1)$ holds.

For Statement 2, suppose that there exist partial words u, v such that $u \neq v$, $u\rho_1 v$, and $(u, v) \notin \rho_2$. Suppose that $v\rho_2 u$. Since $u\rho_1 v$, we have $|u| \leq |v|$, and since $v\rho_2 u$, we have $|v| \leq |u|$. Hence $|u| = |v|$, both $u\rho_1 v$ and $|u| = |v|$ imply $v \subset u$, and both $v\rho_2 u$ and $|v| = |u|$ imply $u \subset v$. We deduce that $u = v$, a contradiction. So $\{u, v\} \in \mathcal{A}(\rho_2)$. As $\mathcal{A}(\rho_2) \subset \mathcal{A}(\rho_1)$, we have $\{u, v\} \in \mathcal{A}(\rho_1)$ which implies that $(u, v) \notin \rho_1$, a contradiction. □

We first define the δ-*relations*. Note that all positive powers of a nonempty word have the same root. For $u, v \in A^+$, $uv = vu$ is equivalent to $\sqrt{u} = \sqrt{v}$. For a nonempty partial word u, let $\mathcal{P}(u)$ denote the set of primitive words $v \in A^+$ such that $u \subset v^n$ for some positive integer n. For $u \in A^+$, we have $\mathcal{P}(u^i) = \mathcal{P}(u) = \{\sqrt{u}\}$, and for each partial word u, we have $\mathcal{P}(u) \subset \mathcal{P}(u^i)$ for all positive powers of u.

For every positive integers i, j and nonempty partial words u, v, define the relation $\delta_{i,j}$ by $u\delta_{i,j}v$ if $\mathcal{P}(u^i) \cap \mathcal{P}(v^j) \neq \emptyset$. In the sequel, $\delta_{1,1}$ is often abbreviated by δ. Note that if $u \subset v$, then $\mathcal{P}(v) \subset \mathcal{P}(u)$ and so $u\delta v$.

The reader can check the following properties of the δ-relations.

on $W(A)$ *satisfying* $a\rho_d\diamond$ *for all* $a \in A$. *That is, if* ρ *is a positive compatible partial ordering on* $W(A)$ *satisfying* $a\rho\diamond$ *for all* $a \in A$, *then* $(\rho_d) \subset (\rho)$.

PROOF The embedding partial ordering ρ_d is clearly compatible on $W(A)$. Now, let ρ be a positive compatible partial ordering on $W(A)$ and let u, v be partial words such that $u\rho_d v$. By induction on $|u| + |v|$, we show that $u\rho v$. If $|u| + |v| = 0$, then $\varepsilon\rho_d\varepsilon$ and $\varepsilon\rho\varepsilon$ since ρ_d and ρ are positive. If $|u| + |v| > 0$ and $u = \varepsilon$, then $\varepsilon\rho_d v$ and $\varepsilon\rho v$ since ρ_d and ρ are positive. If $|u| + |v| > 0$ and $u \neq \varepsilon$, then put $u = au'$ and $v = bv'$ where $a, b \in A \cup \{\diamond\}$. If $a = b$, then $u'\rho_d v'$, and using the inductive hypothesis, we get $u'\rho v'$. Since ρ is compatible, we have $au'\rho av'$ and so $u\rho v$. If $a \neq b$ and $b \neq \diamond$, then $u\rho_d v'$, and thus by the inductive hypothesis, $u\rho v'$. Since ρ is positive, we have $\varepsilon\rho b$ and since ρ is compatible, we have $v'\rho bv'$ and so $v'\rho v$. Since ρ is transitive, we get $u\rho v$ as desired. On the other hand, if $a \neq b$ and $b = \diamond$, then $u'\rho_d v'$, and thus by the inductive hypothesis, $u'\rho v'$. Since $a\rho\diamond$ and ρ is compatible, we have $av'\rho\diamond v'$. Since $u'\rho v'$ and ρ is compatible, we have $au'\rho av'$. Since ρ is transitive, we get $au'\rho\diamond v'$ or $u\rho v$ as desired. □

8.2 Pcodes

In this section, we discuss pcodes of partial words. We start with the full case.

DEFINITION 8.2 *Let X be a nonempty subset of A^+. Then X is called a* **code** *over A if for all integers $m \geq 1, n \geq 1$ and words $u_1, \ldots, u_m, v_1, \ldots, v_n \in X$, the equality*

$$u_1 u_2 \ldots u_m = v_1 v_2 \ldots v_n$$

is trivial, *that is, both $m = n$ and $u_i = v_i$ for $i = 1, \ldots, m$.*

The set $X = \{a, abbbbba, babab, bbbb\}$ is not a code as is seen in Figure 8.1. The word $abbbbbababababbbbba$ can be factorized in two different ways using code words.

FIGURE 8.1: Two distinct factorizations.

In the case of partial words, we define a *pcode* as follows.

DEFINITION 8.3 *Let X be a nonempty subset of $W(A) \setminus \{\varepsilon\}$. Then X is called a* **pcode** *over A if for all integers $m \geq 1, n \geq 1$ and partial words $u_1, \ldots, u_m, v_1, \ldots, v_n \in X$, the compatibility relation*

$$u_1 u_2 \ldots u_m \uparrow v_1 v_2 \ldots v_n$$

is trivial, that is, both $m = n$ and $u_i = v_i$ for $i = 1, \ldots, m$.

Example 8.1
Consider the set $X = \{a \diamond bba, accba\}$ and let $x_1 = a \diamond bba$ and $x_2 = accba$. It is clear that $x_1 \not\uparrow x_2$. This is not sufficient in determining whether or not X is a pcode. However, we can easily check that no *nontrivial* compatibility relation exists since $|x_1| = |x_2|$, and consequently X is a pcode over $\{a, b, c\}$.

Now, the set $Y = \{a \diamond b, aab \diamond bb, \diamond b, ba\}$ is not a pcode. Let $y_1 = a \diamond b, y_2 = aab \diamond bb, y_3 = \diamond b$ and $y_4 = ba$. As seen in Figure 8.2, a nontrivial compatibility relation does exist among the four elements

$$y_1 y_3 y_3 y_4 y_3 \uparrow y_2 y_3 y_1$$

FIGURE 8.2: A nontrivial compatibility relation.

Definition 8.3 has immediate consequences that should be emphasized.

REMARK 8.1

- A nonempty subset of A^+ is a code if and only if it is a pcode.

- A pcode with at least two elements never contains a partial word of the form $\diamond^n$ for any integer n.

- Any nonempty subset of a pcode is a pcode.

- A pcode is always a *pairwise noncompatible* set, but the converse is false. Here a set X is called pairwise noncompatible if $u \not\uparrow v$ for all distinct $u, v \in X$.

- If X is a pcode over A, then $X_\diamond = \{u_\diamond \mid u \in X\}$ is a code over $A \cup \{\diamond\}$. But the converse does not hold. Consider, for instance, the code $X_\diamond = \{a\diamond a, \diamond a \diamond\}$ over $\{a, \diamond\}$. The underlying set X is not a pcode over $\{a\}$ since its elements are compatible. This fact is important, since it justifies the study of pcodes.

▯

The following propositions are natural extensions of the pcode definition.

PROPOSITION 8.3

Let X be a nonempty subset of $W(A) \setminus \{\varepsilon\}$. Then X is a pcode if and only if for every integer $n \geq 1$ and partial words $u_1, \ldots, u_n, v_1, \ldots, v_n \in X$, the condition

$$u_1 u_2 \ldots u_n \uparrow v_1 v_2 \ldots v_n$$

implies $u_i = v_i$ for $i = 1, \ldots, n$.

PROOF If X is a pcode, then clearly the condition holds. Conversely, assume that X satisfies the condition stated in the proposition. Suppose $u_1 u_2 \ldots u_m \uparrow v_1 v_2 \ldots v_n$ for some integers $m \geq 1, n \geq 1$ and partial words $u_1, \ldots, u_m, v_1, \ldots, v_n \in X$. Then

$$u_1 u_2 \ldots u_m v_1 v_2 \ldots v_n \uparrow v_1 v_2 \ldots v_n u_1 u_2 \ldots u_m$$

by multiplication. If $m < n$, then $u_1 = v_1, \ldots, u_m = v_m$ and $\varepsilon \uparrow v_{m+1} \ldots v_n$, which is a contradiction. Similarly, $n < m$ cannot hold. Hence $m = n$ and therefore the condition implies that X is a pcode. ▯

PROPOSITION 8.4

Let X be a nonempty subset of $W(A) \setminus \{\varepsilon\}$. For every $u \in X$, let x_u be a nonempty partial word such that $u \subset x_u$, and let Y be the set $\{x_u \mid u \in X\}$. If X is a pcode, then Y is a pcode.

PROOF Let n be a positive integer and let $x_1, \ldots, x_n, y_1, \ldots, y_n \in Y$ be such that

$$x_1 x_2 \ldots x_n \uparrow y_1 y_2 \ldots y_n$$

For every integer $1 \leq i \leq n$, let $u_i \in X$ be such that $x_{u_i} = x_i$, and let $v_i \in X$ be such that $x_{v_i} = y_i$. Then we have

$$u_1 u_2 \ldots u_n \uparrow v_1 v_2 \ldots v_n$$

since $u_1 u_2 \ldots u_n \subset x_1 x_2 \ldots x_n \subset w$ and $v_1 v_2 \ldots v_n \subset y_1 y_2 \ldots y_n \subset w$ for some w. But since X is a pcode, by Proposition 8.3, $u_i = v_i$ for $i = 1, \ldots, n$. This implies $x_i = x_{u_i} = x_{v_i} = y_i$ for $i = 1, \ldots, n$ showing that Y is a pcode. ▯

The converse of Proposition 8.4 is not true. For example, let $X = \{u, v\}$ where $u = a$ and $v = a \diamond a$. The set $Y = \{a, aba\}$ is a pcode, but X is not a pcode since $u^3 \uparrow v$.

We end this section by introducing some classes of pcodes and their basic properties: the *prefix pcodes*, *suffix pcodes*, *biprefix pcodes*, *uniform pcodes*, and *maximal pcodes*.

DEFINITION 8.4   *Let X be a nonempty subset of $W(A) \setminus \{\varepsilon\}$. Then X is called a* **prefix pcode** *if for all $u, v \in X$,*

$$ux \uparrow v \text{ for some partial word } x \text{ implies } u = v$$

Note that any singleton set is a prefix pcode. Note also that any subset of a prefix pcode is a prefix pcode and hence any intersection of prefix pcodes is also a prefix pcode.

Example 8.2

The set $X = \{a \diamond b, a \diamond\}$ is not a prefix pcode. In this example setting $u = a \diamond, v = a \diamond b$ and $x = b$, we have $ux \uparrow v$. Since X is a pcode, we deduce that a pcode is not necessarily a prefix pcode.

Now, consider $Y = \{a \diamond b, ba\}$ and let $y_1 = a \diamond b$, and $y_2 = ba$. This set is a prefix pcode. No x exists such that $y_2 x \uparrow y_1$. ▯

DEFINITION 8.5   *A set of partial words X is a* **suffix pcode** *if $rev(X)$ is a prefix pcode. A* **biprefix pcode** *is a pcode that is both prefix and suffix.*

DEFINITION 8.6   *If n is a positive integer, then a* largest *pairwise noncompatible set X satisfying $X \subset (A \cup \{\diamond\})^n$ is a biprefix pcode called a* **uniform pcode** *of partial words of length n. By largest we mean that if u is a partial word of length n over A, then there exists $v \in X$ such that $u \uparrow v$.*

DEFINITION 8.7   *A pcode is called a* **maximal pcode** *over A if it is not a proper subset of any other pcode over A.*

It is left as an exercise to show that uniform pcodes over A are maximal over A (see Exercise 8.16). The following proposition holds.

PROPOSITION 8.5
Any pcode X over A is contained in some maximal pcode over A.

8.2.1   The class $\mathcal{F}$

We now consider the following class of binary relations on $W(A)$ partially ordered by inclusion:

$$\mathcal{F} = \{\rho \mid \rho \text{ is a strict binary relation on } W(A) \text{ such that every pcode is an antichain with respect to } \rho\}$$

The class $\mathcal{F}$ is easily seen to be closed under union and intersection. The following proposition gives some closure properties for $\mathcal{F}$.

PROPOSITION 8.6
Let γ be a strict binary relation on $W(A)$ and let $\rho \in \mathcal{F}$. Then the following conditions hold:

1. *If $(\gamma) \subset (\rho)$, then $\gamma \in \mathcal{F}$.*

2. *The membership $\gamma \cap \rho \in \mathcal{F}$ holds.*

PROOF   Statement 1 follows immediately from Proposition 8.1. For Statement 2, since $(\gamma \cap \rho) \subset (\rho)$ and $\gamma \cap \rho$ is strict, then $\gamma \cap \rho \in \mathcal{F}$ follows from Statement 1. ▯

The next proposition implies that $(\delta_{i,j} \cap \rho) \in \mathcal{F}$ for all positive integers i, j and every strict binary relation ρ on $W(A)$.

PROPOSITION 8.7
Let ρ be a strict binary relation on $W(A)$, let X be a nonempty subset of $W(A) \setminus \{\varepsilon\}$, and let i, j be positive integers. If X is a pcode, then X is an $(\delta_{i,j} \cap \rho)$-antichain.

PROOF   Let X be a pcode. The case where X contains only one partial word is trivial. So let $u, v \in X$ be such that $u \neq v$ and $u(\delta_{i,j} \cap \rho)v$. The latter yields $u\delta_{i,j}v$ and by Lemma 8.2(1), $u^i v^j \uparrow v^j u^i$ contradicting the fact that X is a pcode. ▯

The next proposition implies that $\rho_e, \rho_c \in \mathcal{F}$.

PROPOSITION 8.8
Let X be a nonempty subset of $W(A) \setminus \{\varepsilon\}$. If X is a pcode, then X is an ρ_c-antichain (respectively, ρ_e-antichain).

PROOF   Let X be a pcode. The case where X contains only one partial word is trivial. Using Proposition 8.1 and Lemma 8.4, it is enough to show the result for ρ_c. Let $u, v \in X$ be such that $u \neq v$ and $u \rho_c v$. Then $v \subset ux, v \subset xu$ for some $x \in A^*$. If $x = \varepsilon$, then $v \subset u$. This gives $v \uparrow u$ and hence $uv \uparrow vu$. If $x \neq \varepsilon$, then $uv \subset uxu, vu \subset uxu$ and so $uv \uparrow vu$. In either case we get a contradiction with the fact that X is a pcode. Hence X is an antichain with respect to ρ_c. $\Box$

The above proposition does not hold for ρ_o since $X = \{ab^2, ba, ab, b^2a\}$ is an ρ_o-antichain but not a pcode because $(ab^2)(ba) = (ab)(b^2a)$.

The next two propositions relate two-element pcodes with the relation $\bigcup_{\rho \in \mathcal{F}} \rho$.

PROPOSITION 8.9

Let u, v be nonempty partial words such that $|u| < |v|$. Then $u \bigcup_{\rho \in \mathcal{F}} \rho v$ if and only if $\{u, v\}$ is not a pcode.

PROOF   The condition is obviously necessary. To see that the condition is sufficient, suppose that $\{u, v\}$ is not a pcode and let $(u, v) \notin \bigcup_{\rho \in \mathcal{F}} \rho$. Let $\gamma = \{(u, v)\} \cup \bigcup_{\rho \in \mathcal{F}} \rho$. Then $\bigcup_{\rho \in \mathcal{F}} \rho \sqsubset \gamma$ and $\gamma \in \mathcal{F}$, a contradiction. $\Box$

PROPOSITION 8.10

Let $X \subset W(A) \setminus \{\varepsilon\}$ be pairwise noncompatible. Then X is an $\bigcup_{\rho \in \mathcal{F}} \rho$-antichain if and only if for all $u, v \in X$ such that $u \neq v$, $\{u, v\}$ is a pcode.

PROOF   First, suppose that X is an $\bigcup_{\rho \in \mathcal{F}} \rho$-antichain. Let $u, v \in X$ be such that $u \neq v$. Without loss of generality, we can assume that $|u| \leq |v|$. Since X is an $\bigcup_{\rho \in \mathcal{F}} \rho$-antichain, we have $(u, v) \notin \bigcup_{\rho \in \mathcal{F}} \rho$. If $|u| < |v|$, then $\{u, v\}$ is a pcode by Proposition 8.9. If $|u| = |v|$, then $u \not\uparrow v$ since X is pairwise noncompatible. Certainly, in this case, $\{u, v\}$ is a pcode.

Conversely, suppose to the contrary that there exist $u, v \in X$ such that $u \neq v$ and $(u, v) \in \bigcup_{\rho \in \mathcal{F}} \rho$. The set $\{u, v\}$ is a pcode by our assumption. Since $\bigcup_{\rho \in \mathcal{F}} \rho$ is strict, we have $|u| \leq |v|$. If $|u| < |v|$, then $\{u, v\}$ is not a pcode by Proposition 8.9, a contradiction. If $|u| = |v|$, then $v \sqsubset u$ since $\bigcup_{\rho \in \mathcal{F}} \rho$ is strict. So $u \uparrow v$ contradicting the fact that $\{u, v\}$ is a pcode. So X is an $\bigcup_{\rho \in \mathcal{F}} \rho$-antichain. $\Box$

8.2.2   The class $\mathcal{G}$

We now consider the following class of binary relations on $W(A)$ partially ordered by inclusion:

$\mathcal{G} = \{\rho \mid \rho$ is a strict binary relation on $W(A)$ such that every antichain with respect to ρ is a pcode$\}$

The following proposition gives a closure property for $\mathcal{G}$ and immediately implies that $\mathcal{G}$ is closed under union.

PROPOSITION 8.11
Let γ be a strict binary relation on $W(A)$ and let $\rho \in \mathcal{G}$. If $(\rho) \subset (\gamma)$, then $\gamma \in \mathcal{G}$.

PROPOSITION 8.12
Let $u \in A^+, v \in W(A) \setminus \{\varepsilon\}$ be such that $|u| \leq |v|$. If $\{u, v\}$ is an antichain with respect to ρ_o (respectively, $\rho_p, \rho_s, \rho_f, \rho_d, \rho_l$), then $\{u, v\}$ is a pcode.

PROOF   By Proposition 8.1 and Lemma 8.4, it is enough to show the result for ρ_o. Suppose to the contrary that $\{u, v\}$ is not a pcode. Then there exist an integer $n \geq 1$ and partial words $u_1, \ldots, u_n, v_1, \ldots, v_n \in \{u, v\}$ such that

$$u_1 u_2 \ldots u_n \uparrow v_1 v_2 \ldots v_n,$$

and with $|u_1 u_2 \ldots u_n|$ as small as possible contradicting Proposition 8.3. We hence have $u_1 \neq v_1$ and $u_n \neq v_n$. If $n = 1$, then $u \uparrow v$. Since u is full, we get $v \subset u$ and so $u \rho_o v$, which is a contradiction. So we may assume that $n \geq 2$. There are four possibilities: $u_1 = u_n = u, v_1 = v_n = v; u_1 = v_n = u, v_1 = u_n = v; u_1 = v_n = v, v_1 = u_n = u;$ and $u_1 = u_n = v, v_1 = v_n = u$. In all cases, put $u_2 \ldots u_{n-1} = x$ and $v_2 \ldots v_{n-1} = y$. These possibilities can be rewritten as

(1) $uxu \uparrow vyv$

(2) $uxv \uparrow vyu$

(3) $vxu \uparrow uyv$

(4) $vxv \uparrow uyu$

If $|u| = |v|$, for any of the possibilities (1)–(4) we have $u \uparrow v$ which leads to a contradiction. If $|u| < |v|$, for any of the possibilities (1)–(4) there exist nonempty partial words w, w', z, z' such that $v = wz = z'w'$, $w \uparrow u$, and $w' \uparrow u$. The latter two relations give $w \subset u$ and $w' \subset u$ since u is full. There exist $z_1, z_2 \in A^*$ such that $z \subset z_1$ and $z' \subset z_2$. We get $v = wz \subset uz \subset uz_1, v = z'w' \subset z'u \subset z_2 u$ and so $u \rho_o v$, which is a contradiction.   ▯

The converse of the above proposition is not true. For example, the set $X = \{a, aba\}$ is a pcode, but $a \rho_o aba$. The above proposition is not true if u

has a hole. The set $\{u, v\}$ where $u = a\diamond$ and $v = \diamond a$ is an ρ_l-antichain, but $\{u, v\}$ is not a pcode. This latter example shows that $\rho_e, \rho_c, \rho_o, \rho_p, \rho_s, \rho_f, \rho_d$, and ρ_l are not in $\mathcal{G}$.

8.3   Pcodes and monoids

In this section, definitions and some properties of pcodes' generating monoids are given.

For a monoid M, we call a morphism $\varphi : M \to W(A)$ *pinjective* if for all $m, m' \in M$, $\varphi(m) \uparrow \varphi(m')$ implies $m = m'$. The definition of a pcode can be rephrased according to the following proposition.

PROPOSITION 8.13

If a subset X of $W(A)$ is a pcode over A, then a morphism $\varphi : B^* \to W(A)$ which induces a bijection of some alphabet B onto X is pinjective. Conversely, if there exists a pinjective morphism $\varphi : B^* \to W(A)$ such that $X = \varphi(B)$, then X is a pcode over A.

For an alphabet B, a morphism $\varphi : B^* \to W(A)$ which is pinjective and satisfies $X = \varphi(B)$ is called a *pcoding morphism* for X. For any pcode $X \subset W(A)$, the existence of a pcoding morphism for X is straightforward: it suffices to take any bijection of a set B onto X and to extend it to a morphism from B^* into $W(A)$.

We can prove that a set X is a pcode by knowing the submonoid X^* of $W(A)$ it generates. In particular, X is a pcode (respectively, prefix pcode, suffix pcode, biprefix pcode) if and only if X^* is a *pfree* monoid (respectively, *right unitary* monoid, *left unitary* monoid, *biunitary* monoid).

PROPOSITION 8.14

If M is a submonoid of $W(A)$, then the set $X = (M \setminus \{\varepsilon\}) \setminus (M \setminus \{\varepsilon\})^2$ is the unique minimal set that generates M.

We call a submonoid M of $W(A)$ *pfree* if there exists a morphism $\varphi : B^* \to M$ of a free monoid B^* onto M that satisfies

$$\varphi(x) \uparrow \varphi(y) \text{ implies } x = y$$

For instance, for any nonempty partial word u, the submonoid generated by u is pfree.

PROPOSITION 8.15

If M is a pfree submonoid of $W(A)$, then its minimal generating set is a pcode. Conversely, if $X \subset W(A)$ is a pcode, then the submonoid X^* of $W(A)$ is pfree and X is its minimal generating set.

We call the pcode X which generates a pfree submonoid M of $W(A)$ the *base* of M.

Let us give some examples of our definitions.

Example 8.3

The set $X = \{a, \diamond b, a\diamond b\}$ is not a pcode over $\{a, b\}$ since it is not the minimal generating set of X^*.

The set $Y = \{x, y\}$ where $x = \diamond bb$ and $y = abb\diamond$ is the minimal generating set of Y^*, yet Y is not a pcode over $\{a, b\}$ because $xy \uparrow yx$ is a nontrivial compatibility relation over Y. Here Y^* is not pfree. □

Proposition 8.16 gives a characterization of a pfree submonoid of $W(A)$ that does not depend on its base. This proposition can be used to show that a submonoid is pfree (and consequently that its base is a pcode) without knowing its base. We call a submonoid M of $W(A)$ *stable* (in $W(A)$) if for all partial words u, u', v, w with $u \uparrow u'$, the conditions $u, u'w, v \in M$ and $wv \in C(M)$ imply $u = u'$ and $w \in M$.

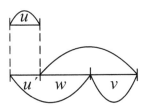

FIGURE 8.3: Representation of stability.

PROPOSITION 8.16

A submonoid M of $W(A)$ is stable if and only if it is pfree.

Note that although the monoid A^* is stable, the monoid $W(A)$ is not stable (and hence not pfree).

Example 8.4

Returning to Example 8.3, the set $Y^* = \{x, y\}^*$ where $x = \diamond bb$ and $y = abb\diamond$ is not pfree, which can be seen by using Proposition 8.16. Indeed, Y^* is not stable by setting $u = \diamond bb$, $u' = abb$, $v = \varepsilon$, and $w = \diamond\diamond bb$ in the definition of stability. $\qquad \Box$

Let M be a submonoid of $W(A)$. Then we call M *right unitary* (in $W(A)$) if for all partial words u, u', v with $u \uparrow u'$, the conditions $u, u'v \in M$ imply $u = u'$ and $v \in M$. Symmetrically, we call M *left unitary* (in $W(A)$) if for all partial words u, u', v with $u \uparrow u'$, the conditions $u, vu' \in M$ imply $u = u'$ and $v \in M$. The submonoid M is *biunitary* if it is both left and right unitary.

PROPOSITION 8.17

Let M be a submonoid of $W(A)$ and let X be its minimal generating set. Then M is right unitary (respectively, left unitary, biunitary) if and only if X is a prefix (respectively, suffix, biprefix) pcode. In particular, a right unitary (left unitary, biunitary) submonoid of $W(A)$ is pfree.

PROPOSITION 8.18

An intersection of pfree submonoids of $W(A)$ is a pfree submonoid of $W(A)$.

If X is a subset of $W(A)$, the set $\mathbf{M}(X)$ of pfree submonoids of $W(A)$ containing X may be empty. If $\mathbf{M}(X) = \emptyset$, then X is pairwise noncompatible. If $\mathbf{M}(X)$ is not empty, then we call the intersection of all elements of $\mathbf{M}(X)$, which is the smallest pfree submonoid of $W(A)$ containing X by Proposition 8.18, the *pfree hull* of X. If X^* is a pfree submonoid of $W(A)$, then X^* coincides with its pfree hull.

PROPOSITION 8.19

Let $X \subset W(A)$ be such that $\mathbf{M}(X) \neq \emptyset$. Let Y be the base of the pfree hull of X. Then

$$Y \subset \{u \mid uy \in X \text{ for some } y \in Y^*\} \cap \{u \mid yu \in X \text{ for some } y \in Y^*\}$$

PROOF   We show that $Y \subset \{u \mid yu \in X \text{ for some } y \in Y^*\}$. Suppose there exists $v \in Y$ such that $v \notin \{u \mid yu \in X \text{ for some } y \in Y^*\}$. Then $X \subset \{\varepsilon\} \cup Y^*(Y \setminus \{v\})$. Let Z be defined by $v^*(Y \setminus \{v\})$. We have $Z^+ = Y^*(Y \setminus \{v\})$, and thus $X \subset Z^*$. Now Z is a pcode. Indeed, a compatibility relation $u_1 u_2 \ldots u_m \uparrow v_1 v_2 \ldots v_n$ where m, n are positive integers and $u_1, \ldots, u_m, v_1, \ldots, v_n \in Z$ can be rewritten as

$$v^{k_1} y_1 v^{k_2} y_2 \ldots v^{k_m} y_m \uparrow v^{l_1} z_1 v^{l_2} z_2 \ldots v^{l_n} z_n$$

with $u_i = v^{k_i} y_i$, $v_j = v^{l_j} z_j$, $y_i \in Y \setminus \{v\}$, $z_j \in Y \setminus \{v\}$, $k_i \geq 0$, $l_j \geq 0$ for $i = 1, \ldots, m$ and $j = 1, \ldots, n$. Since Y is a pcode, we get $k_1 = l_1, y_1 = z_1, k_2 = l_2, y_2 = z_2, \ldots$, and finally $m = n$ and $u_i = v_i$ for $i = 1, \ldots, m$. Thus, the set Z^* is a pfree submonoid of $W(A)$ containing X. But we have $Z^* \sqsubset Y^*$, which contradicts the minimality of the pfree submonoid Y^*. We similarly show that $Y \subset \{u \mid uy \in X \text{ for some } y \in Y^*\}$. □

The following result extends the well-known *Defect Theorem* on words to partial words.

THEOREM 8.1
Let X be a finite subset of $W(A)$ such that $\mathbf{M}(X) \neq \emptyset$. Let Y be the base of the pfree hull of X. If X is not a pcode, then $\|Y\| < \|X\|$.

8.4 Prefix and suffix orderings

In this section, we discuss the prefix and the suffix orderings which we denote by $\preceq_p$ and $\preceq_s$ instead of ρ_p and ρ_s.

A subset X of A^+ is an antichain with respect to $\preceq_p$ if and only if X is a *prefix code*, or if for any $u \in X$, $ux \notin X$ for all $x \in A^+$. We show that with partial words, the antichains with respect to $\preceq_p$ are the *anti-prefix* sets defined as follows.

DEFINITION 8.8   *Let $X \subset W(A) \setminus \{\varepsilon\}$. Then X is **anti-prefix** if for any $u \in X$, the following conditions hold:*

- *If $v \sqsubset u$, then $v \notin X$.*

- *If $v \subset ux$ for some $x \in A^+$, then $v \notin X$.*

It is immediate that a singleton set is anti-prefix and any nonempty subset of an anti-prefix set is anti-prefix. Hence any nonempty intersection of anti-prefix sets is anti-prefix.

PROPOSITION 8.20
Let $X \subset W(A) \setminus \{\varepsilon\}$. Then X is an antichain with respect to $\preceq_p$ if and only if X is anti-prefix.

PROOF   Assume that X is an antichain with respect to $\preceq_p$. Let $u \in X$, and suppose to the contrary that X is not anti-prefix. So either there exists

$v \in X$ with $v \sqsubset u$, or there exist $v \in X$ and $x \in A^+$ such that $v \subset ux$. In either case, we have $u, v \in X$, $u \neq v$, and $u \preceq_p v$ contradicting our assumption. On the other hand, if X is anti-prefix, then suppose to the contrary that there exist $u, v \in X$ with $u \neq v$ and $u \preceq_p v$. Then $v \sqsubset u$ or there exists $x \in A^+$ such that $v \subset ux$. In either case, $v \notin X$ a contradiction. ▯

COROLLARY 8.1
Let $u \in A^+, v \in W(A) \setminus \{\varepsilon\}$ be such that $|u| \leq |v|$. If $\{u, v\}$ is anti-prefix, then $\{u, v\}$ is a pcode.

PROOF   The result follows from Propositions 8.12 and 8.20. ▯

A subset X of A^+ is an antichain with respect to $\preceq_s$ if and only if X is a *suffix code*, or if for any $u \in X$, $xu \notin X$ for all $x \in A^+$.

The family of anti-suffix sets coincides with the family of antichains with respect to $\preceq_s$.

DEFINITION 8.9   *Let $X \subset W(A) \setminus \{\varepsilon\}$. Then X is **anti-suffix** if for any $u \in X$, the following conditions hold:*

- *If $v \sqsubset u$, then $v \notin X$.*

- *If $v \subset xu$ for some $x \in A^+$, then $v \notin X$.*

PROPOSITION 8.21
Let $X \subset W(A) \setminus \{\varepsilon\}$. Then X is an antichain with respect to $\preceq_s$ if and only if X is anti-suffix.

PROOF   The proof is similar to that of Proposition 8.20. ▯

COROLLARY 8.2
Let $u \in A^+, v \in W(A) \setminus \{\varepsilon\}$ be such that $|u| \leq |v|$. If $\{u, v\}$ is anti-suffix, then $\{u, v\}$ is a pcode.

PROOF   The result follows from Propositions 8.12 and 8.21. ▯

We end this section by noticing that there exist anti-prefix (or anti-suffix) sets that are not pcodes. For example, the set $\{u, v\}$ where $u = a \diamond b$ and $v = abbaab$ is both anti-prefix and anti-suffix, but $\{u, v\}$ is not a pcode since $u^2 \uparrow v$.

8.5 Border ordering

In this section, we discuss the border ordering which we denote by $\preceq_o$ instead of ρ_o. Let v be a nonempty partial word. By definition, $\varepsilon \prec_o v$ and $v \not\subset \varepsilon$, and let $N(v)$ be the number of partial words u satisfying $u \prec_o v$ and $v \not\subset u$. For any integer $i \geq 0$, define $\mathcal{O}_i$ as follows:

$$\mathcal{O}_0 = \{\varepsilon\}$$

and for $i \geq 1$,

$$\mathcal{O}_i = \{v \mid v \in W(A) \setminus \{\varepsilon\} \text{ and } N(v) = i\}$$

We are particularly interested in the partial words in $\mathcal{O}_1$. A nonempty partial word v is called *unbordered* if $u \prec_o v$ for some nonempty partial word u implies $v \subset u$. Clearly, v is unbordered if $v \subset ux$ and $v \subset yu$ imply $x = y = \varepsilon$ or $u = \varepsilon$. The fact that v is unbordered means that there exist no nonempty partial words u, x, y satisfying $v \subset ux$ and $v \subset yu$. Note that $\mathcal{O}_1$ is the set of all nonempty unbordered partial words, which is a subset of the primitive partial words (see Chapter 1). From the point of view of the partial order $\preceq_o$, we call the partial words in $\mathcal{O}_1$ *o-primitive*. It is easy to see that $W(A) = \bigcup_{i \geq 0} \mathcal{O}_i$ with $\mathcal{O}_i \cap \mathcal{O}_j = \emptyset$ if $i \neq j$.

PROPOSITION 8.22
Let u be a nonempty partial word such that $0 \notin H(u)$. If $\|A\| \geq 2$, then there exists $v \in A^*$ such that uv is unbordered.

PROOF Let a be the first letter of u, and let $b \in A \setminus \{a\}$. We claim that the partial word $w = uab^{|u|}$ is unbordered. To see this, suppose there exist nonempty partial words x, y, z satisfying $w \subset xy, w \subset zx$. Since $w \subset xy$, the nonempty word x starts with the letter a. Since $w \subset zx$, we have $|x| > |u|$. But then we have $x = x'ab^{|u|}$ for some pword x', and also $x = u'ab^{|x'|}$ for some pword u' satisfying $|u'| = |u|$. Thus $|x'| = |u|$, and hence $w \subset x$, a contradiction. ∎

Let X be a subset of $W(A)$. A partial word u over A is *completable in X* if there exist pwords x, y such that $xuy \in C(X)$. It is equivalent to saying that $W(A)uW(A) \cap C(X) \neq \emptyset$, or, in other words, that $u \in F(C(X))$. The set X is *dense* if all elements of $W(A)$ are completable in X, or equivalently $F(C(X)) = W(A)$. Clearly, each superset of a dense set is dense. The set X is *complete* if X^* is dense. Every dense set is also complete.

The proof that a maximal pcode is complete is based on Proposition 8.23 which describes a method for embedding any pcode in a complete pcode.

PROPOSITION 8.23

Let $X \subset W(A) \setminus \{\varepsilon\}$ be a pcode, let u be an unbordered word over A such that $u \notin F(C(X^*))$, let U be a largest pairwise noncompatible subset of $W(A) \setminus C(W(A)uW(A))$ containing X^*, and let $Y = U \setminus X^*$. Then the set

$$Z = X \cup \{uy_1u \ldots y_n u \mid y_1, \ldots, y_n \in Y \text{ and } n \geq 0\}$$

is a complete pcode.

PROOF First, let us show that the set $V = Uu$ is a prefix pcode. To see this, suppose that $vux \uparrow v'u$ for two partial words $v, v' \in U$ and some pword x. If $|vu| > |v'|$, then $vu \uparrow v'y$ with $u = yz$ for some y, z. We deduce that $yz \uparrow z'y$ for some z'. If $z = \varepsilon$, then $vu \uparrow v'u$ and $v \uparrow v'$. Since U is pairwise noncompatible, we have $v = v'$. If $z \neq \varepsilon$, then since y is full, by Lemma 2.1, there exist words x', y' such that $z' \subset x'y'$, $z \subset y'x'$, and $y \subset (x'y')^n x'$ for some integer $n \geq 0$. But then $u \subset (x'y')^{n+1} x'$, and since u is unbordered, $x' = \varepsilon$. If $n > 0$, u is bordered, and if $n = 0$, we get $y = \varepsilon$ and so $vu \uparrow v'$. This leads to $v' \in C(W(A)uW(A))$, which is a contradiction. Hence $|vu| \leq |v'|$, and $vuy \uparrow v'$ for some y. But then again v' is in $C(W(A)uW(A))$, a contradiction.

Next, we show that Z is a pcode. Assume the contrary and consider a relation

$$u_1 u_2 \ldots u_m \uparrow v_1 v_2 \ldots v_n$$

with $u_1, \ldots, u_m, v_1, \ldots, v_n \in Z$, and $u_1 \neq v_1$. The set X being a pcode, one of these partial words must be in $Z \setminus X$. Assume that one of $u_1, \ldots, u_m$ is in $Z \setminus X$, and let i be the smallest index such that u_i matches $u(Yu)^*$. Since $W(A)uW(A) \cap C(X^*) = \emptyset$, it follows that $W(A)u_iW(A) \cap C(X^*) = \emptyset$. Consequently one of $v_1, \ldots, v_n$ matches $u(Yu)^*$. Let j be the smallest index such that v_j matches $u(Yu)^*$. Then $u_1 \ldots u_{i-1}u, v_1 \ldots v_{j-1}u \in V$, and $u_1 \ldots u_{i-1} = v_1 \ldots v_{j-1}$ since V is a prefix pcode. The set X being a pcode, thus from $u_1 \neq v_1$ it follows that $i = j = 1$. Put

$$u_1 = uy_1u \ldots uy_ku$$
$$v_1 = uy_1'u \ldots uy_l'u$$

with $y_1, \ldots, y_k, y_1', \ldots, y_l' \in Y$. If $|u_1| = |v_1|$, then $u_1 \uparrow v_1$. Since X is a pcode, we get $u_1 = v_1$, a contradiction. So assume that $|u_1| < |v_1|$. Since V is a prefix pcode, the set V^* is right unitary. Since $Y \subset U$, each $y_iu, y_i'u$ is in V. Consequently $y_1 = y_1', \ldots, y_k = y_k'$. Put $w = y_{k+1}'u \ldots uy_l'u$. We have $u_2 \ldots u_m \uparrow wv_2 \ldots v_n$ with $w \in V^*$. The word u is a factor of w, and thus occurs also in $u_2 \ldots u_m$. This shows that one of $u_2, \ldots, u_m$, say u_r, matches $u(Yu)^*$. Suppose r is chosen minimal. Then $u_2 \ldots u_{r-1}u \in V$ and $y_{k+1}'u \in V$, and with the set V being a prefix pcode, we have $y_{k+1}' = u_2 \ldots u_{r-1}$. Thus $y_{k+1}' \in X^*$, a contradiction with the fact that $y_{k+1}' \in Y$.

Last, let us show that Z is complete. Let w be a partial word such that $w \in C(W(A)uW(A))$. Then

$$w \uparrow u_1 u u_2 u \ldots u u_{n-1} u u_n$$

for some positive integer n and some partial words $u_1, \ldots, u_n \in U$. If $w \notin C(W(A)uW(A))$, then $w \in U$ or $w \in C(U)$, and the abovementioned compatibility relation holds. In any case, $uwu \in C(Z^*)$ and so $w \in F(C(Z^*))$. To see this, let $u_{i_1}, u_{i_2}, \ldots, u_{i_k}$ be those u_i's in X^*. Then uwu is compatible with

$$(uu_1 u \ldots u u_{i_1-1} u) u_{i_1} \, (u u_{i_1+1} u \ldots u u_{i_2-1} u) u_{i_2} \ldots u_{i_k} (u u_{i_k+1} u \ldots u u_n u)$$

The parenthesized partial words are in Z and the result follows. ∎

PROPOSITION 8.24

Let $X \subset W(A) \setminus \{\varepsilon\}$ be a pcode. If $u \in A^*$ is an unbordered word such that $u \notin F(C(X^*))$, then the set $Y = X \cup \{u\}$ is a pcode.

PROOF   Let $U = W \setminus C(WuW)$. Then by assumption $X^* \subset U$. Let us first observe the following property of the set $V = Uu$: For all $v, v' \in U$, $v'u \uparrow vux$ for some x implies $v \uparrow v'$. To see this, suppose that $v'u \uparrow vux$ for two partial words v and v' in U and some x. If $|vu| > |v'|$, then $vu \uparrow v'y$ with $u = yz$ for some y, z. We deduce that $yz \uparrow z'y$ for some z'. If $z = \varepsilon$, then $vu \uparrow v'u$ and $v \uparrow v'$. If $z \neq \varepsilon$, then since y is full, there exist words x', y' such that $z' \subset x'y'$, $z \subset y'x'$, and $y \subset (x'y')^n x'$ for some integer $n \geq 0$. But then $u \subset (x'y')^{n+1} x'$, and since u is unbordered, $x' = \varepsilon$. If $n > 0$, u is bordered, and if $n = 0$, we get $y = \varepsilon$ and so $vu \uparrow v'$. This leads to $v' \in C(WuW)$, which is a contradiction. Hence $|vu| \leq |v'|$, and $vuy \uparrow v'$ for some y. But then again v' is in $C(WuW)$, a contradiction.

Now we show that Y is a pcode. Assume the contrary and consider a relation

$$u_1 u_2 \ldots u_m \uparrow v_1 v_2 \ldots v_n$$

with $u_1, \ldots, u_m, v_1, \ldots, v_n \in Y$, and $u_1 \neq v_1$. The set X being a pcode, one of these partial words must be u. Assume that one of $u_1, \ldots, u_m$ is u, and let i be the smallest index such that $u_i = u$. Since $WuW \cap C(X^*) = \emptyset$, it follows that $Wu_iW \cap C(X^*) = \emptyset$. Consequently one of $v_1, \ldots, v_n$ is u. Let j be the smallest index such that $v_j = u$. Then $u_1 \ldots u_{i-1} u, v_1 \ldots v_{j-1} u \in V$ whence $u_1 \ldots u_{i-1} \uparrow v_1 \ldots v_{j-1}$ by the abovementioned property of V. The set X is a pcode, thus from $u_1 \neq v_1$ it follows that $i = j = 1$ leading to a contradiction. ∎

THEOREM 8.2

Let $X \subset W(A) \setminus \{\varepsilon\}$. If X is a maximal pcode, then X is complete.

PROOF   Let $X \subset W(A) \setminus \{\varepsilon\}$ be a maximal pcode that is not complete. If $\|A\| = 1$, then $X = \emptyset$ and X is not maximal. If $\|A\| \geq 2$, consider a partial word u such that $u \notin F(C(X^*))$. We may choose u in A^*. According to Proposition 8.22, there exists a word $v \in A^*$ such that uv is unbordered. We have $uv \notin F(C(X^*))$, and it then follows from Proposition 8.23 that $X \cup \{uv\}$ is a pcode. Thus X is not maximal, a contradiction. $\quad\square$

8.6 Commutative ordering

In this section, we discuss the commutative ordering that we denote by $\preceq_c$ instead of ρ_c.

LEMMA 8.5
Let u, v be nonempty partial words such that v is non $\{|u|, |v| - |u|\}$-special. Then $u \preceq_c v$ if and only if there exists a primitive word z and integers m, n such that $u \subset z^m$ and $v \subset uz^n \subset z^{m+n}, v \subset z^n u \subset z^{m+n}$.

PROOF Let u, v be nonempty partial words such that v is non $\{|u|, |v| - |u|\}$-special. If $u \preceq_c v$, then for some full word x, we have $v \subset xu, v \subset ux$. Let u' be a full word such that $u \subset u'$. If $x = \varepsilon$, then $v \subset u \subset u'$ and there exists a primitive word z and a positive integer m such that $u' = z^m$. Hence $u \subset z^m, v \subset uz^0 \subset z^{m+0}, v \subset z^0 u \subset z^{m+0}$ and the result follows. So we may assume that x is nonempty. We get $v \subset xu', v \subset u'x$ and thus by Lemma 2.5, $xu' = u'x$. There exists a primitive word z and positive integers m, n such that $u' = z^m$ and $x = z^n$ $(z = \sqrt{x})$. This in turn implies that $u \subset u' \subset z^m$ and $v \subset ux = uz^n \subset z^{m+n}, v \subset xu = z^n u \subset z^{m+n}$. $\quad\square$

REMARK 8.2 A subset X of A^+ is an antichain with respect to $\preceq_c$ if and only if X is *anti-commutative*, or if for all $u, v \in X$ satisfying $u \neq v$, we have $uv \neq vu$ (see Exercise 8.25). $\quad\square$

The remark above leads to the following definition.

DEFINITION 8.10 *We call a subset X of $W(A) \backslash \{\varepsilon\}$ **anti-commutative** if for all $u, v \in X$ satisfying $u \neq v$, we have $uv \not\Uparrow vu$.*

Certainly, every pcode is anti-commutative.

PROPOSITION 8.25
Let $X \subset W(A) \setminus \{\varepsilon\}$ be pairwise nonspecial. If X is anti-commutative, then X is an antichain with respect to $\preceq_c$.

PROOF   If X is anti-commutative, then let us show that X is an antichain with respect to $\preceq_c$. Suppose to the contrary that there exist $u, v \in X$ with $u \neq v$ and $u \preceq_c v$. The latter implies that $|u| \leq |v|$. By assumption, v is non $\{|u|, |v| - |u|\}$-special, and by Lemma 8.5, there exists a primitive word z and integers m, n such that $u \subset z^m$ and $v \subset z^{m+n}$. But then $uv \uparrow vu$ contradicting the fact that X is anti-commutative.   ◻

PROPOSITION 8.26

Let $X \subset W(A) \setminus \{\varepsilon\}$. Let $u, v \in X$ be such that u is full, $u \neq v$, and uv is non $\{|u|, |v|\}$-special. If X is an antichain with respect to $\preceq_c$, then $uv \, \slashed{\uparrow} \, vu$.

PROOF   Suppose to the contrary that $uv \uparrow vu$. There exists a full word z such that $uv \subset z$ and $vu \subset z$. Put $z = xy$ where $u \subset x$ and $v \subset y$. We have $uv \subset xy$, and by Exercise 8.26 we also have $uv \subset yx$. Lemma 2.5 implies $xy = yx$, and so x, y are powers of a common word. Say $x = w^m$ and $y = w^n$ for some word w and integers m, n. Since u is full, we have $u = w^m$. If $m = n$, then $v \subset y = w^n = w^m = u$, and so $u \preceq_c v$. For the case $m < n$, we have $v \subset y = w^n = uw^{n-m} = w^{n-m}u$ and thus $u \preceq_c v$. Similarly, we can show that if $m > n$, then $v \preceq_c u$. In all cases, we obtain a contradiction.   ◻

REMARK 8.3   In Proposition 8.26, both the assumptions that u is full and uv is non $\{|u|, |v|\}$-special are needed. Indeed, if we put $X = \{u, v\}$ where $u = a \diamond b$ and $v = aab \diamond ab$, we get that X is an antichain with respect to $\preceq_c$ and that $uv \uparrow vu$. This example is such that u is nonfull and uv is non $\{|u|, |v|\}$-special. Now, if we put $X = \{u, v\}$ where $u = abbaab$ and $v = \diamond\diamond\diamond\diamond$, we get that X is an antichain with respect to $\preceq_c$ and that $uv \uparrow vu$. This example is such that u is full and uv is $\{|u|, |v|\}$-special.   ◻

DEFINITION 8.11   *Let $u', x, y \in A^+, v' \in W(A) \setminus \{\varepsilon\}$ be such that $|x| = |y|$ and $|u'x| = |v'|$. Then the set $\{u, v\}$ where $u = u'x$ and $v = v'y$ (respectively, $u = xu'$ and $v = yv'$) is said to be of **Type 1** (respectively, **Type 2**) if v is not $\{|u|, |x|\}$-special.*

PROPOSITION 8.27

Let u, v be nonempty partial words such that $\{u, v\}$ is of Type 1 or Type 2. Then $uv \, \slashed{\uparrow} \, vu$ if and only if $\{u, v\}$ is a pcode.

PROOF   We prove the result for Type 1 (Type 2 is similar). If $\{u, v\}$ is a pcode, then clearly $uv \, \slashed{\uparrow} \, vu$. Conversely, assume that $\{u, v\}$ is not a pcode and $uv \, \slashed{\uparrow} \, vu$. Then there exist an integer $n \geq 1$ and partial words $u_1, \ldots, u_n, v_1, \ldots, v_n \in \{u, v\}$ such that

$$u_1 u_2 \ldots u_n \uparrow v_1 v_2 \ldots v_n$$

and with $|u_1 u_2 \ldots u_n|$ as small as possible contradicting Proposition 8.3. We hence have $u_1 \neq v_1$ and $u_n \neq v_n$, and we may assume that $n > 2$. There are the four possibilities (1)–(4) as in Proposition 8.12. Since $\{u, v\}$ is of Type 1, there exist nonempty full words u', z_1, z_2 and a nonempty partial word v' such that $|z_1| = |z_2|$, $|u' z_1| = |v'|$, $u = u' z_1$, $v = v' z_2$, and v is not $\{|u|, |z_1|\}$-special. Any possibility gives $v' \uparrow u$. Substituting u by $u' z_1$ and v by $v' z_2$ in (1), (2), (3) and (4) we get

(5) $u' z_1 x u' z_1 \uparrow v' z_2 y v' z_2$

(6) $u' z_1 x v' z_2 \uparrow v' z_2 y u' z_1$

(7) $v' z_2 x u' z_1 \uparrow u' z_1 y v' z_2$

(8) $v' z_2 x v' z_2 \uparrow u' z_1 y u' z_1$

Any possibility implies $z_1 \uparrow z_2$, and hence $z_1 = z_2$ since both z_1 and z_2 are full. So $v = v' z_1$, and hence both u and v end with z_1, and the same is true for both x and y. We deduce that $v \uparrow z_1 u$, and so $v' z_1 \uparrow z_1 u$ and hence $v' z_1 \subset z_1 u$. The fact that $v' \uparrow u$ implies $v' z_1 \subset u z_1$. By Lemma 2.5, we get $u z_1 = z_1 u$ since v is not $\{|u|, |z_1|\}$-special, and u and z_1 are powers of a common word. So $v = v' z_1 \subset u z_1$ is contained in a power of that same common word. But then $uv \uparrow vu$, a contradiction. ▯

Example 8.5

Consider the set $\{abb, ab\diamond a\}$. To determine if the set is of Type 1 or Type 2, at least one word in the set, u, must be full. Let $u = abb$ and $v = ab\diamond a$. For the factorization of v, since $y \in A^+$, $v' = ab\diamond$ and $y = a$. Since $|x| = |y| = 1$, the factorization of $u = abb$ is $(u', x) = (ab, b)$. The word v is not $\{3, 1\}$-special since $\|H(v)\| < 2$, and thus $\{u, v\}$ is of Type 1. Since the example set is of Type 1 and $uv \not\uparrow vu$,

$$ab\diamond a\underline{a}bb \not\uparrow abbab\underline{\diamond}\underline{a}$$

this set is a pcode. ▯

REMARK 8.4   It should be noted that the above proposition is not true in general. Consider the set $\{u, v\}$ where $u = a\diamond b$ and $v = aababb$. The set is not of Type 1 or Type 2, but $uv \not\uparrow vu$,

$$a\diamond b\underline{a}\underline{a}babb \not\uparrow aabab\underline{b}a\diamond b$$

However, this set is not a pcode since a nontrivial compatibility relation does exist,

$$a\diamond ba\diamond baababb \uparrow aababba\diamond ba\diamond b$$

or $u^2v \uparrow vu^2$, or simply $u^2 \uparrow v$. ⬜

We now extend the above proposition further. We assume that $\{u, v\}$ is a set of partial words over an alphabet of size at least two. Otherwise, sets of at least two partial words are obviously nonpcodes.

PROPOSITION 8.28

Let k be an integer satisfying $k > 1$. Let u, v be nonempty partial words such that $|v| = k|u|$ and $\|H(v)\| = 0$. Then $\{u, v\}$ is a pcode if and only if $u^k v \not\uparrow vu^k$.

PROOF If $\{u, v\}$ is a pcode, then clearly $u^k v \not\uparrow vu^k$. Conversely, assume that $\{u, v\}$ is not a pcode and $u^k v \not\uparrow vu^k$. Then there exist $n \geq 1$ and $u_1, \ldots, u_n, v_1, \ldots, v_n \in \{u, v\}$ such that

$$u_1 u_2 \ldots u_n \uparrow v_1 v_2 \ldots v_n \tag{8.1}$$

and with $|u_1 u_2 \ldots u_n|$ as small as possible contradicting Proposition 8.3. We hence have $u_1 \neq v_1$ and $u_n \neq v_n$, and we may assume that $n \geq 2$. There are the four possibilities (1)–(4) of Proposition 8.27. Since $|v| = k|u|$, for any of the possibilities (1)–(4), there exist nonempty pwords $w_1, w_2, \ldots, w_k$ such that $v = w_1 w_2 \ldots w_k$, $|w_1| = |w_2| = \cdots = |w_k| = |u|$, $w_1 \uparrow u$, and $w_k \uparrow u$. The latter two relations give $u \subset w_1$ and $u \subset w_k$ since v is full.

Let us consider the case where $u_1 = u$ and $v_1 = v$ (the other cases are handled similarly).

    Case 1. $u_1 = u_2 = \cdots = u_{k-1} = u$

In this case $w_1 \uparrow u$, $w_2 \uparrow u$, ..., $w_k \uparrow u$, and by multiplication, $u^k v \uparrow vu^k$, contradicting our assumption.

    Case 2. There exists $1 < j < k$ such that $u_1 = u_2 = \cdots = u_{j-1} = u$ and $u_j = v$

Note that each element in $\{w_1, w_2, \ldots, w_{j-1}\}$ is compatible with u. Here, $k = m(j-1) + r$ with $1 \leq r < j$. We get $w_{k-j+2} = w_{k-2j+3} = \cdots =$ element in the set $\{w_1, w_2, \ldots, w_{j-1}\}$, $w_{k-j+3} = w_{k-2j+4} = \cdots =$ element in the set $\{w_1, w_2, \ldots, w_{j-1}\}$, ..., and $w_k = w_{k-j+1} = \cdots =$ element in the set $\{w_1, w_2, \ldots, w_{j-1}\}$. Thus, $w_1 \uparrow u$, $w_2 \uparrow u$, ..., $w_k \uparrow u$. Hence $u^k v \uparrow vu^k$, contradicting our assumption. ⬜

Example 8.6

Let $u = a\diamond ba$ and $v = aabaaabaaabaabba$ so that $|v| = 4|u|$ and $\|H(v)\| = 0$. A nontrivial compatibility relation does exist, $u^4 v \uparrow vu^4$,

$a\diamond baa\diamond baa\diamond baa\diamond baaabaaabaaabaabba \uparrow aabaaabaaabaabbaa\diamond baa\diamond baa\diamond baa\diamond ba$

thus this set is not a pcode.

Now, let $u = ab \diamond ab$ and $v = baababbbababababb$ so that $|v| = 3|u|$ and $\|H(v)\| = 0$. In this case, $u^3 v \, \slashed{\uparrow} \, vu^3$,

$$\underline{ab \diamond abab \diamond abab \diamond ab}\underline{baab}\underline{abb}baababab\underline{b} \, \slashed{\uparrow} \, \underline{baa}\underline{babb}babababab\underline{bab} \diamond \underline{abab} \diamond \underline{abab} \diamond \underline{ab}$$

thus this set is a pcode.   □

We end this section with the following proposition.

PROPOSITION 8.29
Let k be an integer satisfying $k > 1$. Let $u, v, w_1, w_2, \ldots, w_k$ be nonempty partial words such that $v = w_1 w_2 \ldots w_k$, $|w_1| = |w_2| = \cdots = |w_k| = |u|$, $\|H(u)\| = 0$, and $\|H(v)\| = \|H(w_i)\|$ for some $1 \le i \le k$. Then $\{u, v\}$ is a pcode if and only if $uv \, \slashed{\uparrow} \, vu$.

PROOF   We refer the reader to the proof of Proposition 8.28. Any of the possibilities (1)–(4) imply $w_1 \uparrow u$ and $w_k \uparrow u$. The latter two relations give $w_1 \subset u$ and $w_k \subset u$ since u is full. Let us consider the case where $u_1 = u$ and $v_1 = v$ (the other cases are handled similarly).

   Case 1. $u_1 = u_2 = \cdots = u_n = u$
   In this case $w_1 \uparrow u$, $w_2 \uparrow u$, $\ldots$, $w_k \uparrow u$, and thus $uv \uparrow vu$ contradicting our assumption.

   Case 2. There exists $1 < j \le n$ such that $u_1 = u_2 = \cdots = u_{j-1} = u$ and $u_j = v$
   Here we consider the cases where $j \ge k$ and $j < k$. Note that each element in the set $\{w_1, w_2, \ldots, w_{j-1}\}$ is compatible with u. If $j \ge k$, then $w_1 \uparrow u$, $w_2 \uparrow u$, $\ldots$, $w_k \uparrow u$, and thus $uv \uparrow vu$ contradicting our assumption. Now, if $j < k$, then $k = m(j-1) + r$ and $i = m'(j-1) + r'$ with $1 \le r < j$ and $1 \le r' < j$. We get

$$w_i \subset w_{i-j+1} = w_{i-2j+2} = \cdots = w_{r'} \tag{8.2}$$

We also get

$$w_i \subset w_{i+j-1} = w_{i+2j-2} = \cdots = w_{k-j+1-r+r'} \tag{8.3}$$

if $r' > r$, and

$$w_i \subset w_{i+j-1} = w_{i+2j-2} = \cdots = w_{k-r+r'} \tag{8.4}$$

if $r' \le r$. Moreover, if $1 \le i' \le k$ and $i' \not\equiv r' \bmod j - 1$, then $w_{i'} = u$. We consider the following three cases.

   Case 2.1. $j \le i \le k - j + 1$
   In this case, the compatibility relation 8.1 yields $w_1 = w_2 = \cdots = w_{j-1} = u = w_{k-j+2} = \cdots = w_{k-1} = w_k$. The relations 8.2, 8.3 and 8.4 imply that $v = u^{i-1} w_i u^{k-i}$ with $w_i \subset u$. Hence $uv \uparrow vu$, contradicting our assumption.

Case 2.2. $1 \leq i < j$

Here $i = r'$ and $w_i \subset u$. Consider the case where $r' > r$ (the case where $r' \leq r$ is handled similarly). Referring to relations 8.1, 8.2 and 8.3, we have $w_{k-j+1-r+r'} \uparrow u$ or $w_{k-j+1-r+r'} \uparrow w_s$ where $s \not\equiv r' \bmod j - 1$. In either case, $w_{k-j+1-r+r'} \uparrow u$ and since u is full, we get $w_{k-j+1-r+r'} \subset u$. Again $uv \uparrow vu$, contradicting our assumption.

Case 2.3. $k - j + 2 \leq i \leq k$

This case is symmetric to Case 2.2. ▯

8.7   Circular pcodes

In this section, we start by defining the *circular codes* and then extending them to the *circular pcodes*.

DEFINITION 8.12   *Let X be a nonempty subset of A^+. Then X is called a **circular code** if for all integers $m \geq 1, n \geq 1$, words $u_1, \ldots, u_m, v_1, \ldots, v_n \in X$, and $r \in A^*$ and $s \in A^+$, the conditions*

$$su_2 \ldots u_m r = v_1 v_2 \ldots v_n$$
$$u_1 = rs$$

imply $m = n$, $r = \varepsilon$, and $u_i = v_i$ for $i = 1, \ldots, m$.

DEFINITION 8.13   *Let X be a subset of $W(A) \setminus \{\varepsilon\}$. Then X is called a **circular pcode** over A if for all integers $m \geq 1, n \geq 1$, partial words $u_1, \ldots, u_m, v_1, \ldots, v_n \in X$, and $r \in W(A)$ and $s \in W(A) \setminus \{\varepsilon\}$, the conditions*

$$su_2 u_3 \ldots u_m r \uparrow v_1 v_2 \ldots v_n$$
$$u_1 \subset rs$$

imply $m = n$, $r = \varepsilon$, and $u_i = v_i$ for $i = 1, \ldots, m$.

Figure 8.4 depicts the circular pcode concept.

It is clear from the definition that a subset X of A^+ is a circular code if and only if it is a circular pcode. A circular pcode is a pcode, and any subset of a circular pcode is also a circular pcode.

Circular pcodes turn out to have numerous interesting properties. We start by two propositions.

PROPOSITION 8.30

Let $X \subset W(A) \setminus \{\varepsilon\}$. If X is a circular pcode, then X does not contain two distinct conjugate partial words.

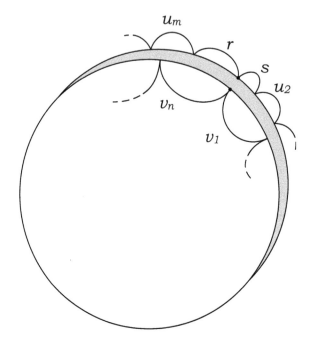

FIGURE 8.4:   A circular pcode.

PROPOSITION 8.31

Let $X \subset W(A) \setminus \{\varepsilon\}$ be a circular pcode. If $u \in X$, then u is primitive.

We will characterize in various ways the submonoids generated by circular pcodes.

DEFINITION 8.14

- *A submonoid M of $W(A)$ is called* **pure** *if for each partial word u and integer $n \geq 1$, the conditions $u_1 \ldots u_n \in M$ and $u_i \subset u$ for all $i = 1, \ldots, n$ imply $u_1 = u_2 = \cdots = u_n$ and $u_i \in M$ for all $i = 1, \ldots, n$.*

- *A submonoid M of $W(A)$ is called* **very pure** *if for all partial words u, v, u', v' satisfying $|v'| = |v|$ and $|u'| = |u|$, the conditions $vu \uparrow v'u'$, $uv \in M$, and $v'u' \in M$ imply $u = u'$ and $u, v \in M$.*

Note that a very pure monoid is pure.

PROPOSITION 8.32

A submonoid M of $W(A)$ is very pure if and only if its minimal generating set is a circular pcode.

PROOF   Let M be a very pure submonoid of $W(A)$. Let u, u', v, w be partial words with $u \uparrow u'$, $u, u'w, v \in M$, and $wv \in C(M)$. We have $(vu')w = v(u'w) \in M$ and $w(vu') = (wv)u' \in C(M)$. This implies $u = u'$ and $w \in M$. Thus M is stable, hence M is pfree by Proposition 8.16. Let X be its base. Assume that there exist positive integers m, n, partial words $u_1, \ldots, u_m, v_1, \ldots, v_n \in X$, and $r \in W(A)$ and $s \in W(A) \setminus \{\varepsilon\}$ such that

$$su_2u_3 \ldots u_m r \uparrow v_1 v_2 \ldots v_n$$
$$u_1 \subset rs$$

Put $u_1 = r's'$ where $|r'| = |r|$ and $|s'| = |s|$. Put $u = s'$ and $v = u_2 \ldots u_m r'$. Then $vu \in M$ and by weakening $uv \in C(M)$. Since M is very pure, $u, v \in M$. Since $u_2 \ldots u_m, u_2 \ldots u_m r', s', r's' \in M$, the stability of M implies that $r' \in M$. From $r's' \in X$, it follows that $r' = \varepsilon$ (and $r = \varepsilon$). By weakening, $u_1 u_2 \ldots u_m \uparrow v_1 v_2 \ldots v_n$. Since X is a pcode by Proposition 8.15, this implies $m = n$ and $u_i = v_i$ for $i = 1, \ldots, m$.

Conversely, let X be the minimal generating set for M and assume that X is a circular pcode (here $M = X^*$). To show that M is very pure, consider partial words u, v such that $uv \in M$ and $vu \in C(M)$. The latter implies that $vu \uparrow v'u'$ with u', v' satisfying $v'u' \in M$, $|v'| = |v|$, and $|u'| = |u|$. If $u = \varepsilon$ or $v = \varepsilon$, then $u = u'$ and $u, v \in M$. If $u \neq \varepsilon$ and $v \neq \varepsilon$, then put

$$uv = u_1 u_2 \ldots u_m$$
$$vu \uparrow v_1 v_2 \ldots v_n$$

with $u_1, \ldots, u_m, v_1, \ldots, v_n \in X$. There exists an integer i, $1 \leq i \leq m$, such that

$$u = u_1 u_2 \ldots u_{i-1} r$$
$$v = su_{i+1} \ldots u_m$$

where $u_i = rs$, $r \in W(A)$, and $s \in W(A) \setminus \{\varepsilon\}$. Then

$$su_{i+1} \ldots u_m u_1 u_2 \ldots u_{i-1} r \uparrow v_1 v_2 \ldots v_n$$

Since X is a circular pcode, this implies $m = n$, $r = \varepsilon$, and $u_i = v_1, u_{i+1} = v_2, \ldots, u_m = v_{n-i+1}, u_1 = v_{n-i+2}, u_2 = v_{n-i+3}, \ldots, u_{i-1} = v_n$. Thus $u = u_1 u_2 \ldots u_{i-1} = v_{n-i+2} v_{n-i+3} \ldots v_n = u'$ and $u, v \in M$, showing that M is very pure. $\quad\Box$

We now give a characterization of circular pcodes in terms of the conjugacy operation that was defined in Chapter 2. We start with a definition.

DEFINITION 8.15   Let $X \subset W(A) \setminus \{\varepsilon\}$ be a pcode. Two partial words $u, v \in X^*$ are called **X-conjugate** if there exist $x, y \in X^*$ such that $u = xy$, $v = yx$.

Two partial words in X^* which are X–conjugate are obviously conjugate.

PROPOSITION 8.33

Let $X \subset W(A) \setminus \{\varepsilon\}$ be a pcode. The following conditions are equivalent:

1. *The set X is a circular pcode.*

2. *The monoid X^* is pure, and any two partial words in X^* which are conjugate are also X-conjugate.*

PROOF   We first show that Condition 1 implies Condition 2. Since X^* is very pure, it is pure. Next, let $u, v \in X^*$ be conjugate partial words. Then $u \subset xy, v \subset yx$ for some pwords x, y. Put $u = x'y'$ where $|x'| = |x|$ and $|y'| = |y|$. Also, put $v = y''x''$ where $|y''| = |y|$ and $|x''| = |x|$. Since $x' \subset x$ and $y' \subset y$, we get $y'x' \subset yx$. The latter and the fact that $y''x'' \subset yx$ imply that $y'x' \uparrow y''x''$. We get the two conditions $x'y' \in X^*$ and $y'x' \in C(X^*)$. Since X^* is very pure, $x' = x''$ and $x', y' \in X^*$. With a similar reasoning, we can deduce that $y' = y''$. So $u = x'y', v = y'x'$ with $x', y' \in X^*$, showing that u, v are X-conjugate.

Now, we show that Condition 2 implies Condition 1. Let u, v be partial words such that $uv \in X^*$ and $vu \in C(X^*)$. The latter implies that $vu \uparrow v'u'$ with u', v' satisfying $v'u' \in X^*$, $|v'| = |v|$ and $|u'| = |u|$. If $u = \varepsilon$, then $u = u'$ and $u, v \in X^*$. If $u \neq \varepsilon$ and $v = \varepsilon$, then $u, u', v \in X^*$ and $u \uparrow u'$. Since X is a pcode, this yields $u = u'$. If $u \neq \varepsilon$ and $v \neq \varepsilon$, then by definition, there exists a primitive word x and a positive integer n such that $vu \subset x^n$ and $v'u' \subset x^n$. We get words r, s and integers p, q such that $x = rs$, $u \subset sx^q$, $v \subset x^p r$, and $p + q + 1 = n$. Put $y = sr$ (x being primitive, y is primitive as well). Since $v'u' \subset x^n$ and $uv \subset y^n$, write $v'u' = x_1x_2 \ldots x_n$ and $uv = y_1y_2 \ldots y_n$ where $|x_1| = |x_2| = \cdots = |x_n|$, $|y_1| = |y_2| = \cdots = |y_n|$. Since X^* is pure, we have $x_1 = x_2 = \cdots = x_n$, $y_1 = y_2 = \cdots = y_n$, and $x_1, \ldots, x_n, y_1, \ldots, y_n \in X^*$. Thus $v'u' = (x')^n$ and $uv = (y')^n$ with $x', y' \in X^*$. Since $x' \subset x \subset rs, y' \subset y \subset sr$, we get that x', y' are conjugate and thus X-conjugate. So there exist $r', s' \in X^*$ such that $x' = r's'$ and $y' = s'r'$. Thus $u = s'(x')^q$, $u' = s'(x')^q$, and $v = (x')^p r'$ showing that $u = u'$ and $u, v \in X^*$.   ☐

The reader is invited to prove the following result which is an analogue of Theorem 9.2.

PROPOSITION 8.34

Let $X \subset W(A) \setminus \{\varepsilon\}$ be a circular pcode. If X is maximal as a circular pcode, then X is complete.

Exercises

8.1 $\boxed{\text{s}}$ Prove Lemma 8.1.

8.2 Prove Lemma 8.2.

8.3 Complete the proof of Lemma 8.3.

8.4 Show that the set $Y = \{a\diamond b, \diamond b, baba\}$ is not a pcode.

8.5 Give a necessary and sufficient condition for a nonempty set X satisfying $X \subset \{a, \diamond\}^*$ to be a pcode over $\{a\}$.

8.6 Is $X = \{a\diamond b, \diamond cb\}$ a pcode? Why or why not?

8.7 Let u, v be nonempty partial words such that $|u| = |v|$. Prove that $uv \not\uparrow vu$ if and only if $\{u, v\}$ is a pcode (actually $u \not\uparrow v$ if and only if $\{u, v\}$ is a pcode).

8.8 $\boxed{\text{s}}$ Is the pcode $X = \{a\diamond b, abbab\}$ over $\{a, b\}$ complete?

8.9 Prove that a prefix pcode is a pcode.

8.10 $\boxed{\text{s}}$ Consider the two-element set $\{a\diamond b, aababbaababb\}$. Show that it is not a pcode by using Proposition 8.28.

8.11 Let k, l be integers satisfying $1 \le k \le l$. Let $u, v, w, w_1, \ldots, w_k$ be nonempty partial words such that $u = w_1 w_2 \ldots w_k$, $v = w^l$, and $|w_1| = |w_2| = \cdots = |w_k| = |w|$. Prove that $\{u, v\}$ is a pcode if and only if $uv \not\uparrow vu$.

8.12 $\boxed{\text{s}}$ Let u, v be nonempty partial words such that $|u| = 2|v|$ and $\|H(v)\| = 0$. Prove that $uv \not\uparrow vu$ if and only if $\{u, v\}$ is a pcode. Are the following sets pcodes?

- $\{aba, a\diamond\diamond bab\}$
- $\{b\diamond abb\diamond, bba\}$

8.13 Let X and Y be pcodes over A. Prove that if $X^* = Y^*$, then $X = Y$.

8.14 $\boxed{\text{s}}$ Prove Proposition 8.30.

8.15 Prove Proposition 8.31.

Challenging exercises

8.16 ⬚s Show that a uniform pcode X over A is maximal over A.

8.17 ⬚s Prove Proposition 8.13 and deduce the following corollaries:

1. Let $\varphi : W(A) \to W(C)$ be a pinjective morphism. If X is a pcode over A, then $\varphi(X)$ is a pcode over C. Similarly, if $\varphi : A^* \to W(C)$ is a pinjective morphism and X is a code over A, then $\varphi(X)$ is a pcode over C.

2. If $X \subset W(A)$ is a pcode over A, then X^n is a pcode over A for all positive integers n.

8.18 ⬚s Prove Proposition 8.14.

8.19 Prove Proposition 8.15.

8.20 ⬚s Prove Proposition 8.16.

8.21 Prove Proposition 8.17.

8.22 Prove Proposition 8.18.

8.23 Prove Theorem 8.1.

8.24 ⬚s Let u, v be nonempty partial words such that $|v| = 2|u|$. Prove that $u^2 v \not\uparrow vu^2$ if and only if $\{u, v\}$ is a pcode. Illustrate this with the following three sets $\{u, v\}$ where:

1. $u = bab\diamond$ and $v = baa\diamond\diamond bab$
2. $u = a\diamond b$ and $v = aa\diamond abb$
3. $u = ab\diamond$ and $v = \diamond b\diamond abb$

8.25 Prove that a subset X of A^+ is an antichain with respect to $\preceq_c$ if and only if X is *anti-commutative*, or if for all $u, v \in X$ satisfying $u \neq v$, we have $uv \neq vu$.

8.26 Prove Proposition 8.34.

Programming exercises

8.27 Write a program to find out whether or not a two-element set $\{u, v\}$ is of Type 1 or Type 2.

8.28 Design an applet that takes as input a two-element set $\{u, v\}$ and computes the compatibility relations $u^k v \uparrow vu^k$ for suitably bounded k.

Website

A World Wide Web server interface at

$$\texttt{http://www.uncg.edu/mat/pcode}$$

has been established for automated use of a program that discovers if a nonempty finite set of partial words is or is not a pcode (the algorithm will be discussed in Chapter 9). The program also discovers a nontrivial compatibility relation in case the set is not a pcode. In addition, in such cases where the given set X is a two-element nonpcode, the program outputs whether or not X is of Type 1 or Type 2.

Bibliographic notes

The theory of codes of words is exposited in [12]. Pcodes were introduced by Blanchet-Sadri [16] and further studied by Blanchet-Sadri and Moorefield [39]. We refer the reader to [132] for an exposition of some of the results of this chapter in the framework of full words.

The Defect Theorem for partial words was proposed by Leupold in [103].

The class $\mathcal{F}$ was considered in [133] for strict positive binary relations on A^* and the class $\mathcal{G}$ for strict positive binary relations on A^*. Theoretical aspects of the embedding ordering on A^* can be found in [51, 86, 93, 106]. An algorithmic aspect of the embedding ordering is motivated by molecular biology. The problem is to find, for a given set $X = \{u_1, \ldots, u_n\}$ of words, a shortest word v such that $u_i \rho_d v$ for all i. This problem is referred to as the shortest common supersequence problem which is known to be NP-complete [129].

Proposition 8.22, Proposition 8.23 and Theorem 8.2 extend results of [12] to partial words.

Exercise 8.25 is from [94].

Chapter 9

Deciding the Pcode Property

In Section 9.1, we give an analog of the Sardinas and Patterson algorithm for testing whether or not a given finite set of partial words is a pcode. In Section 9.2, we adapt a technique of Head and Weber related to dominoes to show that the pcode property is decidable.

9.1 First algorithm

In this section, we describe a first algorithm for deciding the pcode property. A subset X of $W(A) \setminus \{\varepsilon\}$ containing two distinct compatible partial words is obviously not a pcode. Recall that X is pairwise noncompatible if no distinct partial words $u, v \in X$ satisfy $u \uparrow v$.

We build a sequence of sets as follows: Let U_1 be the set

- $\{x \mid x \neq \varepsilon \text{ and there exist } u \in X \text{ and } u'x \in X \text{ such that } u \uparrow u'\}$

and for $i \geq 1$, let U_{i+1} be the union of the two sets

- $\{x \mid \text{there exist } u \in X \text{ and } u'x \in U_i \text{ such that } u \uparrow u'\}$

- $\{x \mid \text{there exist } u \in U_i \text{ and } u'x \in X \text{ such that } u \uparrow u'\}$

REMARK 9.1   To obtain the U_1-set, compare each element in the set X with every longer element in X to determine if x exists. For any given u, v in X satisfying $|u| < |v|$, if $v = u'x$ with $u \uparrow u'$, then $x \in U_1$. This can be pictured as follows:

$$u \quad \in X$$
$$\uparrow$$
$$u' \underline{x} \in X$$

To obtain the U_{i+1}-sets where $i \geq 1$: First, compare each element in the set X with every nonshorter element in the set U_i to determine if x exists. For any given $u \in X$ and $v \in U_i$ satisfying $|u| \leq |v|$, if $v = u'x$ with $u \uparrow u'$, then $x \in U_{i+1}$. This can be pictured as follows:

$$u \quad \in X$$
$$\uparrow$$
$$u' \; \underline{x} \in U_i$$

Second, compare each element in the set U_i with every nonshorter element in the set X to determine if x exists. For any given $u \in U_i$ and $v \in X$ satisfying $|u| \le |v|$, if $v = u'x$ with $u \uparrow u'$, then $x \in U_{i+1}$. This can be pictured as follows:

$$u \quad \in U_i$$
$$\uparrow$$
$$u' \; \underline{x} \in X$$

$\square$

LEMMA 9.1

Let $X \subset W(A) \setminus \{\varepsilon\}$. For all $n \ge 1$ and $k \in \{1, \dots, n\}$, we have $\varepsilon \in U_n$ if and only if there exist a partial word $x \in U_k$ and integers $i, j \ge 0$ such that $xX^i \cap C(X^j) \ne \emptyset$ and $i + j + k = n$.

PROOF   We prove the statement for all n by descending induction on k. Assume first that $k = n$. If $\varepsilon \in U_n$, then the condition is satisfied with $x = \varepsilon$ and $i = j = 0$. Conversely, if the condition is satisfied, then $i = j = 0$ and $x = \varepsilon$ and $\varepsilon \in U_n$.

Now, let $n > k \ge 1$, and suppose that the equivalence holds for $n, n - 1, \dots, k+1$. If $\varepsilon \in U_n$, then by the inductive hypothesis, there exist a partial word $x \in U_{k+1}$ and integers $i, j \ge 0$ such that $xX^i \cap C(X^j) \ne \emptyset$ and $i + j + (k+1) = n$. Thus there exist partial words $u_1, \dots, u_i, v_1, \dots, v_j \in X$ such that

$$xu_1 \dots u_i \uparrow v_1 \dots v_j$$

Now $x \in U_{k+1}$, and there are two cases: Either there exists $u \in C(X)$ such that $ux \in U_k$, or there exists $y \in U_k$ and a partial word y' such that $y \uparrow y'$ and $y'x \in X$. In the first case, we have $u \uparrow u'$ for some $u' \in X$ and

$$uxu_1 \dots u_i \uparrow u'v_1 \dots v_j$$

Consequently, there exist a partial word $ux \in U_k$ and integers $i, j + 1 \ge 0$ such that $uxX^i \cap C(X^{j+1}) \ne \emptyset$ and $i + (j+1) + k = n$, and the condition is satisfied. In the second case, we have

$$y'xu_1 \dots u_i \uparrow yv_1 \dots v_j$$

Consequently, there exist a partial word $y \in U_k$ and integers $j, i + 1 \ge 0$ such that $yX^j \cap C(X^{i+1}) \ne \emptyset$ and $j + (i+1) + k = n$, and the condition is satisfied.

Conversely, assume that there exist a partial word $x \in U_k$ and integers $i, j \ge 0$ such that $xX^i \cap C(X^j) \ne \emptyset$ and $i + j + k = n$. Then

$$xu_1 \ldots u_i \uparrow v_1 \ldots v_j$$

for some $u_1, \ldots, u_i, v_1, \ldots, v_j \in X$. If $j = 0$, then $i = 0$ and $k = n$. If $j > 0$, then we consider two cases:

Case 1. $|x| \geq |v_1|$

If $x = v_1' y$ for some pword y and some v_1' satisfying $v_1' \uparrow v_1$, then $y \in U_{k+1}$ and $y u_1 \ldots u_i \uparrow v_2 \ldots v_j$. Thus $y X^i \cap C(X^{j-1}) \neq \emptyset$ and by the inductive hypothesis $\varepsilon \in U_n$.

Case 2. $|x| < |v_1|$

If $v_1 = x' y$ for some nonempty pword y and some x' satisfying $x \uparrow x'$, then $y \in U_{k+1}$ and $u_1 \ldots u_i \uparrow y v_2 \ldots v_j$. Thus $y X^{j-1} \cap C(X^i) \neq \emptyset$ and by the inductive hypothesis $\varepsilon \in U_n$. $\quad\square$

Note that if X is a finite set, then $\{U_n \mid n \geq 1\}$ is finite (this is because each U_n contains only suffixes of partial words in X). The next theorem provides an algorithm for testing whether or not a finite set is a pcode.

THEOREM 9.1
Let $X \subset W(A) \setminus \{\varepsilon\}$ be pairwise noncompatible. The set X is a pcode if and only if none of the sets U_n contains the empty word.

PROOF If X is not a pcode, then there exists a compatibility relation

$$u_1 u_2 \ldots u_m \uparrow v_1 v_2 \ldots v_n$$

where m, n are positive integers, $u_1 \neq v_1$, and $u_1, \ldots, u_m, v_1, \ldots, v_n \in X$. Assume first that $|u_1| = |v_1|$. Then $u_1 \uparrow v_1$, a contradiction since X is pairwise noncompatible. Now assume that $|u_1| > |v_1|$. Then $u_1 = v_1' x$ for some pword x and some v_1' satisfying $v_1' \uparrow v_1$. But then $x \in U_1$ and $x X^{m-1} \cap C(X^{n-1}) \neq \emptyset$. By Lemma 9.1, $\varepsilon \in U_{m+n-1}$.

If X is a pcode and $\varepsilon \in U_n$, then put $k = 1$ in Lemma 9.1. There exist $x \in U_1$ and integers $i, j \geq 0$ such that $i + j = n - 1$ and $x X^i \cap C(X^j) \neq \emptyset$. Since $x \in U_1$, we have $v = ux$ for some $u \in C(X), v \in X$. Furthermore, $u \neq v$ since $x \neq \varepsilon$. Since $u \in C(X)$, there exists $u' \in X$ such that $u \uparrow u'$. It follows from $uxX^i \cap uC(X^j) \neq \emptyset$ that $vX^i \cap C(u'X^j) \neq \emptyset$, showing that X is not a pcode. $\quad\square$

Example 9.1
Consider the pairwise noncompatible set $Y = \{a \diamond b, \diamond a, abba, bba\}$. The computations that follow will show that

$U_1 = \{a, b\}$,

$U_2 = \{\diamond b, a, ba, bba\}$,

$U_3 = \{\varepsilon, \diamond b, a, b, ba, bba\}$,

$U_4 = \{\varepsilon, \diamond a, \diamond b, a, a \diamond b, abba, b, ba, bba\},$

$$\vdots$$

Since $\varepsilon \in U_3$, this set is not a pcode.

- For U_1: First, consider $u = a \diamond b$. In this case, $a \diamond b \uparrow ab\underline{ba}$ and therefore $x = a$. Now, consider $u = \diamond a$. Here, $\diamond a \uparrow a \diamond \underline{b}$ and therefore $x = b$. No other choices of u are successful and $U_1 = \{a, b\}$.

- For U_2: The first set is empty since every $u \in Y$ is greater in length than every word in U_1. However, comparing U_1 with Y produces a nonempty set. First, comparing $u = a$ generates the following:

$a \uparrow a\underline{\diamond b}$	$\diamond b$
$a \uparrow \diamond \underline{a}$	a
$a \uparrow a\underline{bba}$	bba

Now, comparing $u = b$ generates the following:

$b \uparrow \diamond \underline{a}$	a
$b \uparrow b\underline{ba}$	ba

- For U_3: First, compare the set Y with U_2:

$bba \uparrow bba$	ε
$\diamond a \uparrow ba$	ε

Now, compare U_2 with Y:

$\diamond b \uparrow a\diamond \underline{b}$	b
$\diamond b \uparrow ab\underline{ba}$	ba
$\diamond b \uparrow bb\underline{a}$	a
$a \uparrow a\underline{\diamond b}$	$\diamond b$
$a \uparrow \diamond \underline{a}$	a
$a \uparrow a\underline{bba}$	bba
$ba \uparrow \diamond a$	ε
$bba \uparrow bba$	ε

- For U_4: First, compare the set Y with U_3:

$\diamond a \uparrow ba$	ε
$bba \uparrow bba$	ε

Now, compare U_3 with Y:

$\varepsilon \uparrow a\diamond b$	$a\diamond b$
$\varepsilon \uparrow \diamond a$	$\diamond a$
$\varepsilon \uparrow abba$	$abba$
$\varepsilon \uparrow bba$	bba
$\diamond b \uparrow a\diamond \underline{b}$	b
$\diamond b \uparrow ab\underline{ba}$	ba
$\diamond b \uparrow bb\underline{a}$	a
$a \uparrow a\diamond \underline{b}$	$\diamond b$
$a \uparrow \diamond \underline{a}$	a
$a \uparrow a\underline{bba}$	bba
$b \uparrow \diamond \underline{a}$	a
$b \uparrow b\underline{ba}$	ba
$ba \uparrow \diamond a$	ε
$bba \uparrow bba$	ε

Since $\varepsilon \in U_3$, the algorithm may stop at this point since it is determined that Y is not a pcode. The set U_4 was computed as an exercise to further the example.

□

We now discuss how to use the algorithm to discover some nontrivial compatibility relations for nonpcodes.

Example 9.2

Consider the set $Z = \{u_1, u_2, u_3, u_4\}$ where $u_1 = \diamond b$, $u_2 = a\diamond b$, $u_3 = aa\diamond bba$, and $u_4 = ba$. The generated sets are as follows:

$U_1 = \{b, bba\}$,

$U_2 = \{a, b\}$,

$U_3 = \{\diamond b, a, a\diamond bba, b\}$,

$U_4 = \{\varepsilon, \diamond b, a, a\diamond bba, b, ba, bba\}$,

$U_5 = \{\varepsilon, \diamond b, a, a\diamond bba, b, ba, bba, a\diamond b, aa\diamond bba\} = U_6 = \cdots$.

Since $\varepsilon \in U_4$, the set Z is not a pcode.

The following list of compatibilities are derived from the sets generated by the algorithm:

$$
\begin{array}{ccc}
u_2 \underline{bba} & \uparrow & u_3 \\
u_2 u_1 \underline{a} & \uparrow & u_3 \\
u_2 u_1 u_1 & \uparrow & u_3 \underline{b} \\
u_2 u_1 u_1 \underline{a} & \uparrow & u_3 u_4 \\
u_2 u_1 u_1 u_2 & \uparrow & u_3 u_4 \diamond \underline{b} \\
u_2 u_1 u_1 u_2 & \uparrow & u_3 u_4 u_1
\end{array}
$$

Thus, a nontrivial compatibility relation for the set Z is $u_2u_1u_1u_2 \uparrow u_3u_4u_1$, or $a\diamond b\diamond b\diamond ba\diamond b \uparrow aa\diamond bbaba\diamond b$.

The discovery of the nontrivial compatibility relation in the above example requires a more detailed explanation. In fact, deriving the U_i-sets with the algorithm, several possible nontrivial compatibility relations emerge.

In obtaining U_1, since $\diamond b \uparrow a\diamond \underline{b}$, $b \in U_1$. This relation may be translated as follows. Since $u_1 = \diamond b$ and $u_2 = a\diamond b$, the original relation may be rewritten as $u_1b \uparrow u_2$. Note that the additional $b \in U_1$ is concatenated to u_1 to make a word compatible with $a\diamond b$. In the same manner, the relation $a\diamond b \uparrow aa\diamond \underline{bba}$ is rewritten as $u_2bba \uparrow u_3$.

When comparing the elements of the set Z with the elements of the set U_1, the set U_2 yields the relation $\diamond b \uparrow bb\underline{a}$, or $u_1 \uparrow bb\underline{a}$. Due to this relation, $a \in U_2$. The compatibility relation from the previous set U_1 that generated $bba \in U_1$ is $u_2bba \uparrow u_3$. In this case, $u_1 = \diamond b$ replaces the bb of $\underline{bba}$ which leaves only the extraneous $a \in U_2$. Therefore, the new relation is $u_2u_1a \uparrow u_3$.

Comparing the elements of the set U_1 with the elements of the set Z requires alternate handling. The relation $b \uparrow \diamond\underline{b}$ generates $b \in U_2$. The compatibility relation from the previous set U_1 that generated b is $u_1b \uparrow u_2$. In this case, $u_1 = \diamond b$ replaces the b on the left side of the relation. However, the extraneous $b \in U_2$ is concatenated to the right side of the relation. Therefore, the new relation is $u_1u_1 \uparrow u_2b$. In the same manner, the relation $b \uparrow b\underline{a}$ where $a \in U_2$ is translated as $u_1u_4 \uparrow u_2a$ since $u_4 = ba$.

The following lists illustrate the compatibility relations that unfold as the set U_3 is determined by the algorithm. For clarity, relations generating the same partial word that becomes an element of U_3 are grouped together. Note that multiple possibilities may exist for a single relation as derived from the algorithm. The original relation from the algorithm is on the left, and the respective translation is on the right.

The following relations allow $b \in U_3$:

$$a \uparrow \diamond\underline{b} \quad\quad u_2u_1u_1 \uparrow u_3b$$
$$a \uparrow \diamond\underline{b} \quad\quad u_1u_4b \uparrow u_2u_1$$

The following relations allow $\diamond b \in U_3$:

$$a \uparrow a\underline{\diamond b} \quad\quad u_2u_1u_2 \uparrow u_3\diamond b$$
$$a \uparrow a\underline{\diamond b} \quad\quad u_1u_4\diamond b \uparrow u_2u_2$$

The following relations allow $a\diamond bba \in U_3$:

$$a \uparrow a\underline{a\diamond bba} \quad\quad u_2u_1u_3 \uparrow u_3a\diamond bba$$
$$a \uparrow a\underline{a\diamond bba} \quad\quad u_1u_4a\diamond bba \uparrow u_2u_3$$

The following relation allows $a \in U_3$:

$$b \uparrow b\underline{a} \quad\quad u_1u_1a \uparrow u_2u_4$$

The relations from U_4 are presented in a similar manner. Observe that once $\varepsilon \in U_4$ two different nontrivial compatibility relations emerge, $u_2u_1u_2 \uparrow u_3u_1$ and $u_1u_4u_1 \uparrow u_2u_2$, thus proving the example set Z is not a pcode.

The following relations allow $\varepsilon \in U_4$:

$$\diamond b \uparrow \diamond b\underline{\varepsilon} \quad u_2u_1u_2 \uparrow u_3u_1$$
$$\diamond b \uparrow \diamond b\underline{\varepsilon} \quad u_1u_4u_1 \uparrow u_2u_2$$

The following relations allow $bba \in U_4$:

$$\diamond b \uparrow a\diamond\underline{bba} \quad u_2u_1u_3 \ \uparrow u_3u_1 bba$$
$$\diamond b \uparrow a\diamond\underline{bba} \quad u_1u_4u_1 bba \uparrow \quad u_2u_3$$

The following relation allows $a\diamond bba \in U_4$:

$$a \uparrow a\underline{a}\diamond\underline{bba} \quad u_1u_1u_3 \uparrow u_2u_4 a\diamond bba$$

The following relations allow $ba \in U_4$:

$$a\diamond b \uparrow a\diamond b\underline{ba} \quad u_2u_1u_3 \ \uparrow u_3u_2 ba$$
$$a\diamond b \uparrow a\diamond b\underline{ba} \quad u_1u_4u_2 ba \uparrow \quad u_2u_3$$

The following relations allow $b \in U_4$:

$$a \uparrow \diamond\underline{b} \quad u_1u_1u_1 \ \uparrow u_2u_4 b$$
$$b \uparrow \diamond\underline{b} \quad u_2u_1u_1 b \uparrow \ u_3u_1$$
$$b \uparrow \diamond\underline{b} \quad u_1u_4u_1 \ \uparrow u_2u_1 b$$

The following relation allows $\diamond b \in U_4$:

$$a \uparrow a\diamond\underline{b} \quad u_1u_1u_2 \uparrow u_2u_4\diamond b$$

The following relations allow $a \in U_4$:

$$b \uparrow b\underline{a} \quad u_2u_1u_1 a \uparrow \ u_3u_4$$
$$b \uparrow b\underline{a} \quad u_1u_4u_4 \uparrow u_2u_1 a$$

Due to $\varepsilon \in U_4$, the algorithm may stop at this point. However, to demonstrate additional possible nontrivial compatibility relations satisfied by Z, the sets U_5 and U_6 will be briefly examined. With each additional iteration of the algorithm, new nontrivial compatibility relations emerge.

All relations generating the set U_5 will not be delved into, instead only the relations pertaining to the elements ε and $\diamond b$ will be apprised. Observe that since $\varepsilon \in U_5$ another nontrivial compatibility relation emerges, $u_1u_1u_2 \uparrow u_2u_4u_1$, thus furthering the fact that the example set X is not a pcode.

The following relation allows $\varepsilon \in U_5$:

$$\diamond b \uparrow \diamond b\underline{\varepsilon} \quad u_1u_1u_2 \uparrow u_2u_4u_1$$

The following relations allow $\diamond b \in U_5$:

$$a \uparrow a\diamond\underline{b} \qquad u_2 u_1 u_1 u_2 \uparrow u_3 u_4 \diamond b$$
$$a \uparrow a\diamond\underline{b} \qquad u_1 u_4 u_4 \diamond b \uparrow u_2 u_1 u_2$$

Subsequent sets generated by the algorithm are equivalent to the set U_5. However, additional nontrivial compatibility relations become apparent.

The following relations allow $\varepsilon \in U_6$:

$$\diamond b \uparrow \diamond b\underline{\varepsilon} \qquad u_2 u_1 u_1 u_2 \uparrow u_3 u_4 u_1$$
$$\diamond b \uparrow \diamond b\underline{\varepsilon} \qquad u_1 u_4 u_4 u_1 \uparrow u_2 u_1 u_2$$

Therefore, since $\varepsilon \in U_6$, two additional nontrivial compatibility relations are derived, $u_2 u_1 u_1 u_2 \uparrow u_3 u_4 u_1$ and $u_1 u_4 u_4 u_1 \uparrow u_2 u_1 u_2$, thus furthering the fact that the example set Z is not a pcode. ◻

9.2   Second algorithm

In this section, we describe a second algorithm for deciding the pcode property. It is based on a domino technique that we start investigating for full words and then extending it to include partial words.

9.2.1   Domino technique on words

Let X be a nonempty finite subset of A^+. For $\alpha, \beta \in X^*$ satisfying $\alpha = \beta$, put $\alpha = \alpha_1\alpha_2\ldots\alpha_m$, $\beta = \beta_1\beta_2\ldots\beta_n$ for some $\alpha_1,\ldots,\alpha_m,\beta_1,\ldots,\beta_n \in X$. The relation $\alpha = \beta$ is trivial if $m = n$ and $\alpha_1 = \beta_1,\ldots,\alpha_m = \beta_m$, and the relation $\alpha = \beta$ is *factorizable* if there exist $\alpha',\alpha'',\beta',\beta'' \in X^+$ such that $\alpha = \alpha'\alpha''$, $\beta = \beta'\beta''$, $\alpha' = \beta'$, and $\alpha'' = \beta''$.

Example 9.3
Consider the set

$$Y = \{u_1, u_2, u_3, u_4\}$$

over $\{a,b\}$ where

$$u_1 = a, u_2 = abbbbba, u_3 = babab, \text{ and } u_4 = bbbb$$

Setting

$$\alpha = (a)(bbbb)(babab)(abbbbba) = u_1\, u_4\, u_3\, u_2$$
$$\beta = (abbbbba)(babab)(bbbb)(a) = u_2\, u_3\, u_4\, u_1$$

the relation $\alpha = \beta$ is seen to be nontrivial and nonfactorizable. ◻

In order to study the relations satisfied by the set X, we define the *simplified domino graph* and the *domino function* of X.

DEFINITION 9.1 *Let X be a nonempty finite subset of A^+. Let $G = (V, E)$ be the directed graph with vertex set*

$$V = \{open, close, \tbinom{u}{\varepsilon}, \tbinom{\varepsilon}{u} \mid u \in P(X) \setminus \{\varepsilon\}\},$$

and with edge set $E = E_1 \cup E_2 \cup E_3 \cup E_4$ where

$$E_1 = \{(open, \tbinom{\varepsilon}{u})) \mid u \in X\},$$

$$E_2 = \{((\tbinom{u}{\varepsilon}), close) \mid u \in X\},$$

$$E_3 = \{((\tbinom{u}{\varepsilon}), \tbinom{uv}{\varepsilon})), ((\tbinom{\varepsilon}{u}), \tbinom{\varepsilon}{uv})) \mid v \in X\},$$

$$E_4 = \{((\tbinom{u}{\varepsilon}), \tbinom{\varepsilon}{v})), ((\tbinom{\varepsilon}{u}), \tbinom{v}{\varepsilon})) \mid uv \in X\}.$$

*The **simplified domino graph associated with X**, denoted by $G(X)$, is the directed graph $G' = (V', E')$ where V' consists of open, close and those vertices v in V such that there exists a path from open to close that goes through v, and E' consists of those edges e in E such that there exists a path from open to close going through e.*

*The **domino function associated with X** is the mapping d from E to $\{\tbinom{u}{\varepsilon}, \tbinom{\varepsilon}{u} \mid u \in X\}$ defined on*

E_1 by $(open, \tbinom{\varepsilon}{u}) \mapsto \tbinom{u}{\varepsilon}$,

E_2 by $((\tbinom{u}{\varepsilon}), close) \mapsto \tbinom{u}{\varepsilon}$,

E_3 by $((\tbinom{u}{\varepsilon}), \tbinom{uv}{\varepsilon})) \mapsto \tbinom{\varepsilon}{v}$ and $((\tbinom{\varepsilon}{u}), \tbinom{\varepsilon}{uv})) \mapsto \tbinom{v}{\varepsilon}$,

E_4 by $((\tbinom{u}{\varepsilon}), \tbinom{\varepsilon}{v})) \mapsto \tbinom{uv}{\varepsilon}$ and $((\tbinom{\varepsilon}{u}), \tbinom{v}{\varepsilon})) \mapsto \tbinom{\varepsilon}{uv}$.

*The **domino** associated with an edge e of E is defined as $d(e) = \binom{d_1(e)}{d_2(e)}$ where $d_1(e)$ denotes the top component and $d_2(e)$ the bottom one.*

Example 9.4

Returning to the set Y of Example 9.3, we see that there are fifteen elements in $P(Y) \setminus \{\varepsilon\}$ which are:

$$a, ab, abb, abbb, abbbb, abbbbb, abbbbba, b, ba, bab, baba, babab, bb, bbb, bbbb$$

The set of vertices V includes *open*, *close* and thirty other elements such as $\binom{a}{\varepsilon}$ and $\binom{\varepsilon}{bab}$. The set of edges E consists of several edges split into the four sets E_1, E_2, E_3 and E_4:

- The set E_1 has four edges including $e_1 = (open, \binom{\varepsilon}{a})$ since $a = u_1 \in Y$. The domino function d associated with Y maps e_1 to $d(e_1) = \binom{u_1}{\varepsilon}$ where $d_1(e_1) = u_1$ and $d_2(e_1) = \varepsilon$.

- The set E_2 has four edges including $e_2 = (\binom{babab}{\varepsilon}, close)$ since $babab = u_3 \in Y$. This edge is mapped to $\binom{u_3}{\varepsilon}$.

- The set E_3 has several edges including $e_3 = \left(\binom{a}{\varepsilon}, \binom{abbbb}{\varepsilon}\right)$. Here, both a and $abbbb$ belong to $P(Y) \setminus \{\varepsilon\}$, and $bbbb = u_4 \in Y$. The edge e_3 is mapped to $\binom{\varepsilon}{u_4}$.

- The set E_4 has several edges including $e_4 = \left(\binom{bab}{\varepsilon}, \binom{\varepsilon}{ab}\right)$. Here, both bab and ab belong to $P(Y) \setminus \{\varepsilon\}$, and $(bab)(ab) = u_3 \in Y$. The edge e_4 is mapped to $\binom{u_3}{\varepsilon}$.

The simplified domino graph and the domino function associated with Y are as in Figure 9.1 where the domino $d(e)$ associated with an edge e is represented as the label of e. Note that, for instance, the vertex $v = \binom{bb}{\varepsilon}$ is not in $G(Y)$

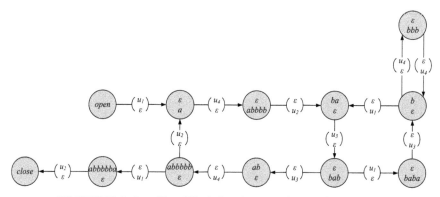

FIGURE 9.1:   Simplified domino graph and function of $Y = \{u_1, u_2, u_3, u_4\}$ where $u_1 = a$, $u_2 = abbbba$, $u_3 = babab$, and $u_4 = bbbb$.

since there is no path from *open* to *close* going through v, and the edge $e = (open, \binom{\varepsilon}{bbbb}))$ is not in $G(Y)$ since there is no path from *open* to *close* going through e.   ▯

     The function d induces mappings d_1 and d_2 from E to $X \cup \{\varepsilon\}$ also called domino functions. If $p = e_1 e_2 \ldots e_i$ is a path in G, then $d(e_1)d(e_2) \ldots d(e_i)$ (respectively, $d_1(e_1)d_1(e_2) \ldots d_1(e_i)$, $d_2(e_1)d_2(e_2) \ldots d_2(e_i)$) is denoted by $d(p)$ (respectively, $d_1(p)$, $d_2(p)$).

     There are many paths starting at *open* and ending at *close* in Figure 9.1. They include the path

$$q = open, \binom{\varepsilon}{a}, \binom{\varepsilon}{abbbb}, \binom{ba}{\varepsilon}, \binom{\varepsilon}{bab}, \binom{ab}{\varepsilon}, \binom{abbbb}{\varepsilon}, \binom{abbbba}{\varepsilon}, close$$

with associated domino sequence

$$d(q) = \binom{u_1}{\varepsilon}\binom{u_4}{\varepsilon}\binom{\varepsilon}{u_2}\binom{u_3}{\varepsilon}\binom{\varepsilon}{u_3}\binom{\varepsilon}{u_4}\binom{\varepsilon}{u_1}\binom{u_2}{\varepsilon} = \binom{u_1 u_4 u_3 u_2}{u_2 u_3 u_4 u_1}$$

If we look at the two sequences built by $d(q)$, $d_1(q) = u_1 u_4 u_3 u_2$ and $d_2(q) = u_2 u_3 u_4 u_1$, then we see that $d_1(q) = d_2(q)$ is the nontrivial nonfactorizable relation over Y that was discussed in Example 9.3.

A path p in G from *open* to some vertex $\binom{u}{\varepsilon}$ (respectively, $\binom{\varepsilon}{u}$) is trying to find two decodings of the same message over X into codewords beginning with distinct codewords. The decodings obtained so far are $d_1(p)$ and $d_2(p)$. The word u in A^* denotes the backlog of the first (respectively, second) decoding as against the second (respectively, first) one. In the example of Figure 9.1, the path

$$r = open, \binom{\varepsilon}{a}, \binom{\varepsilon}{abbbb}, \binom{ba}{\varepsilon}$$

from *open* to $\binom{ba}{\varepsilon}$ is such that

$$d_1(r) = a\, b\, b\, b\, b$$
$$d_2(r) = a\, b\, b\, b\, b\, b\, a$$

and we notice that ba is the backlog of $d_1(r)$ as against $d_2(r)$.

The next proposition illustrates how the paths from *open* to *close* in $G(X)$ correspond to nontrivial nonfactorizable relations satisfied by X. It is stated without proof as we will prove a more general result in the next section.

PROPOSITION 9.1
Let X be a nonempty finite subset of A^+. For $\alpha, \beta \in X^*$, $\alpha = \beta$ is a nontrivial nonfactorizable relation if and only if there exists a path p in $G(X)$ from open to close such that $d(p) = \binom{\alpha}{\beta}$ or $d(p) = \binom{\beta}{\alpha}$.

Whether or not X is a code can be determined by looking at its simplified domino graph $G(X)$. More specifically, the following result holds.

THEOREM 9.2
Let X be a nonempty finite subset of A^+. Then X is a code if and only if there is no path in $G(X)$ from open to close.

We have already noted that there exist paths from *open* to *close* in the simplified domino graph $G(Y)$ of Figure 9.1, showing that Y is not a code.

9.2.2 Domino technique on partial words

Let X be a nonempty finite subset of $W(A) \setminus \{\varepsilon\}$. For $\alpha, \beta \in X^*$ satisfying $\alpha \uparrow \beta$, put

$$\alpha = \alpha_1 \alpha_2 \dots \alpha_m \text{ and } \beta = \beta_1 \beta_2 \dots \beta_n$$

for some $\alpha_1, \dots, \alpha_m, \beta_1, \dots, \beta_n \in X$. The relation $\alpha \uparrow \beta$ is trivial if $m = n$ and $\alpha_1 = \beta_1, \dots, \alpha_m = \beta_m$, and that it is *factorizable* if there exist $\alpha', \alpha'', \beta', \beta'' \in X^+$ such that $\alpha = \alpha'\alpha''$, $\beta = \beta'\beta''$, $\alpha' \uparrow \beta'$, and $\alpha'' \uparrow \beta''$.

Example 9.5
Consider the set

$$Z = \{u_1, u_2, u_3, u_4\}$$

over $\{a, b\}$ where

$$u_1 = a\diamond b,\ u_2 = aab\diamond bb,\ u_3 = \diamond b,\ \text{and}\ u_4 = ba$$

Setting

$$\alpha = (a\diamond b)(\diamond b)(\diamond b)(ba)(\diamond b) = u_1\ u_3\ u_3\ u_4\ u_3$$
$$\beta = \ \ (aab\diamond bb)(\diamond b)(a\diamond b)\ \ = u_2\ u_3\ u_1$$

the relation $\alpha \uparrow \beta$ is seen to be nontrivial and nonfactorizable.   ∎

In order to study the compatibility relations

$$\alpha_1\alpha_2 \ldots \alpha_m \uparrow \beta_1\beta_2 \ldots \beta_n$$

where $\alpha_1, \ldots, \alpha_m, \beta_1, \ldots, \beta_n \in X$, we extend the domino technique on words of Section 9.2.1.

DEFINITION 9.2   *Let X be a nonempty finite subset of $W(A) \setminus \{\varepsilon\}$. Let $G = (V, E)$ be the directed graph with vertex set*

$$V = \{open, close, \binom{u}{\varepsilon}, \binom{\varepsilon}{u} \mid u \in C(P(X)) \setminus \{\varepsilon\}\},$$

and with edge set $E = E_1 \cup E_2 \cup E_3 \cup E_4$ where

$E_1 = \{(open, \binom{\varepsilon}{u}) \mid u \in X\},$

$E_2 = \{(\binom{u}{\varepsilon}, close), (\binom{\varepsilon}{u}, close) \mid u \in C(X)\},$

$E_3 = \{(\binom{u}{\varepsilon}, \binom{uv}{\varepsilon}), (\binom{\varepsilon}{u}, \binom{\varepsilon}{uv}) \mid v \in X\},$

$E_4 = \{(\binom{u}{\varepsilon}, \binom{\varepsilon}{v}), (\binom{\varepsilon}{u}, \binom{v}{\varepsilon}) \mid w = u'v, u \uparrow u', w \in X\}.$

*The **simplified domino graph associated with X**, denoted by $G(X)$, is the directed graph $G' = (V', E')$ where V' consists of open, close and those vertices v in V such that there exists a path from open to close that goes through v, and E' consists of those edges e in E such that there exists a path from open to close going through e.*

*The **domino function associated with X** is the mapping d from E to the set of nonempty subsets of $\{\binom{u}{\varepsilon}, \binom{\varepsilon}{u} \mid u \in X\}$ defined on*

E_1 *by* $(open, \binom{\varepsilon}{u}) \mapsto \{\binom{u}{\varepsilon}\},$

E_2 *by* $(\binom{u}{\varepsilon}, close) \mapsto \{\binom{v}{\varepsilon} \mid u \uparrow v \text{ and } v \in X\}$ *and* $(\binom{\varepsilon}{u}, close) \mapsto \{\binom{\varepsilon}{v} \mid u \uparrow v$ *and* $v \in X\},$

E_3 by $\left(\binom{u}{\varepsilon}, \binom{uv}{\varepsilon}\right) \mapsto \{\binom{\varepsilon}{v}\}$ and $\left(\binom{\varepsilon}{u}, \binom{\varepsilon}{uv}\right) \mapsto \{\binom{v}{\varepsilon}\}$,

E_4 by $\left(\binom{u}{\varepsilon}, \binom{\varepsilon}{v}\right) \mapsto \{\binom{w}{\varepsilon} \mid w = u'v, u \uparrow u', \text{ and } w \in X\}$ and $\left(\binom{\varepsilon}{u}, \binom{v}{\varepsilon}\right) \mapsto \{\binom{\varepsilon}{w} \mid w = u'v, u \uparrow u', \text{ and } w \in X\}$.

The domino set *associated with an edge* e of E is the set $d(e)$.

Example 9.6

Returning to the set Z of Example 9.5, we see that there are twelve elements in $P(Z) \setminus \{\varepsilon\}$ which are:

$$a, a\diamond, a\diamond b, aa, aab, aab\diamond, aab\diamond b, aab\diamond bb, \diamond, \diamond b, b, ba$$

and many more elements in $C(P(Z)) \setminus \{\varepsilon\}$ that include $\diamond\diamond, \diamond a, \ldots.$. The set of vertices V includes *open*, *close* and several other elements such as $\binom{a}{\varepsilon}$ and $\binom{\varepsilon}{a\diamond b\diamond}$. The set of edges E consists of several edges split into the four sets E_1, E_2, E_3 and E_4:

- The set E_1 has four edges including $e_1 = (open, \binom{\varepsilon}{a\diamond b}))$ since $a\diamond b = u_1 \in Z$. The domino function d associated with Z maps e_1 to $d(e_1) = \{\binom{u_1}{\varepsilon}\}$.

- The set E_2 has several edges including $e_2 = (\binom{\diamond\diamond}{\varepsilon}, close)$ since $\diamond\diamond \uparrow u_3$ and thus $\diamond\diamond \in C(Z)$. This edge is mapped to $\{\binom{u_3}{\varepsilon}, \binom{u_4}{\varepsilon}\}$.

- The set E_3 has several edges including $e_3 = (\binom{\diamond\diamond b}{\varepsilon}, \binom{\diamond\diamond b\diamond b}{\varepsilon})$. Here, both $\diamond\diamond b$ and $\diamond\diamond b\diamond b$ belong to $C(P(Z)) \setminus \{\varepsilon\}$, and $\diamond b = u_3 \in Z$. The edge e_3 is mapped to $\{\binom{\varepsilon}{u_3}\}$.

- The set E_4 has several edges including $e_4 = (\binom{\diamond\diamond b}{\varepsilon}, \binom{\varepsilon}{\diamond bb})$. Here, both $u = \diamond\diamond b$ and $v = \diamond bb$ belong to $C(P(Z)) \setminus \{\varepsilon\}$. Setting $u' = aab$, we get $w = (aab)(\diamond bb) = u'v$ with $u \uparrow u'$ and $w = u_2 \in Z$. The edge e_4 is mapped to $\{\binom{u_2}{\varepsilon}\}$.

Parts of the simplified domino graph and the domino function associated with Z are displayed in Figures 9.2, 9.3 and 9.4 where the domino set $d(e)$ associated with an edge e is represented as the label of e. Since the domino sets in $G(Z)$ are all singletons, the domino set $\{\binom{u_1}{\varepsilon}\}$ say has been abbreviated by $\binom{u_1}{\varepsilon}$. The reader is invited to complete the graph and discover that, for instance, the vertex $v = \binom{aab}{\varepsilon}$ is not in $G(Z)$ since there is no path from *open* to *close* going through v, and the edge $e = (open, \binom{\varepsilon}{aabb}))$ is not in $G(Z)$ since there is no path from *open* to *close* going through e. ⬜

If $p = e_1 e_2 \ldots e_i$ is a path in G, the set

$$d(e_1)d(e_2)\ldots d(e_i) = \{x_1 x_2 \ldots x_i \mid x_1 \in d(e_1), x_2 \in d(e_2), \ldots, x_i \in d(e_i)\}$$

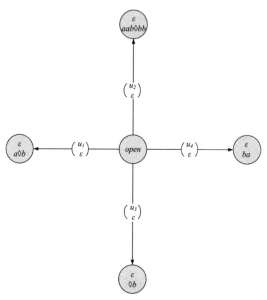

FIGURE 9.2: Neighborhood of vertex *open* in $G(Z)$ where $Z = \{u_1, u_2, u_3, u_4\}$ with $u_1 = a\diamond b$, $u_2 = aab\diamond bb$, $u_3 = \diamond b$, and $u_4 = ba$.

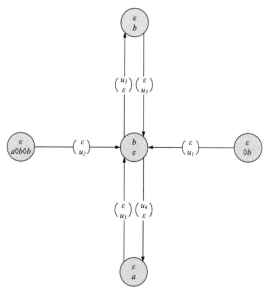

FIGURE 9.3: Neighborhood of vertex $\binom{b}{\varepsilon}$ in $G(Z)$.

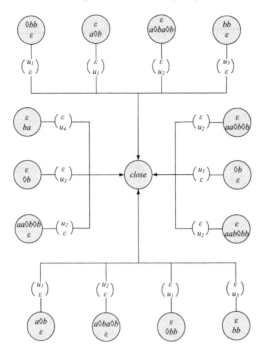

FIGURE 9.4: Neighborhood of vertex *close* in $G(Z)$.

is denoted by $d(p)$. For $x = \binom{y_1}{z_1}\binom{y_2}{z_2}\ldots\binom{y_i}{z_i}$ in $d(p)$, we abbreviate $y_1 y_2 \ldots y_i$ by $above(x)$ and $z_1 z_2 \ldots z_i$ by $below(x)$. We will also write $x = \binom{above(x)}{below(x)}$. Note that $above(x), below(x)$ are in X^*.

There are many paths of length at least three starting at *open* and ending at *close* in $G(Z)$. Such a path, which appears in Figure 9.5, shows how a domino sequence x associated with its edges leads to a nontrivial nonfactorizable compatibility relation of the form $above(x) \uparrow below(x)$. The sequence of labels

$$\binom{u_1}{\varepsilon}\binom{u_3}{\varepsilon}\binom{\varepsilon}{u_2}\binom{u_3}{\varepsilon}\binom{\varepsilon}{u_3}\binom{u_4}{\varepsilon}\binom{u_3}{\varepsilon}\binom{\varepsilon}{u_1}$$

is in $d(q)$ and note that $u_1 u_3 u_3 u_4 u_3 \uparrow u_2 u_3 u_1$ is a nontrivial nonfactorizable compatibility relation over Z which was already discussed in Example 9.5.

A path p in $G(X)$ from *open* to some vertex $\binom{u}{\varepsilon}$ is trying to find a nontrivial compatibility relation over X. The factorizations obtained so far for a particular $x \in d(p)$ are $above(x)$ and $below(x)$. More precisely, if $above(x) = \alpha_1 \alpha_2 \ldots \alpha_m$ and $below(x) = \beta_1 \beta_2 \ldots \beta_n$, then $\alpha_1 \neq \beta_1$ and $\alpha_1 \alpha_2 \ldots \alpha_m u \uparrow \beta_1 \beta_2 \ldots \beta_n$ and u is a suffix of $\beta_1 \beta_2 \ldots \beta_n$. The partial word u denotes the backlog of the first factorization as against the second one. Similarly, if p is from *open* to some vertex $\binom{\varepsilon}{u}$, then $\alpha_1 \neq \beta_1$ and $\alpha_1 \alpha_2 \ldots \alpha_m \uparrow$

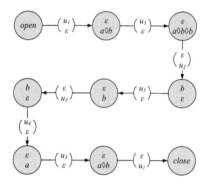

FIGURE 9.5:   A path q of length at least three from *open* to *close* in $G(Z)$.

$\beta_1\beta_2\ldots\beta_n u$ and u is a suffix of $\alpha_1\alpha_2\ldots\alpha_m$. In this case, u denotes the backlog of the second factorization as against the first one.

In the sequel, in order to simplify the notation, we identify both *open* and *close* with $\binom{\varepsilon}{\varepsilon}$.

LEMMA 9.2
Let X be a nonempty finite subset of $W(A) \setminus \{\varepsilon\}$.

1. *If $u \in C(P(X))$ and there exists a path p in $G(X)$ from open to $\binom{u}{\varepsilon}$ (respectively, $\binom{\varepsilon}{u}$), then $d(p)$ consists of elements of the form $\binom{\alpha}{\beta u}$ (respectively, $\binom{\alpha u}{\beta}$) for some $\alpha, \beta \in W(A)$ satisfying $\alpha \uparrow \beta$.*

2. *If there exists a path p in $G(X)$ from open to close such that $\binom{\alpha}{\beta} \in d(p)$, then $\alpha \uparrow \beta$ is a nonfactorizable compatibility relation satisfied by X. Moreover, if p is of length at least 3, then $\alpha \uparrow \beta$ is nontrivial.*

PROOF   First, Statement 1 follows by induction. The only path of length 1 from *open* is an E_1-edge of the form $(open, \binom{\varepsilon}{u})$ for some $u \in X$. Here, $d(p) = \{\binom{u}{\varepsilon}\}$ and the result follows with $\alpha = \beta = \varepsilon$. Now, consider the path $q = pe$ where p is a path from *open* to $\binom{u}{\varepsilon}$ and e is an edge from $\binom{u}{\varepsilon}$. By the inductive hypothesis, $d(p)$ consists of elements of the form $\binom{\alpha}{\beta u}$ for some $\alpha, \beta \in W(A)$ satisfying $\alpha \uparrow \beta$. For $e = (\binom{u}{\varepsilon}, close) \in E_2$, $d(pe) = d(p)d(e)$ consists of elements of the form $\binom{\alpha}{\beta u}\binom{v}{\varepsilon} = \binom{\alpha v}{\beta u} = \binom{\alpha'}{\beta'}$ where $u \uparrow v$ and $v \in X$. For $e = (\binom{u}{\varepsilon}, \binom{uv}{\varepsilon}) \in E_3$, $d(pe)$ consists of elements of the form $\binom{\alpha}{\beta u}\binom{\varepsilon}{v} = \binom{\alpha}{\beta uv} = \binom{\alpha}{\beta' uv}$ where $v \in X$. Finally, for $e = (\binom{u}{\varepsilon}, \binom{\varepsilon}{v}) \in E_4$, $d(pe)$ consists of elements of the form $\binom{\alpha}{\beta u}\binom{w}{\varepsilon} = \binom{\alpha w}{\beta u} = \binom{\alpha u' v}{\beta u} = \binom{\alpha'}{\beta'}$ where $w = u'v$, $u \uparrow u'$, and $w \in X$. In any case, the result follows with some

$\alpha', \beta' \in W(A)$ satisfying $\alpha' \uparrow \beta'$. The result follows similarly when p is a path from *open* to $\binom{\varepsilon}{u}$ and e is an edge from $\binom{\varepsilon}{u}$.

Second, let us show that Statement 2 holds. If there exists a path p from *open* to *close* such that $\binom{\alpha}{\beta} \in d(p)$, then by Statement 1, $\alpha \uparrow \beta$ since *close* = $\binom{\varepsilon}{\varepsilon}$. But by the definition of $d(p)$, we have $\alpha, \beta \in X^*$ and thus $\alpha \uparrow \beta$ is a compatibility relation satisfied by X. □

The next lemma shows how to obtain the path corresponding to a nontrivial nonfactorizable compatibility relation. First, we need some definitions.

For two partial words $\alpha, \beta \in W(A)$, we write $\alpha \preceq \beta$ if $\alpha \in C(P(\beta))$ where $P(\beta)$ is the set of all prefixes of β, and $\alpha \prec \beta$ if $\alpha \preceq \beta$ and $\alpha \not\vdash \beta$.

Let $\alpha, \beta \in X^*$, and put $\alpha = \alpha_1\alpha_2\ldots\alpha_m$ and $\beta = \beta_1\beta_2\ldots\beta_n$. We say that $\binom{\alpha}{\beta}$ has a *proper prefix compatibility relation* if there exist $\alpha', \beta' \in X^+$ such that α' is a prefix of α, β' is a prefix of β, $\binom{\alpha}{\beta} \neq \binom{\alpha'}{\beta'}$, and $\alpha' \uparrow \beta'$ is a compatibility relation. Note that a nonfactorizable compatibility relation $\alpha \uparrow \beta$ is such that $\binom{\alpha}{\beta}$ has no proper prefix compatibility relation. We say that $\binom{\alpha}{\beta}$ has the *nppcr* property if the following three conditions hold:

(*i*) $\alpha \preceq \beta$ and the suffix γ of β satisfying $\beta \uparrow \alpha\gamma$ belongs to $C(P(X))$, or $\beta \preceq \alpha$ and the suffix γ of α satisfying $\alpha \uparrow \beta\gamma$ belongs to $C(P(X))$.

(*ii*) $\binom{\alpha}{\beta}$ has no proper prefix compatiblity relation.

(*iii*) If $n > 0$, then $m > 0$ and $|\alpha_1| < |\beta_1|$.

LEMMA 9.3
Let X be a nonempty finite subset of $W(A) \setminus \{\varepsilon\}$.

    1. *Let $\alpha, \beta \in X^*$ be such that there exists a path p in $G(X)$ from open to $v_1 \in V$ with $\binom{\alpha}{\beta} \in d(p)$.*

        (a) *If $v_1 = \binom{u}{\varepsilon}$ and $v \in X$ is such that $uv \in C(P(X))$, then there exist $v_2 \in V$ and a path q from open to v_2 such that $\binom{\alpha}{\beta v} \in d(q)$.*

        (b) *If $v_1 = \binom{u}{\varepsilon}$ and $w = u'v \in X$ is such that $u \uparrow u'$ and $v \in C(P(X))$, then there exist $v_2 \in V$ and a path q from open to v_2 such that $\binom{\alpha w}{\beta} \in d(q)$.*

        (c) *If $v_1 = \binom{\varepsilon}{u}$ and $v \in X$ is such that $uv \in C(P(X))$, then there exist $v_2 \in V$ and a path q from open to v_2 such that $\binom{\alpha v}{\beta} \in d(q)$.*

        (d) *If $v_1 = \binom{\varepsilon}{u}$ and $w = u'v \in X$ is such that $u \uparrow u'$ and $v \in C(P(X))$, then there exist $v_2 \in V$ and a path q from open to v_2 such that $\binom{\alpha}{\beta w} \in d(q)$.*

    2. *Let $\alpha, \beta \in X^*$ be such that $\binom{\alpha}{\beta}$ has the nppcr property. Then there exist $v \in V$ and a path p in $G(X)$ from open to v such that $\binom{\alpha}{\beta} \in d(p)$.*

3. Let $\alpha, \beta \in X^*$ be such that $\alpha \uparrow \beta$ is a nontrivial nonfactorizable compatibility relation. Then there exists a path p in $G(X)$ from open to close such that $\binom{\alpha}{\beta} \in d(p)$ or $\binom{\beta}{\alpha} \in d(p)$.

PROOF Cases (a) and (c) of Statement 1 lead to edges in E_3, and Cases (b) and (d) lead to edges in E_2 or E_4 depending on whether $v = \varepsilon$ or $v \neq \varepsilon$. Let us consider Case (b) (the other cases are left as exercises for the reader). If $v \neq \varepsilon$, then put $v_2 = \binom{\varepsilon}{v}$ and $e = (\binom{u}{\varepsilon}, \binom{\varepsilon}{v}) \in E_4$. Here, $\binom{\alpha}{\beta}\binom{w}{\varepsilon} = \binom{\alpha w}{\beta} \in d(p)d(e) = d(q)$. On the other hand, if $v = \varepsilon$, then $u' = w$ and take $v_2 = close$ and $e = (\binom{u}{\varepsilon}, close) \in E_2$. Here, $\binom{\alpha}{\beta}\binom{w}{\varepsilon} = \binom{\alpha w}{\beta} \in d(p)d(e) = d(q)$.

For Statement 2, the proof is by induction on $m + n$ where $\alpha = \alpha_1 \alpha_2 \ldots \alpha_m$ and $\beta = \beta_1 \beta_2 \ldots \beta_n$. If $m + n = 1$, then by the *nppcr* property, we must have $m = 1$ and $n = 0$. Thus, $\alpha = \alpha_1$ and $\beta = \varepsilon$. Let $v = \binom{\varepsilon}{\alpha_1}$ and p be the path consisting of the edge $e = (open, v) \in E_1$. Then $\binom{\alpha}{\beta} = \binom{\alpha_1}{\varepsilon} \in d(e) = d(p)$.

If $m + n > 1$, then $m > 0$ by the *nppcr* property. So let $\alpha' = \alpha_1 \alpha_2 \ldots \alpha_{m-1}$, and whenever $n > 0$, let $\beta' = \beta_1 \beta_2 \ldots \beta_{n-1}$. Note that when $\alpha \prec \beta$, we have $n > 0$ and β' is defined. Moreover, by the *nppcr* property, we have $\alpha \not\vee \beta'$ and $\alpha' \not\vee \beta$. So we consider the following cases:

- If $\alpha \prec \beta$ and $\alpha \prec \beta'$, then use the inductive hypothesis on $\binom{\alpha}{\beta'}$ and Statement 1(a).

- If $\alpha \prec \beta$ and $\beta' \prec \alpha$, then use the inductive hypothesis on $\binom{\alpha}{\beta'}$ and Statement 1(d).

- If $\beta \preceq \alpha$ and $\alpha' \prec \beta$, then use the inductive hypothesis on $\binom{\alpha'}{\beta}$ and Statement 1(b).

- If $\beta \preceq \alpha$ and $\beta \prec \alpha'$, then use the inductive hypothesis on $\binom{\alpha'}{\beta}$ and Statement 1(c).

Let us consider the third case (the other cases are similar). If $\beta \preceq \alpha$ and $\alpha' \prec \beta$, then put $w = \alpha_m$. Since $\alpha' \prec \beta$, let u be the suffix of β such that $\beta \uparrow \alpha'u$. The latter and the fact that $\beta \preceq \alpha$ imply that $w = u'v \in X$ with $u \uparrow u'$. Since $\beta \preceq \alpha$, the suffix v of α satisfying $\alpha \uparrow \beta v$ belongs to $C(P(X))$. We have $u \in C(P(X))$, and so $\binom{\alpha'}{\beta}$ has the *nppcr* property. By the inductive hypothesis, there exist $v_1 \in V$ and a path q from *open* to v_1 such that $\binom{\alpha'}{\beta} \in d(q)$. By Lemma 9.2(1), $v_1 = \binom{u}{\varepsilon}$. So by Statement (1)(b), there exist $v_2 \in V$ and a path p from *open* to v_2 such that $\binom{\alpha}{\beta} = \binom{\alpha'w}{\beta} \in d(p)$.

For Statement 3, we first note that if α, β are distinct compatible elements of X, then the path $p = e_1 e_2$ in $G(X)$ where $e_1 = (open, \binom{\varepsilon}{\alpha})$ and $e_2 = (\binom{\varepsilon}{\alpha}, close)$ is such that $\binom{\alpha}{\beta} \in d(p)$. Otherwise, since $\alpha \uparrow \beta$ is a compatibility relation satisfied by X, Condition (i) of *nppcr* is satisfied. Since it is nonfactorizable, Condition (ii) is satisfied. Finally, since it is nontrivial and nonfactorizable, one of $\binom{\alpha}{\beta}$ and $\binom{\beta}{\alpha}$, say the first, satisfies Condition (iii). Hence $\binom{\alpha}{\beta}$ has the *nppcr* property. By Statement 2, there exist $v \in V$ and a path p from *open* to v such that $\binom{\alpha}{\beta} \in d(p)$. By Lemma 9.2(1), we must have $v = close$. ☐

Whether or not a pairwise noncompatible set X is a pcode can be determined by looking at its simplified domino graph $G(X)$ as stated in the next theorem.

THEOREM 9.3
Let X be a nonempty finite subset of $W(A)\backslash\{\varepsilon\}$ that is pairwise noncompatible. Then X is a pcode if and only if there is no path of length at least 3 in $G(X)$ from open to close.

PROOF   The above two lemmas illustrate how the paths of length at least 3 from *open* to *close* in $G(X)$ correspond to nontrivial nonfactorizable compatibility relations satisfied by X. Indeed, for $\alpha, \beta \in X^*$, $\alpha \uparrow \beta$ is a nontrivial nonfactorizable compatibility relation if and only if there exists a path p of length at least 3 in $G(X)$ from *open* to *close* such that $\binom{\alpha}{\beta} \in d(p)$ or $\binom{\beta}{\alpha} \in d(p)$. ☐

We have already noted that there exist paths of length at least three from *open* to *close* in $G(Z)$ (see Figure 9.5), showing that Z is not a pcode.

Exercises

9.1 Is $X = \{a \diamond b, bbb, \diamond ab\}$ pairwise noncompatible?

9.2 ⟨s⟩ Consider the set $X = \{a \diamond b, \diamond baaa, abba\}$. Carry the computations of the U_i-sets on X as is done in Example 9.1. Is X a pcode?

9.3 Repeat Exercise 9.2 for $X = \{a \diamond, aab, b \diamond b\}$.

9.4 What can be said about X if $U_i = \emptyset$ for some $i \geq 1$? What is U_{i+1} then?

9.5 Show that if X is a finite set, then $\{U_n \mid n \geq 1\}$ is finite.

9.6 ⟨s⟩ Show that the first algorithm ends immediately for prefix pcodes.

9.7 Build the simplified domino graph and its associated domino function for the set $X = \{aaabba, abb, ba, bb\}$. Is X a code or not?

9.8 Build the simplified domino graph of the set $X = \{a \diamond b, a \diamond\}$. Conclude that X is a pcode over $\{a, b\}$.

9.9 Classify the edges in Figure 9.1 as E_1-, E_2-, E_3- or E_4-edges.

9.10 $\boxed{\text{s}}$ Explain why E_1-, E_2-, and E_3-edges cannot be bidirectional.

Challenging exercises

9.11 Use the first algorithm to discover some nontrivial compatibility relation for the set $Z = \{a, abbbbba, babab, bbbb\}$ as was done in Example 9.2.

9.12 $\boxed{\text{s}}$ Classify the following sets as pcodes or nonpcodes:

- $X_1 = \{bba, bc \diamond\}$
- $X_2 = \{aabb \diamond\diamond, ba \diamond, bb \diamond c\}$
- $X_3 = \{ac \diamond, bb\}$
- $X_4 = \{\diamond b, a \diamond b, aa \diamond bba\}$
- $X_5 = \{ab \diamond, abb \diamond, abbb \diamond\}$

9.13 Give the neighborhood of the vertex *open* in $G(X)$ when

$$X = \{\diamond b, a \diamond b, aa \diamond bba, ba\}$$

9.14 Repeat Exercise 9.13 for the vertex *close*.

9.15 If e is an E_4-edge, then what is a necessary condition in order for e to be bidirectional?

9.16 Build the simplified domino graph of $X = \{a \diamond b, \diamond b, baba\}$. What can you conclude?

9.17 Build the simplified domino graph and the domino function associated with $X = \{a \diamond b, \diamond a, abba, bba\}$. Then

1. Give an example of a path p of length at least three from *open* to *close* with its associated domino sequence.
2. From the path p of your answer to 1, extract a nontrivial nonfactorizable relation.

9.18 $\boxed{\text{H}}$ Build the simplified domino graph of $X = \{a \diamond b, aab \diamond bb, \diamond b, ba\}$ that was started in Figures 9.2, 9.3 and 9.4.

9.19 Prove Case (d) of Statement 1 of Lemma 9.3.

9.20 Prove the first case of Statement 2 of Lemma 9.3.

Programming exercises

9.21 Implement the algorithm of Section 9.1 for full words. Run your program on the set Y of Example 9.3.

9.22 Design an applet that receives as input a nonempty finite subset $X \subset A^+$ and an element $u \in P(X) \setminus \{\varepsilon\}$, and outputs all the E_1-, E_2-, E_3- and E_4-edges leaving $\binom{u}{\varepsilon}$ in the graph $G = (V, E)$ of Definition 9.1.

9.23 Repeat Exercise 9.22 for an input defined as a nonempty finite subset $X \subset W(A) \setminus \{\varepsilon\}$ and an element $u \in C(P(X)) \setminus \{\varepsilon\}$, and an output defined as all the E_1-, E_2-, E_3- and E_4-edges leaving $\binom{u}{\varepsilon}$ in the graph $G = (V, E)$ of Definition 9.2.

9.24 Write a program that when given as input a nonempty finite subset $X \subset A^+$ computes the simplifed domino graph of X, $G(X)$. Run your program on the set of Example 9.3.

9.25 Repeat Exercise 9.24 for an input defined as a nonempty finite subset $X \subset W(A) \setminus \{\varepsilon\}$. Run your program on the set of Example 9.5.

Website

A World Wide Web server interface at

$$\texttt{http://www.uncg.edu/mat/pcode}$$

has been established for automated use of programs related to the first algorithm discussed in Section 9.1.

Bibliographic notes

Sardinas and Patterson developed an algorithm to test whether or not a set of full words is a code [127]. The partial word adaptation from Section 9.1 of their algorithm is from Blanchet-Sadri and Moorefield [39].

The domino technique on full words from Section 9.2.1 is from Head and Weber [92]. In order to study the relations satisfied by X, Guzmán suggested to look at the simplified domino graph and the domino function of X [85]. Proposition 9.1 is from Guzmán [85]. That approach was further considered in [13] for instance. The simplified domino graph of X is a subgraph of the Head and Weber's domino graph of X defined in [92].

The domino technique on partial words from Section 9.2.2 is from Blanchet-Sadri [16].

Part V

FURTHER TOPICS

Chapter 10

Equations on Partial Words

As we saw in Chapter 2, some of the most basic properties of words, like the conjugacy and the commutativity, can be expressed as solutions of word equations. Recall that two words x and y are conjugate if there exist words u and v such that $x = uv$ and $y = vu$. The latter is equivalent to the existence of a word z satisfying $xz = zy$ in which case there exist words u, v such that $x = uv$, $y = vu$, and $z = (uv)^k u$ for some nonnegative integer k. And two words x and y commute, namely $xy = yx$, if and only if x and y are powers of the same word, that is, there exists a word z such that $x = z^k$ and $y = z^l$ for some integers k and l.

Another equation of interest is $x^m = y^n$. It turns out that if x and y are words, then $x^m = y^n$ for some positive integers m, n if and only if there exists a word z such that $x = z^k$ and $y = z^l$ for some integers k and l. Yet, another interesting equation is $x^m y^n = z^p$ which has only periodic solutions in a free monoid, that is, if $x^m y^n = z^p$ holds with integers $m, n, p \geq 2$, then there exists a word w such that x, y and z are powers of w.

In this chapter, we pursue our investigation of equations on partial words. In Section 10.1, we give a result that gives the structure of partial words satisfying the equation $x^m \uparrow y^n$, which provides the conditions for when x and y are contained in powers of a common word. In Section 10.2, we solve the equation $x^2 \uparrow y^m z$. This result is a first step for solving the equation $x^m y^n \uparrow z^p$ in Section 10.3.

10.1 The equation $x^m \uparrow y^n$

In this section, we investigate the equation $x^m \uparrow y^n$ on partial words. When dealing with partial words x and y, if there exists a pword z such that $x \subset z^k$ and $y \subset z^l$ for some integers k, l, then $x^m \uparrow y^n$ for some positive integers m, n. Indeed, by the multiplication rule, $x^l \subset z^{kl}$ and $y^k \subset z^{kl}$, showing that $x^l \uparrow y^k$. For the converse, it is beneficial to define the following manipulation of a partial word x. For a positive integer p and an integer $0 \leq i < p$, define

$$x\begin{bmatrix} i \\ p \end{bmatrix} = x(i)x(i+p)x(i+2p)\ldots x(i+jp)$$

where j is the largest nonnegative integer such that $i + jp < |x|$. We shall call this *the ith residual word of x modulo p*.

The following two lemmas provide equivalent conditions for periodicity and weak periodicity.

LEMMA 10.1

A partial word x is p-periodic if and only if $x\begin{bmatrix} i \\ p \end{bmatrix}$ is 1-periodic for all $0 \le i < p$.

LEMMA 10.2

A partial word x is weakly p-periodic if and only if $x\begin{bmatrix} i \\ p \end{bmatrix}$ is weakly 1-periodic for all $0 \le i < p$.

Using the multiplication and the simplification rules, we can demonstrate the following lemma. Consequently, if $x^{m'} \uparrow y^{n'}$ and $\gcd(m', n') \ne 1$, then $x^m \uparrow y^n$ where $m = m'/\gcd(m', n')$ and $n = n'/\gcd(m', n')$. And therefore the assumption that $\gcd(m, n) = 1$ may be made without losing generality.

LEMMA 10.3

Let x, y be partial words and let m, n and p be positive integers. Then $x^m \uparrow y^n$ if and only if $x^{mp} \uparrow y^{np}$.

LEMMA 10.4

Let x, y be partial words and let m, n be positive integers such that $x^m \uparrow y^n$ with $\gcd(m, n) = 1$. Call $|x|/n = |y|/m = p$. If there exists an integer i such that $0 \le i < p$ and $x\begin{bmatrix} i \\ p \end{bmatrix}$ is not 1-periodic, then $D(y\begin{bmatrix} i \\ p \end{bmatrix})$ is empty.

PROOF   Assume that there is an integer i such that $0 \le i < p$ and $x\begin{bmatrix} i \\ p \end{bmatrix}$ is not 1-periodic. Then for some j and k such that $i + jp$ and $i + kp$ are in the domain of x,

$$x(i + jp) \ne x(i + kp)$$

Now assume that $D(y\begin{bmatrix} i \\ p \end{bmatrix})$ is not empty, that is, there is a constant l such that

$$y(i + (l + j)p) = x(i + jp)$$

and hence

$$y(i + (l + j)p) \uparrow x((i + jp + l'|y|) \bmod |x|)$$

for all l'. Now we make the claim that there exists an l' such that

$$(i + jp + l'|y|) \equiv (i + kp) \bmod |x| \qquad (10.1)$$

Since $|y| = mp$ and $|x| = np$, (10.1) becomes

$$(j + l'm)p \equiv kp \bmod np$$

which may be reduced to

$$k - j \equiv l'm \bmod n$$

Since $\gcd(m, n) = 1$, such an l' exists that satisfies our claim. Therefore

$$x(i + kp) = x((i + jp + l'|y|) \bmod |x|) \uparrow y(i + (l + j)p) = x(i + jp) \qquad (10.2)$$

but we assumed earlier that $x(i + jp) \neq x(i + kp)$ and that $i + jp$ and $i + kp$ were both in the domain of x. Therefore the compatibility relation in (10.2) is a contradiction. ☐

LEMMA 10.5
Let x be a partial word, let m, p be positive integers, and let i be an integer such that $0 \leq i < p$. Then the relation

$$x^m \begin{bmatrix} i \\ p \end{bmatrix} = x \begin{bmatrix} i \\ p \end{bmatrix} x \begin{bmatrix} (i - |x|) \bmod p \\ p \end{bmatrix} \cdots x \begin{bmatrix} (i - (m - 1)|x|) \bmod p \\ p \end{bmatrix}$$

holds.

PROOF The proof is by induction on m. Consider the case of $m = 2$. Note that

$$x^2 \begin{bmatrix} i \\ p \end{bmatrix} = x \begin{bmatrix} i \\ p \end{bmatrix} y$$

for some partial word y. Let k be the largest nonnegative integer such that $i + kp < |x|$. Then

$$y(0) = x(j)$$

where $j = (i + (k + 1)p) \bmod |x|$. Therefore $i - j + (k + 1)p = |x|$ by the definition of k and so

$$j = (i - |x|) \bmod p$$

Hence

$$y = x \begin{bmatrix} j \\ p \end{bmatrix} = x \begin{bmatrix} (i - |x|) \bmod p \\ p \end{bmatrix}$$

and the basis follows. Assume the relation holds for $m \leq n$. Then $x^{n+1} \begin{bmatrix} i \\ p \end{bmatrix}$ is equal to

$$x^n \begin{bmatrix} i \\ p \end{bmatrix} x \begin{bmatrix} (i - n|x|) \bmod p \\ p \end{bmatrix}$$

which in turn equals

$$x \begin{bmatrix} i \\ p \end{bmatrix} x \begin{bmatrix} (i - |x|) \bmod p \\ p \end{bmatrix} \ldots x \begin{bmatrix} (i - (n-1)|x|) \bmod p \\ p \end{bmatrix} x \begin{bmatrix} (i - n|x|) \bmod p \\ p \end{bmatrix}$$

The result follows. ▯

The following concept of a "good pair" of partial words is basic in this section.

DEFINITION 10.1   *Let x, y be partial words and let m, n be positive integers such that $x^m \uparrow y^n$ with $\gcd(m, n) = 1$. If for all $i \in H(x)$ the word*

$$y^n \begin{bmatrix} i \\ |x| \end{bmatrix} = y^n(i) y^n(i + |x|) \ldots y^n(i + (m-1)|x|)$$

is 1-periodic and for all $i \in H(y)$ the word

$$x^m \begin{bmatrix} i \\ |y| \end{bmatrix} = x^m(i) x^m(i + |y|) \ldots x^m(i + (n-1)|y|)$$

*is 1-periodic, then the pair (x, y) is called a **good pair**.*

Let us illustrate the concept of good pair with a few examples.

Example 10.1
Consider $x = a\diamond babba\diamond babba\diamond\diamond$ of length $|x| = 15$ and $y = ab\diamond a\diamond b$ of length $|y| = 6$ that satisfy $x^2 \uparrow y^5$. For all $i \in H(x)$, the word

$$y^5(i) y^5(i + |x|)$$

is 1-periodic since for $i = 1$, we get $b\diamond$; for $i = 7$, we get $b\diamond$; for $i = 13$, we get $b\diamond$; and for $i = 14$, we get $\diamond b$. Similarly, for all $i \in H(y)$, the word

$$x^2(i) x^2(i + |y|) x^2(i + 2|y|) x^2(i + 3|y|) x^2(i + 4|y|)$$

is 1-periodic since for $i = 2$, we get $bb\diamond bb$ and for $i = 4$, we get $bb\diamond\diamond\diamond$. Thus (x, y) qualifies as a good pair.

Now, consider $x = a\diamond b$ of length $|x| = 3$ and $y = acbadb$ of length $|y| = 6$ which satisfy $x^2 \uparrow y^1$. Here (x, y) is not a good pair since $y^1(1) y^1(1 + |x|) = y(1) y(4) = cd$ is not 1-periodic. ▯

We now state the "good pair" theorem.

THEOREM 10.1

Let x, y be partial words and let m, n be positive integers such that $x^m \uparrow y^n$ with $\gcd(m, n) = 1$. If (x, y) is a good pair, then there exists a partial word z such that $x \subset z^k$ and $y \subset z^l$ for some integers k, l.

PROOF Since $\gcd(m, n) = 1$, there exists an integer p such that $\frac{|x|}{n} = \frac{|y|}{m} = p$. Now assume there exists an integer i such that $0 \le i < p$ and $x\begin{bmatrix} i \\ p \end{bmatrix}$ is not 1-periodic. Then by Lemma 10.4, $i + jp \in H(y)$ for $0 \le j < m$ which by the assumption that (x, y) is a good pair implies that $x^m\begin{bmatrix} i + jp \\ |y| \end{bmatrix}$ must be 1-periodic for any choice of j. Note that $|y| = mp$ and similarly $|x| = np$. Therefore by Lemma 10.5, $x^m\begin{bmatrix} i + jp \\ mp \end{bmatrix}$ is equal to

$$x\begin{bmatrix} i + jp \\ mp \end{bmatrix} x\begin{bmatrix} (i + jp - |x|) \bmod mp \\ mp \end{bmatrix} \cdots x\begin{bmatrix} (i + jp - (m-1)|x|) \bmod mp \\ mp \end{bmatrix}$$

When l is chosen so that $0 \le l < m$, we have

$$i + jp - l|x| = i + (j - ln)p$$

For $0 \le j < m$, we claim that

$$\{(j - ln) \bmod m \mid 0 \le l < m\} = \{0, 1, \ldots, m-1\}$$

Indeed, assuming there exist $0 \le l_1 < l_2 < m$ such that

$$(j - l_1 n) \equiv (j - l_2 n) \bmod m$$

we get that m divides $(l_1 - l_2)n$, and since $\gcd(m, n) = 1$, that m divides $(l_1 - l_2)$, whence $l_1 = l_2$. So there exist $j_0, j_1, \ldots, j_{m-1}$ such that $j_0 = j$ and $\{j_0, j_1, \ldots, j_{m-1}\} = \{0, 1, \ldots, m-1\}$ and

$$x^m\begin{bmatrix} i + jp \\ mp \end{bmatrix} = x\begin{bmatrix} i + j_0 p \\ mp \end{bmatrix} x\begin{bmatrix} i + j_1 p \\ mp \end{bmatrix} \cdots x\begin{bmatrix} i + j_{m-1}p \\ mp \end{bmatrix}$$

Since $x^m\begin{bmatrix} i + jp \\ mp \end{bmatrix}$ is 1-periodic, there exists a letter a such that for all $0 \le k < m$,

$$x\begin{bmatrix} i + j_k p \\ mp \end{bmatrix} \subset a^{m_{j_k}}$$

for some integer m_{j_k}. This contradicts our assumption that there is an i for which $x\begin{bmatrix} i \\ p \end{bmatrix}$ is not 1-periodic (here $x\begin{bmatrix} i \\ p \end{bmatrix} = x(i)x(i+p)\ldots x(i+(n-1)p) \subset a^n$). Therefore $x\begin{bmatrix} i \\ p \end{bmatrix}$ is 1-periodic for all $0 \le i < p$. By the equivalent condition for periodicity, this implies that x is p-periodic. The same argument holds for y, and since $x^m \uparrow y^n$, the result that there exists a word z of length p such that $x \subset z^n$ and $y \subset z^m$ is proven. $\square$

We illustrate Theorem 10.1 with the following example.

Example 10.2
Given $x = a\diamond babba\diamond babba\diamond\diamond$ of length $|x| = 15$ and $y = ab\diamond a\diamond b$ of length $|y| = 6$, the alignment of x^2 and y^5 may be observed with the depiction in Figure 10.1.[1] We can check that $x^2 \uparrow y^5$ and we saw in Example 10.1 that (x, y) is a good pair. Here $x \subset z^5$ and $y \subset z^2$ with $z = abb$. $\square$

$x^m =$ a^b abb a^b abb a^^ a^b abb a^b abb a^^
$y^n =$ ab^ a^b ab^ a^b ab^ a^b ab^ a^b ab^ a^b

FIGURE 10.1: An example of the good pair equation.

REMARK 10.1 The compatibility relation $x^2 = (a\diamond b)^2 \uparrow (acbadb)^1 = y^1$ shows that the assumption of (x, y) being a good pair is necessary in Theorem 10.1. It was noticed in Example 10.1 that (x, y) is not a good pair, and we can check that there exists no partial word z as desired. $\square$

COROLLARY 10.1
Let x and y be primitive partial words such that (x, y) is a good pair. If $x^m \uparrow y^n$ for some positive integers m and n, then $x \uparrow y$.

PROOF Suppose to the contrary that $x \not\uparrow y$. Since (x, y) is a good pair, there exists a word z such that $x \subset z^k$ and $y \subset z^l$ for some integers k, l. Since $x \not\uparrow y$, we get $k \neq l$. But then x or y is not primitive, a contradiction. $\square$

REMARK 10.2 Note that if both x and y are full words such that $x^m = y^n$ for some positive integers m and n, then (x, y) is a good pair. Corollary 10.1 hence implies that if x, y are primitive full words satisfying $x^m = y^n$ for some positive integers m and n, then $x = y$. $\square$

We conclude this section by further investigating the equation $x^2 \uparrow y^m$ on partial words where m is a positive integer. The proof is left as an exercise

---

[1]This graphic and the other that follows were generated using a C++ applet on one of the author's websites, mentioned in the Website Section at the end of this chapter.

for the reader.

PROPOSITION 10.1
Let x, y be partial words. Then $x^2 \uparrow y^m$ for some positive integer m if and only if there exist $u, v, u_0, v_0, \ldots, u_{m-1}, v_{m-1}$ such that $y = uv$,

$$x = (u_0 v_0) \ldots (u_{n-1} v_{n-1}) u_n = v_n (u_{n+1} v_{n+1}) \ldots (u_{m-1} v_{m-1})$$

where $0 \leq n < m$, $u \uparrow u_i$ and $v \uparrow v_i$ for all $0 \leq i < m$, and where one of the following holds:

- *$m = 2n$ and $u = \varepsilon$.*

- *$m = 2n + 1$ and $|u| = |v|$.*

10.2 The equation $x^2 \uparrow y^m z$

In this section, we investigate the equation $x^2 \uparrow y^m z$ on partial words where it is assumed that m is a positive integer and z is a prefix of y. The equation $x^2 \uparrow y^m z$ will play a crucial role in the solution of the equation $x^m y^n \uparrow z^p$ discussed in the next section.

Consider the compatibility relation

$$(a \diamond\diamond a)^2 \uparrow (aab)^2 aa$$

where $x = a \diamond\diamond a$, $y = aab$ and $z = aa$. We say that the triple (x, y, z) is a "nontrivial" solution of the equation $x^2 \uparrow y^2 z$. More formally, we have the following definition.

DEFINITION 10.2   *Let m be a positive integer. A triple (x, y, z) satisfying $x^2 \uparrow y^m z$ with z a prefix of y is called a **trivial solution** if x, y, z are contained in powers of a common word (or there exists a word w such that $x \subset w^{k_1}$, $y \subset w^{k_2}$, and $z \subset w^{k_3}$ for some integers k_1, k_2, k_3).*

Obviously, if x, y, z are contained in powers of a common word, then the equation $x^2 \uparrow y^m z$ may have a solution for some m. In order to characterize all other solutions, we need the concept of a "good triple" of partial words.

DEFINITION 10.3   *Let x, y, z be partial words such that z is a proper prefix of y. Then (x, y, z) is a **good triple** if for some positive integer m there exist partial words $u, v, u_0, v_0, \ldots, u_{m-1}, v_{m-1}, z_x$ such that $u \neq \varepsilon$, $v \neq \varepsilon$,*

$$y = uv,$$

$$x = (u_0v_0)\ldots(u_{n-1}v_{n-1})u_n \qquad (10.3)$$
$$= v_n(u_{n+1}v_{n+1})\ldots(u_{m-1}v_{m-1})z_x \qquad (10.4)$$

where $0 \le n < m$, $u \uparrow u_i$ and $v \uparrow v_i$ for all $0 \le i < m$, $z \uparrow z_x$, and where one of the following holds:

- $m = 2n$, $|u| < |v|$, and there exist partial words u', u'_n such that $z_x = u'u_n$, $z = uu'_n$, $u \uparrow u'$ and $u_n \uparrow u'_n$.

- $m = 2n+1$, $|u| > |v|$, and there exist partial words v'_{2n} and z'_x such that $u_n = v_{2n}z_x$, $u = v'_{2n}z'_x$, $v_{2n} \uparrow v'_{2n}$ and $z_x \uparrow z'_x$.

Let us give an example before we characterize all solutions of the equation $x^2 \uparrow y^m z$.

Example 10.3

Let $x = abca\diamond c\diamond abca$, $y = abcaa\diamond c$ and $z = a$. Here z is a proper prefix of y. Set $m = 3$. We can decompose x into a factor of length $|y| = 7$ with a remaining factor of length 4:

$$x = (abca\diamond c\diamond)abca$$

Then we split the factor of length 7 into a first factor of length 4 and a second factor of length 3:

$$(abca)(\diamond c\diamond)(abca) = u_0v_0u_1$$

Here $n = 1$, and so $m = 2n + 1$. Decomposing x starting with a block of length 3 instead leads to

$$(abc)(a\diamond c\diamond)(abc)(a) = v_1u_2v_2z_x$$

Note that all the u's have length 4 and all the v's have length 3. The pword y can then be split into a 4-length piece followed by a 3-length piece: $y = (abca)(a\diamond c) = uv$. We can check that $u \uparrow u_i$ and $v \uparrow v_i$ for all $0 \le i < 3$ and that $z \uparrow z_x$. Moreover, setting $v'_2 = abc$ and $z'_x = a$, we have $u_1 = v_2z_x$, $u = v'_2z'_x$, $v_2 \uparrow v'_2$ and $z_x \uparrow z'_x$. Thus the triple (x, y, z) qualifies as a good triple. ∎

We now state the "good triple" theorem.

THEOREM 10.2

Let x, y, z be partial words such that z is a proper prefix of y. Then $x^2 \uparrow y^m z$ for some positive integer m if and only if (x, y, z) is a good triple.

PROOF   Note that if the conditions hold, then trivially $x^2 \uparrow y^m z$ for some positive integer m. If $x^2 \uparrow y^m z$ for some positive integer m, then there exist partial words u, v and an integer n such that $y = uv$, $x \uparrow (uv)^n u$ and $x \uparrow v(uv)^{m-n-1} z$. Thus $|x| = n(|u|+|v|)+|u| = (m-n-1)(|u|+|v|)+|v|+|z|$ which clearly shows

$$|z| = (2n - m + 2)|u| + (2n - m)|v| \tag{10.5}$$

This determines a relationship between m and n. There are two cases to consider which correspond to assumptions on $|u|$ and $|v|$. Under the assumption $|u| = |v|$ we see that z must be either empty or equal to y which is a contradiction. If we assume $|u| < |v|$, then (10.5) shows $|z| = 2|u|$, and if we assume $|u| > |v|$, then $|z| = |u| - |v|$. Now note that x^2 may be factored in the following way:

$$x^2 = (u_0 v_0) \ldots (u_{n-1} v_{n-1})(u_n v_n)(u_{n+1} v_{n+1}) \ldots (u_{m-1} v_{m-1}) z_x$$

Here $u_i \uparrow u$ and $v_i \uparrow v$ and $z_x \uparrow z$. From this it is clear that (10.3) and (10.4) are satisfied.

Note that $u \neq \varepsilon$ (otherwise $|u| < |v|$, in which case $|z| = 2|u| = 0$), and also $v \neq \varepsilon$ (otherwise, $|u| > |v|$, in which case $|z| = |u| - |v| = |y|$). First assume $|u| < |v|$, equivalently $|z| = 2|u|$ and $m = 2n$. Note that the suffix of length $|u|$ of z_x must be u_n and therefore is compatible with u. The prefix of length $|u|$ of z must be u itself since z is a prefix of y. Thus $z_x = u' u_n$ and $z = uu'_n$ where $u \uparrow u'$ and $u_n \uparrow u'_n$ which is one of our assertions. Now assume $|u| > |v|$, that is $|z| = |u| - |v|$ and $m = 2n + 1$. Note by cancellation that $u_n = v_{2n} z_x$. Since $u_n \uparrow u$, we can rewrite u as $v'_{2n} z'_x$ where $v_{2n} \uparrow v'_{2n}$ and $z_x \uparrow z'_x$, which is our other assertion. ▯

Example 10.4
Returning to Example 10.3 where $x = abca \diamond c \diamond abca$, $y = abcaa \diamond c$ and $z = a$, we can check that $x^2 \uparrow y^3 z$. Figure 10.2 shows the decomposition of x, y, and z according to Definition 10.3. ▯

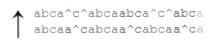

```
abca^c^abcaabca^c^abca
abcaa^cabcaa^cabcaa^ca
```

FIGURE 10.2:   An example of the good triple equation.

We end this section with two corollaries.

COROLLARY 10.2
Let x, y be partial words such that $|x| \geq |y| > 0$ and let z be a prefix of y. Assume that $x^2 \uparrow y^m z$ for some positive integer m.

Referring to the notation of Theorem 10.2 (when $z \neq \varepsilon$ and $z \neq y$) or referring to the notation of Proposition 10.1 (otherwise), both $w \uparrow uv$ and $w \uparrow vu$ hold where w denotes the prefix of length $|y|$ of x.

Moreover, u and v are contained in powers of a common word if $(z = \varepsilon$ and $m = 2n)$ or $(z = y$ and $m+1 = 2n)$. This is also true if any of the following six conditions hold with $u \neq \varepsilon$ and $v \neq \varepsilon$:

1. y is full and w has at most one hole.

2. y is full and w is not $\{|u|, |v|\}$-special.

3. w is full and y has at most one hole.

4. w is full, and either $(|u| \leq |v|$ and uv is not $(|u|, |v|)$-special) or $(|v| \leq |u|$ and vu is not $(|v|, |u|)$-special).

5. $uv \uparrow vu$ and y has at most one hole.

6. $uv \uparrow vu$, and either $(|u| \leq |v|$ and uv is not $(|u|, |v|)$-special) or $(|v| \leq |u|$ and vu is not $(|v|, |u|)$-special).

PROOF   We show the result when z is a proper prefix of y (or when $z \neq \varepsilon$ and $z \neq y$). This part of the proof refers to the notation of Theorem 10.2. If $m > n+1$, then from the fact that $y = uv$ and $x \uparrow y^n u$ and $x \uparrow vy^{m-n-1}z$, we get $w \uparrow uv$ and $w \uparrow vu$. If on the other hand $m = n+1$, then $x \uparrow yu$ and $x \uparrow vz$. It follows that $|u| < |z|$ and we also get $w \uparrow uv$ and $w \uparrow vu$.

For Statement 1, since u, v are full, we get $w \subset uv$ and $w \subset vu$ and by Lemma 10.5, $uv = vu$ and u, v are powers of a common word.

For Statement 2, the result follows similarly since $uv = vu$ by Lemma 2.4.

For Statement 3, we get $uv \uparrow vu$. By Theorem 2.5, u and v are contained in powers of a common word.

Statement 4 follows similarly as Statement 3 using Theorem 2.6. Statement 5 follows similarly as Statement 3, and Statement 6 as Statement 4.   □

COROLLARY 10.3
Let x, y, z be words such that z is a prefix of y. If x, y are primitive and $x^2 = y^m z$ for some integer $m \geq 2$, then $x = y$.

10.3   The equation $x^m y^n \uparrow z^p$

Certainly, if there exist a word w such that

$$x \subset w^{np} \text{ and } y \subset w^{mp} \text{ and } z \subset w^{2mn}$$

then

$$x^m y^m \subset w^{2mnp}$$
$$z^p \quad \subset w^{2mnp}$$

and the equation $x^m y^n \uparrow z^p$ has a "trivial" solution.

However, there may be "nontrivial" solutions as is seen with the compatibility relation

$$(a \diamond b)^2 (b \diamond a)^2 \uparrow (abba)^3$$

There is no common word w such that all $a \diamond b$, $b \diamond a$ and $abba$ are contained in powers of w.

In this section, we give the structure of all the solutions of the equation $x^m y^n \uparrow z^p$ when $m \geq 2, n \geq 2$ and $p \geq 4$. We will reduce the number of cases in proving the main Theorem 10.3 by using the following lemma.

LEMMA 10.6
Let x, y, z be partial words and let m, n, p be positive integers. If $x^m y^n \uparrow z^p$, then $(rev(y))^n (rev(x))^m \uparrow (rev(z))^p$.

We start by defining two types of solutions.

DEFINITION 10.4
*There exists a partial word w such that x, y, z are contained in powers of w. We call such solutions the **trivial** or **Type 1** solutions.*

DEFINITION 10.5
*The partial words x, y, z satisfy $x \uparrow z$ and $y \uparrow z$. We call such solutions the **Type 2** solutions.*

It is an easy exercise to check that if z is full, then Type 2 solutions are Type 1 solutions.

THEOREM 10.3
Let x, y, z be primitive partial words such that (x, z) and (y, z) are good pairs. Let m, n, p be integers such that $m \geq 2, n \geq 2$ and $p \geq 4$. Then the equation $x^m y^n \uparrow z^p$ has only solutions of Type 1 or Type 2 unless one of the following holds:

- *$x^2 \uparrow z^k z'$ for some integer $k \geq 2$ and nonempty prefix z' of z.*

- *$z^2 \uparrow x^l x'$ for some integer $l \geq 2$ and nonempty prefix x' of x.*

PROOF By Lemma 10.6, we need only examine the case when $|x^m| \geq |y^n|$. Now assume $x^m y^n \uparrow z^p$ has some solution that is not of Type 1 or Type 2.

Our assumption on the lengths of x^m and y^n implies that $|x^m| \geq |z^2|$ and therefore either $|x^2| \geq |z^2|$ or $|x^2| < |z^2|$. If $|x^2| \geq |z^2|$, then $x^2 \uparrow z^k z'$ for some integer $k \geq 2$ and prefix z' of z. And if $|x^2| < |z^2|$, then $z^2 \uparrow x^l x'$ for some integer $l \geq 2$ and prefix x' of x.

Consider the case where $z' = \varepsilon$ (the case $x' = \varepsilon$ is similar). Corollary 10.1 implies that $x \uparrow z$. From $x^m y^n \uparrow z^p$ and $x \uparrow z$, using the simplification rule, we get $y^n \uparrow z^{p-m}$. Using Corollary 10.1 again, we have $y \uparrow z$. Hence this case forms Type 2 solutions. ☐

In the case of full words, there only exist the Type 1 solutions.

COROLLARY 10.4
Let x, y, z be words and let m, n, p be integers such that $m \geq 2, n \geq 2$ and $p \geq 4$. Then the equation $x^m y^n = z^p$ has no nontrivial solutions.

PROOF We prove the result when x, y, z are primitive (the nonprimitive case is left to the reader). As in the proof of Theorem 10.3, we need only examine the case when $|x^m| \geq |y^n|$. This assumption leads to either $x^2 = z^k z'$ for some integer $k \geq 2$ and prefix z' of z, or $z^2 = x^l x'$ for some integer $l \geq 2$ and prefix x' of x. In the first case, we have $x = z$ by Corollary 10.3, and from the equation $x^m y^n = z^p$, we get $y^n = z^{p-m}$. The latter implies that y and z are powers of a common word, and the solution is trivial. The second case is analogous. ☐

Exercises

10.1 Show that for any partial word x and positive integers m, p such that $|x|$ is divisible by p,

$$x^m\begin{bmatrix}i\\p\end{bmatrix} = (x\begin{bmatrix}i\\p\end{bmatrix})^m$$

where $0 \leq i < p$.

10.2 ⬛s Give the decompositions of $x = \diamond bbab\diamond$ and $y = a\diamond b$ that satisfy Proposition 10.1.

10.3 Is $(a\diamond ba\diamond\diamond bb\diamond\diamond abb, ab\diamond\diamond a, ab\diamond)$ a good triple?

10.4 Let $x = \diamond bb\diamond b\diamond$, $y = ab\diamond\diamond b$ and $z = ab$. Display the alignment of x^2 and $y^m z$ for some m. What can you conclude about the triple (x, y, z)?

10.5 Show that $(a \diamond ba \diamond \diamond \diamond bb \diamond \diamond \diamond bb, ab \diamond \diamond a, ab \diamond)$ is a good triple according to Definition 10.3. Highlight the factorizations of x, y and z as is done in Example 10.3.

10.6 Prove Lemma 10.6.

10.7 Show that if z is full, then Type 2 solutions of the equation $x^m y^n \uparrow z^p$ are also Type 1 solutions.

10.8 ⊞ Prove the nonprimitive case of Corollary 10.4.

10.9 Let x, y be full words. Prove that $x^m = y^n$ for some positive integers m, n if and only if there exists a word z such that $x = z^k$ and $y = z^l$ for some integers k and l.

10.10 Check that the following are solutions of the equation $x^m y^n \uparrow z^p$ for some suitable values of m, n and p:

- $x = abc$, $y = abd$ and $z = ab \diamond$ is a solution of $x^2 y^2 \uparrow z^4$,
- $x = ab \diamond$, $y = a \diamond c$ and $z = \diamond bcabc$ is a solution of $x^4 y^4 \uparrow z^4$.

Which type of solutions are they?

10.11 ⓢ Find integers m, n and p for which

$$x = abca \diamond c \diamond abca, \; y = bc \diamond aac \text{ and } z = abcaa \diamond c$$

is a solution of $x^m y^n \uparrow z^p$. Repeat for

$$x = abca \diamond \diamond c, \; y = a \text{ and } z = abc \diamond bc \diamond \diamond bca$$

Challenging exercises

10.12 ⓢ Prove Proposition 10.1.

10.13 What does it mean for a triple of full words (x, y, z) to be a good triple?

10.14 Show that Corollary 10.3 does not hold when $m = 1$.

10.15 Prove Corollary 10.3.

10.16 Show Corollary 10.2 when $z = \varepsilon$ or $z = y$.

10.17 ⓢ Let x, y, z be partial words such that z is a prefix of y. Assume that x, y are primitive and that $x^2 \uparrow y^m z$ for some integer $m \geq 2$. Show that if x has at most one hole and y is full, then $x \uparrow y$.

10.18 Show that Exercise 10.17 does not hold when $m = 1$.

10.19 [H] Prove that Exercise 10.17 does not hold when x is full and y has one hole.

10.20 [S] Find a nontrivial solution to the equation $x^2 y^2 \uparrow z^3$.

10.21 Show the case of $x' = \varepsilon$ in the proof of Theorem 10.3.

Programming exercises

10.22 Write a program that takes as input two partial words x, y and outputs two positive integers m, n such that $x^m \uparrow y^n$ whenever they exist, in which case your program should create an alignment of x^m and y^n.

10.23 Write a program to check whether or not a pair (x, y) of partial words is a good pair. Run you program on the pairs

- $(acbadb, a \diamond b)$
- $(ab \diamond a \diamond b, a \diamond babba \diamond babba \diamond \diamond)$

10.24 Design an applet that when given a good pair (x, y) as input, outputs a pword z and integers k, l such that $x \subset z^k$ and $y \subset z^l$.

10.25 Design an algorithm that discovers the decomposition of a pair of partial words x, y according to Proposition 10.1 in case $x^2 \uparrow y^m$ for some positive integer m.

10.26 Implement the decomposition of x, y and z as described in Definition 10.3 in case (x, y, z) is a good triple. Run your program on

- $(a \diamond ba \diamond \diamond \diamond bb \diamond \diamond \diamond bb, ab \diamond \diamond a, ab \diamond)$
- $(a \diamond ba \diamond \diamond \diamond bb \diamond \diamond abb, ab \diamond \diamond a, ab \diamond)$

Website

A World Wide Web server interface at

$$\texttt{http://www.uncg.edu/mat/research/equations}$$

has been established for automated use of programs related to the equations discussed in this chapter. In particular, one of the programs takes as input a good pair (x, y) of partial words according to Definition 10.1, and outputs a partial word z and integers k, l such that $x \subset z^k$ and $y \subset z^l$ (this program implements the good pair Theorem 10.1.) Another program takes as input a triple (x, y, z) of partial words such that z is a proper prefix of y, and outputs an integer m such that $x^2 \uparrow y^m z$, if such m exists, and shows the decomposition of x, y and z according to Definition 10.3 (this program implements the good triple Theorem 10.2).

Bibliographic notes

An important topic in algorithmic combinatorics on words is the *satisfiability problem for equations on words*, that is, the problem to decide whether or not a given equation on the free monoid has a solution. The problem was proposed in 1954 by Markov [113] and remained open until 1977 when Makanin answered it positively [110]. However, Makanin's algorithm is one of the most complicated algorithms ever presented and has at least exponential space complexity [107]. Rather recently, Plandowski showed, with a completely new algorithm, that the problem is actually in polynomial space [120] and [121]. However, the structure of the solutions cannot be found using Makanin's algorithm. Even for rather short instances of equations, for which the existence of solutions may be easily established, the structure of the solutions may be very difficult to describe.

For integers $m \geq 2, n \geq 2$ and $p \geq 2$, the equation $x^m y^n = z^p$ possesses a solution in a free group only when x, y and z are each a power of a common element. This result, which received a lot of attention, was first proved by Lyndon and Schützenberger for free groups [109]. Their proof implied the case for free monoids since every free monoid can be embedded in a free group. Direct proofs for free monoids appear in [51, 52, 89]. Corollaries 10.3 and 10.4 are from Chu and Town [52].

The results on partial words of Sections 10.1, 10.2 and 10.3 are from Blanchet-Sadri, Blair and Lewis [20].

Chapter 11

Correlations of Partial Words

In this chapter, we study the combinatorics of possible sets of periods and weak periods of partial words. In Section 11.1, we introduce the notions of binary and ternary correlations, which are binary and ternary vectors indicating the periods and weak periods of partial words. In Section 11.2, we characterize precisely which of these vectors represent the period and weak period sets of partial words and prove that all *valid* correlations may be taken over the binary alphabet. In Section 11.3, we show that the sets of all such vectors of a given length form distributive lattices under suitably defined partial orderings. In Section 11.4, we show that there is a well defined minimal set of generators, which we call an *irreducible* set of periods, for any binary correlation of length n and demonstrate in Section 11.5 that these generating sets are the so-called *primitive* subsets of $\{1, 2, ..., n-1\}$. Finally, we investigate the number of partial word correlations of length n.

11.1 Binary and ternary correlations

The leading concept in this chapter is that of "correlation" which we first define for full words.

DEFINITION 11.1 *Let u be a (full) word and let v be the binary vector of length $|u|$ for which $v_0 = 1$ and*

$$v_i = \begin{cases} 1 & \text{if } i \in \mathcal{P}(u) \\ 0 & \text{otherwise} \end{cases} \tag{11.1}$$

*We call v the **correlation** of u.*

Example 11.1
The word *abbababbab* has periods 5 and 8 (and 10) and thus has correlation 1000010010. ⬚

This representation gives a useful and concise way of representing period sets of strings of a given length over a finite alphabet.

Among the possible 2^n binary vectors of length n, only a proper subset are *valid* correlations. Indeed, the vector 100001001000 is not the correlation of any word, a fact implied by Theorem 3.1 because any word having periods 5 and 8 and length at least $5 + 8 - \gcd(5, 8)$ has also $\gcd(5, 8) = 1$ as period. Valid correlations will also be called *full word correlations*.

When $p \in \mathcal{P}'(u) \setminus \mathcal{P}(u)$ we say that the partial word u has a *strictly* weak period of p. We now extend the definition of a "correlation" of a full word to incorporate the difference between strictly weak periods and strong periods, a difference which does not occur in the case of full words.

DEFINITION 11.2

- The **binary correlation** of a partial word u satisfying $\mathcal{P}'(u) = \mathcal{P}(u)$ is the binary vector v of length $|u|$ such that $v_0 = 1$ and

$$
v_i = \begin{cases} 1 & \text{if } i \in \mathcal{P}(u) \\ 0 & \text{otherwise} \end{cases} \tag{11.2}
$$

- The **ternary correlation** of a partial word u is the ternary vector v of length $|u|$ such that $v_0 = 1$ and

$$
v_i = \begin{cases} 1 & \text{if } i \in \mathcal{P}(u) \\ 2 & \text{if } i \in \mathcal{P}'(u) \setminus \mathcal{P}(u) \\ 0 & \text{otherwise} \end{cases} \tag{11.3}
$$

We will say that a ternary vector v of length n is a *valid* ternary correlation provided that there exists a partial word u of length n over an alphabet A such that v is the ternary correlation of u. We define

$$
\mathcal{P}(v) = \{i \mid 0 < i < n \text{ and } v_i = 1\} \cup \{n\}
$$

and

$$
\mathcal{P}'(v) = \{i \mid 0 < i < n \text{ and } v_i > 0\} \cup \{n\}
$$

as the period set and the weak period set of v respectively. Valid binary correlations and related terminology are defined similarly. *Valid* binary or ternary correlations will also be called *partial word correlations*. When $i \in \mathcal{P}(v) \setminus \{n\}$, we say that i is a *nontrivial* period of v. Similarly, when $i \in \mathcal{P}'(v) \setminus \{n\}$, we say that i is a *nontrivial* weak period of v.

Example 11.2

The partial word $abca\diamond cadca$ has periods 9 (and 10) and strictly weak period 3. Thus its binary correlation vector is 1000000001 and its ternary correlation vector is 1002000001. $\square$

11.2   Characterizations of correlations

We now state one of the results that motivates this chapter. It gives a complete characterization of the possible period sets of full words of arbitrary length. We first need a couple of definitions.

DEFINITION 11.3   *A binary vector v of length n is said to satisfy the* **forward propagation rule** *provided that for all $0 \leq p < q < n$ such that $v_p = v_q = 1$ we have that $v_{p+i(q-p)} = 1$ for all integers i satisfying $2 \leq i < \frac{n-p}{q-p}$.*

DEFINITION 11.4   *A binary vector v of length n is said to satisfy the* **backward propagation rule** *provided that for all $0 \leq p < q < \min(n, 2p)$ such that $v_p = v_q = 1$ and $v_{2p-q} = 0$ we have that $v_{p-i(q-p)} = 0$ for all integers i satisfying $2 \leq i \leq \min(\lfloor \frac{p}{q-p} \rfloor, \lfloor \frac{n-p}{q-p} \rfloor)$.*

Example 11.3
The vector $v = 100001001000$ of length $n = 12$ does not satisfy the forward propagation rule since $p = 5$ and $q = 8$ satisfy $v_p = v_q = 1$ but $v_{p+i(q-p)} = v_{11} \neq 1$ when $i = 2$. ▯

THEOREM 11.1
For correlation v of length n the following are equivalent:

1. *There exists a word over the binary alphabet with correlation v.*

2. *There exists a word over some alphabet with correlation v.*

3. *The correlation v satisfies the forward and backward propagation rules.*

REMARK 11.1   As a corollary, we obtain that for any word u over an alphabet A, there exists a binary word v of length $|u|$ such that $\mathcal{P}(v) = \mathcal{P}(u)$, a result that was stated in Chapter 5 (see Theorem 5.1). ▯

Referring to Example 11.3, notice that if a twelve-letter word has periods 5 and 8, then it must also have period 11. Some periods are implied by other periods because of the forward propagation rule.

We begin the process of characterizing the partial word correlations by recording the next lemma which formalizes the relationship between partial words and the words that are compatible with them.

LEMMA 11.1

Let u be a partial word over an alphabet A. Then

$$\mathcal{P}(u) = \bigcup_{w \in C(u) \cap A^*} \mathcal{P}(w)$$

PROOF   Consider first a period p of u. This implies that for each $0 \le i < p$ the partial word $u_{i,p} = u(i)u(i+p)u(i+2p)\ldots$ is 1-periodic, say with letter $c_i \in A$ (if $u_{i,p}$ is a string of $\diamond$'s, then c_i can be chosen as any letter in A). Letting $|u| = mp + r$ for $0 \le r < p$, we see that $u \subset (c_0 c_1 \cdots c_{p-1})^m c_0 c_1 \cdots c_{r-1} = w$. The full word w has period p and is compatible with u.

In the other direction, let w be a full word with period p that is compatible with u. Then $w(i) = u(i)$ for all $i \in D(u)$. If $0 \le i, j < |u|$ with $i \equiv j \bmod p$, we have that $w(i) = w(j)$ by the definition of periodicity. But then if $i, j \in D(u)$ with $i \equiv j \bmod p$, we have that $u(i) = w(i) = w(j) = u(j)$ and thus p is a period of u. ∎

Example 11.4

Consider the partial word $u = abca\diamond cabca$ over the alphabet $A = \{a, b, c\}$. Then $\mathcal{P}(u) = \{3, 6, 9, 10\} = \mathcal{P}(w_1) \cup \mathcal{P}(w_2) \cup \mathcal{P}(w_3)$ where $w_1 = abcaacabca$, $w_2 = abcabcabca$, $w_3 = abcaccabca$ are the words w satisfying $w \in C(u) \cap A^*$. ∎

The characterization of partial word correlations relies on the following concept.

DEFINITION 11.5   *Let u and v be partial words of equal length. The* **greatest lower bound of u and v** *is the partial word $u \wedge v$, where*

$$(u \wedge v) \subset u \text{ and } (u \wedge v) \subset v, \text{ and}$$
$$\text{if } w \subset u \text{ and } w \subset v, \text{ then } w \subset (u \wedge v)$$

Example 11.5

If $u = ab\diamond cdef\diamond\diamond gh$ and $v = acbcdef\diamond fhh$, then we can use the following table to calculate $u \wedge v$:

$$
\begin{array}{rl}
u & = a\ b\ \diamond\ c\ d\ e\ f\ \diamond\ \diamond\ g\ h \\
v & = a\ c\ b\ c\ d\ e\ f\ \diamond\ f\ h\ h \\
\hline
u \wedge v & = a\ \diamond\ \diamond\ c\ d\ e\ f\ \diamond\ \diamond\ \diamond\ h
\end{array}
$$

That is, whenever the partial words differ, we put a $\diamond$. Whenever they are the same, we put their common symbol. ∎

One property we notice immediately about the greatest lower bound and which we leave as an exercise for the reader is the following.

LEMMA 11.2
If u, v are partial words of equal length over an alphabet A, then $\mathcal{P}(u) \cup \mathcal{P}(v) \subset \mathcal{P}(u \wedge v)$ and $\mathcal{P}'(u) \cup \mathcal{P}'(v) \subset \mathcal{P}'(u \wedge v)$.

We are now ready to state the first part of the characterization theorem.

THEOREM 11.2
Let n be a positive integer. Then for any finite collection $u_1, u_2, \ldots, u_k$ of full words of length n over an alphabet A, there exists a partial word w of length n over the binary alphabet with $\mathcal{P}(w) = \mathcal{P}'(w) = \mathcal{P}(u_1) \cup \mathcal{P}(u_2) \cup \cdots \cup \mathcal{P}(u_k)$.

PROOF   The case $k = 1$ follows from Theorem 5.1 and so we assume that $k \geq 2$.

For all integers $p > 0$, define $\langle p \rangle_n$ to be the set of positive integers less than n which are multiples of p. Then

$$\bigcup_{j=1}^{k} \mathcal{P}(u_j) \setminus \{n\} = \bigcup_{p \in P} \langle p \rangle_n$$

for some $P \subset \{1, \ldots, n-1\}$. Thus for all $1 \leq j \leq k$, we assume that $\mathcal{P}(u_j) = \langle p_j \rangle_n \cup \{n\}$ for some $0 < p_j < n$.

With these assumptions, we move on to the case when $k = 2$. For notational clarity we set $u = u_1$, $v = u_2$, $\mathcal{P}(u) \setminus \{n\} = \langle p \rangle_n$ and $\mathcal{P}(v) \setminus \{n\} = \langle q \rangle_n$ for some $0 < p < q < n$. Define

$$w_p = \begin{cases} (ab^{p-1})^m ab^{r-1} & \text{if } r > 0 \\ (ab^{p-1})^m & \text{if } r = 0 \end{cases} \tag{11.4}$$

where $n = mp + r$ with $0 \leq r < p$. Similarly define w_q. Obviously $\mathcal{P}(w_p) = \langle p \rangle_n \cup \{n\}$ and $\mathcal{P}(w_q) = \langle q \rangle_n \cup \{n\}$. Then we claim that $\mathcal{P}(w_p \wedge w_q) = \mathcal{P}(w_p) \cup \mathcal{P}(w_q)$.

By Lemma 11.2 we have $\mathcal{P}(w_p) \cup \mathcal{P}(w_q) \subset \mathcal{P}(w_p \wedge w_q)$. In the other direction, consider $\xi \in \mathcal{P}(w_p \wedge w_q)$. Assume that $\xi \notin \mathcal{P}(w_p) \cup \mathcal{P}(w_q)$. Then by definition we have that neither p nor q divides ξ. Now the first letter of $w_p \wedge w_q$ is a as both w_p and w_q begin with a. Then for all i divisible by ξ we have that $(w_p \wedge w_q)(i)$ is either a or $\diamond$. But both the symbols a and $\diamond$ can appear only where a appears in either w_p or w_q. These occur precisely at the positions j where $p|j$ or $q|j$ respectively. As neither p nor q divides ξ we have that $(w_p \wedge w_q)(\xi) = b$, a contradiction.

Moreover, we see that $w_p \wedge w_q$ has no strictly weak periods. Assume the contrary and let $\xi \in \mathcal{P}'(w_p \wedge w_q) \setminus \mathcal{P}(w_p \wedge w_q)$. Then there exist $i, j \in D(w_p \wedge w_q)$ such that $i \equiv j \bmod \xi$ and $(w_p \wedge w_q)(i) = a$ and $(w_p \wedge w_q)(j) = b$, and for all $0 \leq l < n$ such that $l \equiv i \bmod \xi$ and l is strictly between i and j we have

$l \in H(\omega_p \wedge \omega_q)$. Let l be such that $|i - l|$ is minimized, that is, if $i < j$ then l is minimal and if $i > j$ then l is maximal:

$$
\begin{array}{cc}
i\ j \\
l \to \diamond\ \diamond \\
\vdots\ \ \vdots \\
\diamond\ \diamond \leftarrow l \\
j\ i
\end{array}
$$

This minimal distance is obviously ξ. Then p and q divide i and at least one of them divides l. But we see that only one of p and q divides l, for if both did then $(\omega_p \wedge \omega_q)(l) = a \neq \diamond$. Without loss of generality let $p|l$. But as $p|i$ and $p|l$, we have p divides $|i - l| = \xi$. Then since ω_p is p-periodic, we have that $\omega_p(i') = \omega_p(i) = a$ for all $i' \equiv i \bmod p$. But $j \equiv i \bmod \xi$ and $p|\xi$, so $j \equiv i \bmod p$. Therefore, $\omega_p(j) = a$ and thus $(\omega_p \wedge \omega_q)(j) \neq b$, a contradiction.

Now let $k > 2$ and let $\{p_1, \ldots, p_k\} \subset \{1, \ldots, n-1\}$ be the periods such that $\mathcal{P}(u_j) = \langle p_j \rangle_n \cup \{n\}$. We claim that $\mathcal{P}(\omega_{p_1} \wedge \cdots \wedge \omega_{p_k}) = \mathcal{P}(\omega_{p_1}) \cup \cdots \cup \mathcal{P}(\omega_{p_k})$. But we see the same proof applies. Specifically, $\omega_{p_j}(0) = a$ for all $1 \leq j \leq k$. Moreover, we see that $(\omega_{p_1} \wedge \cdots \wedge \omega_{p_k})(\xi)$ is a or $\diamond$ if and only if $\omega_{p_j}(\xi) = a$ for some $1 \leq j \leq k$. But $\omega_{p_j}(\xi) = a$ if and only if $p_j|\xi$. Thus, if $\xi \in \mathcal{P}(\omega_{p_1} \wedge \cdots \wedge \omega_{p_k}) \setminus \{n\}$ then $p_j|\xi$ for some j, that is, $\xi \in \langle p_j \rangle_n = \mathcal{P}(\omega_{p_j}) \setminus \{n\}$ for some j. The proof of the nonexistence of strictly weak periods translates easily as well. ☐

Theorem 11.2 tells us that every union of the period sets of full words over any alphabet is the period set of a binary partial word. But Lemma 11.1 tells us that the period set of any partial word u over an alphabet A (including the binary alphabet) is the union of the period sets of all full words over A compatible with u. Thus, we have a bijection between these sets which we record as the following corollary.

COROLLARY 11.1
The set of binary correlations of partial words of length n over the binary alphabet is precisely the set of unions of correlations of full words of length n over all nonempty alphabets.

In light of Lemma 11.1, the following corollary is essentially a rephrasing of the previous one:

$$u \text{ is a partial word over } A$$

$$\Downarrow\quad \text{Lemma } 11.1$$

$$\mathcal{P}(u) = \mathcal{P}(u_1) \cup \mathcal{P}(u_2) \cup \cdots \cup \mathcal{P}(u_k)$$

where $u_1, \ldots, u_k$ are the full words over A compatible with u

↓ Theorem 11.2

There exists a partial word v over $\{a,b\}$ with
$$\mathcal{P}(v) = \mathcal{P}'(v) = \mathcal{P}(u_1) \cup \mathcal{P}(u_2) \cup \cdots \cup \mathcal{P}(u_k) = \mathcal{P}(u)$$

COROLLARY 11.2
The set of binary correlations of partial words over an alphabet A with $\|A\| \geq 2$ is the same as the set of binary correlations of partial words over the binary alphabet. Phrased differently, if u is a partial word over an alphabet A, then there exists a binary partial word v of length $|u|$ such that $\mathcal{P}(v) = \mathcal{P}(u)$.

Theorem 11.2 and Corollaries 11.1 and 11.2 give us characterizations of binary correlations of partial words over an arbitrary alphabet. They do not mention at all, though, the effect of strictly weak periods. The second part of the characterization theorem completely characterizes the partial word ternary correlations.

THEOREM 11.3
A ternary vector v of length n is the ternary correlation of a partial word of length n over an alphabet A if and only if $v_0 = 1$ and the following two conditions hold:

1. *If $v_p = 1$, then $v_{ip} = 1$ for all $0 \leq i < \frac{n}{p}$.*

2. *If $v_p = 2$, then $v_{ip} = 0$ for some $2 \leq i < \frac{n}{p}$.*

PROOF Corollaries 11.1 and 11.2 imply the result for the case when $\mathcal{P}'(v) \setminus \mathcal{P}(v) = \emptyset$. For the opposite case, let v satisfy the above conditions along with the assumption that n is at least 3 since the cases of one-letter and two-letter partial words are trivial by simple enumeration considering all possible renamings of letters. So we may now define

$$u = (\bigwedge_{p>0|v_p=1} \omega_p) \wedge (\bigwedge_{p|v_p=2} \psi_p) \tag{11.5}$$

where ω_p is as in Equality 11.4 and

$$\psi_p = ab^{p-1} \diamond b^{n-p-1} \tag{11.6}$$

with distinct letters a, b. We claim that u is a partial word with correlation v.

Set $P = \{p \mid p > 0 \text{ and } v_p = 1\} \cup \{n\}$ and $Q = \{p \mid v_p = 2\}$. By the proof of Theorem 11.2, $P \subset \mathcal{P}(u)$. We show the reverse inclusion by contradiction. Let p be a period of u that is not in P. Since $u(0) = a$, by the definition of periodicity, $u(ip) = a$ or $u(ip) = \diamond$ for all positive integers i. But $u(ip) \neq a$

since $\psi_q(j) \neq a$ for all $q \in Q$ and $j > 0$. So $u(ip) = \diamond$ for all positive i. Specifically, $u(p) = \diamond$ and so p is either in P or Q, but we assumed that $p \notin P$. Then $v_p = 2$ and by our assumptions, there exists an integer i such that $2 \leq i < \frac{n}{p}$ and $v_{ip} = 0$ (or $ip \notin P \cup Q$). But this means by construction that $u(ip) = b$, a contradiction.

Since this gives that $P = \mathcal{P}(u)$ and we have that $P \cup Q \subset \mathcal{P}'(u)$ it suffices to show that if $p \in \mathcal{P}'(u) \backslash \mathcal{P}(u)$ then $p \in Q$. So assume that $p \in \mathcal{P}'(u) \backslash \mathcal{P}(u)$. We have that some $u_i = u(i)u(i+p)u(i+2p) \dots$ contains both a and b. But the only possible location of a is 0, so we may write this as $u(0) = a$, $u(jp) = \diamond$, and $u(kp) = b$ for some $k \geq 2$ and $0 < j < k$. But notice then that u does not have period p, and so $p \notin P$. Thus, since $u(p) = \diamond$, we have that $p \in Q$ and have thus completed this direction of the proof.

Now consider the other direction, that is, if we are given a partial word u with correlation v, then v satisfies the conditions. By Lemma 11.1 we have that Condition 1 must be met. So it suffices to show that Condition 2 holds.

If $v_p = 2$, then there must exist some $0 \leq i < p$ such that two distinct letters a, b appear in u_i. Assume without loss of generality that a appears before b. Let k be a position with letter a in u_i and k' be a position with letter b in u_i, that is, $u(i + kp) = a$ and $u(i + k'p) = b$. Then u is neither $(k' - k)p$-periodic nor $(k' - k)p$-strictly weak periodic, or in other words, $v_{(k'-k)p} = 0$. Thus v satisfies Condition 2 and the result follows. □

Example 11.6
The ternary vector $v_1 = 102000101$ is a valid ternary correlation since it satisfies both Conditions 1 and 2 of Theorem 11.3. However, the vector $v_2 = 102010101$ violates Condition 2 and therefore is not a valid ternary correlation. □

To emphasize the algorithm described in the proof of Theorem 11.3, we record it as follows.

ALGORITHM 11.1
For $n \geq 3$ and $0 < p < n$, let $n = mp + r$ where $0 \leq r < p$. Then define

$$
\omega_p = \begin{cases} (ab^{p-1})^m & \text{if } r = 0 \\ (ab^{p-1})^m ab^{r-1} & \text{if } r > 0 \end{cases}
$$

$$
\psi_p = ab^{p-1} \diamond b^{n-p-1}
$$

Then given a valid ternary correlation v of length n, the partial word

$$
\left(\bigwedge_{p>0|v_p=1} \omega_p \right) \wedge \left(\bigwedge_{p|v_p=2} \psi_p \right)
$$

has ternary correlation v.

Example 11.7
Given $v = 1020000101$, then $ab\diamond bbbb\diamond b\diamond$ has correlation v as is seen by the following computations:

$$\omega_7 = a\;b\;b\;b\;b\;b\;b\;a\;b\;b$$
$$\omega_9 = a\;b\;b\;b\;b\;b\;b\;b\;b\;a$$
$$\psi_2 = a\;b\;\diamond\;b\;b\;b\;b\;b\;b\;b$$
$$\overline{a\;b\;\diamond\;b\;b\;b\;b\;\diamond\;b\;\diamond}$$

▯

In analogy to Corollary 11.2, we record the following.

COROLLARY 11.3
The set of ternary correlations of partial words over an alphabet A with $\|A\| \geq 2$ is the same as the set of ternary correlations of partial words over the binary alphabet. Phrased differently, if u is a partial word over an alphabet A, then there exists a binary partial word v of length $|u|$ such that $\mathcal{P}(v) = \mathcal{P}(u)$ and $\mathcal{P}'(v) = \mathcal{P}'(u)$.

We end this section with the following two remarks.

REMARK 11.2   Note that this corollary was shown true in the case of one hole in Chapter 5 (see Theorem 5.3). ▯

REMARK 11.3   Note that this corollary stipulates that the alphabet A must contain at least two letters. Otherwise, the only possible correlation of length n is 1^n. ▯

11.3   Distributive lattices

Having completely characterized the set of full word correlations of length n as well as the sets of binary and ternary partial word correlations of length n and having shown that all such correlations may be taken as over the binary alphabet, we give these sets names.

We will denote by Γ_n the set of all correlations of full words of length n. We will also denote by Δ_n (respectively, Δ'_n) the set of all partial word binary (respectively, ternary) correlations of length n.

In this section, we study some structural properties of the above mentioned sets. We show that Γ_n, Δ_n and Δ'_n are all lattices under inclusion (suitably defined in the case of Δ'_n). Moreover both Δ_n and Δ'_n are distributive. We start by defining the "distributive lattice" concept.

Let ρ be a binary relation defined on an arbitrary set S, that is, $\rho \subset S \times S$. We recall from Chapter 8, that instead of denoting $(u,v) \in \rho$, we often write $u\rho v$. There, a reflexive, antisymmetric, and transitive relation ρ defined on S was called a *partial ordering*, and (S,ρ) was called a *partially ordered set* or *poset*.

Some examples of posets include the following.

Example 11.8

- Given a set S, the pair $(2^S, \subset)$, where $2^S = \{X \mid X \subset S\}$ is the power set of S and $\subset$ is standard inclusion, is a poset.

- Let $A^n_\diamond$ be the set of partial words of length n over the alphabet A where $A_\diamond = A \cup \{\diamond\}$. Then the pair $(A^n_\diamond, \subset)$, where $\subset$ denotes the "containment," is a poset.

$\square$

If ρ is a partial ordering on a finite set S, we can construct a "Hasse diagram" for ρ on S by drawing a line segment from u up to v if $u, v \in S$ with $u\rho v$ and, most importantly, if there is no other element $w \in S$ such that $u\rho w$ and $w\rho v$ (so there is nothing "in between" u and v). If we adopt the convention of reading the diagram from bottom to top, then it is not necessary to direct any edges.

DEFINITION 11.6   *If (S,ρ) is a poset, then an element $u \in S$ is called a* **maximal element** *of S if for all $w \in S$, $w \neq u$ implies $(u,w) \notin \rho$. Similarly, an element $u \in S$ is called a* **minimal element** *of S if for all $w \in S$, $w \neq u$ implies $(w,u) \notin \rho$. An element $u \in S$ is called a* **null element** *if $u\rho w$ for all $w \in S$. Finally, an element $u \in S$ is called a* **universal element** *if $w\rho u$ for all $w \in S$.*

DEFINITION 11.7

- *A* **join semilattice** *is a poset (S,ρ) such that for all $u,v \in S$, there exists an element $(u \vee v) \in S$, called the* **join** *of u and v, such that $u\rho(u \vee v)$ and $v\rho(u \vee v)$ and for all w with $u\rho w$ and $v\rho w$ we have that $(u \vee v)\rho w$.*

- *A* **meet semilattice** *is a poset (S,ρ) such that for all $u,v \in S$, there exists an element $(u \wedge v) \in S$, called the* **meet** *of u and v, such that*

$(u \wedge v)\rho u$ and $(u \wedge v)\rho v$ and for all w with $w\rho u$ and $w\rho v$ we have that $w\rho(u \wedge v)$.

- A **lattice** *is a poset* (S, ρ) *that is both a join and a meet semilattice.*

We give some examples.

Example 11.9

- The poset $(2^S, \subset)$ is a lattice where the join $\vee$ is set union $\cup$ and the meet $\wedge$ is set intersection $\cap$. It has a null element, $\emptyset$, and a universal element, S.

- The poset $(A^n_\diamond, \subset)$ is a meet semilattice. Recall from Definition 11.5 that for any two partial words u, v over A of length n we have that $u \wedge v$, the greatest lower bound of u and v, is the maximal word which is contained in both u and v. It is represented in Figure 11.1 for $n = 3$ and $A = \{a, b\}$.

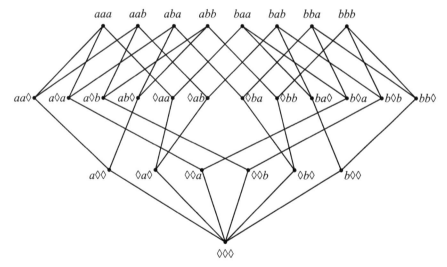

FIGURE 11.1: Meet semilattice $(A^3_\diamond, \subset)$ where $A = \{a, b\}$.

DEFINITION 11.8 *A lattice* (S, ρ) *is called* **distributive** *if for all*

$u, v, w \in S$, the following two equalities hold:

$$u \wedge (v \vee w) = (u \wedge v) \vee (u \wedge w)$$
$$u \vee (v \wedge w) = (u \vee v) \wedge (u \vee w)$$

Example 11.10
The lattice $(2^S, \subset)$ is distributive since the familiar distributive laws of sets hold:

$$X \cap (Y \cup Z) = (X \cap Y) \cup (X \cap Z)$$
$$X \cup (Y \cap Z) = (X \cup Y) \cap (X \cup Z)$$

for all subsets X, Y, Z of S.   ⬜

We now state the Jordan-Dedekind condition.

DEFINITION 11.9   *Let (S, ρ) be a poset. A nonempty subset X of S is called a* **chain** *if for all distinct $u, v \in X$ we have that either $u\rho v$ or $v\rho u$. The* **length** *of a chain is its number of elements minus one. A chain X is called* **maximal** *provided that for all $u, v \in X$ with $u\rho v$ and $w \in S$, if $u\rho w$ and $w\rho v$ then $w \in X$. The poset (S, ρ) is said to satisfy the* **Jordan-Dedekind condition** *if all maximal chains between two elements of S are of equal length.*

Returning to the poset $(A_\diamond^3, \subset)$ of Figure 11.1, it is easy to check that it satisfies the Jordan-Dedekind condition. For instance, all maximal chains between $\diamond\diamond\diamond$ and abb have length 3. They are

$$X_1 = \{\diamond\diamond\diamond, \diamond\diamond b, a\diamond b, abb\}$$
$$X_2 = \{\diamond\diamond\diamond, a\diamond\diamond, ab\diamond, abb\}$$
$$X_3 = \{\diamond\diamond\diamond, \diamond b\diamond, \diamond bb, abb\}$$

REMARK 11.4   If a poset violates the Jordan-Dedekind condition then the poset is not distributive.   ⬜

In the next two sections, we will give partial word counterparts to the following theorem which is left as an exercise. For $u, v \in \Gamma_n$, define $u \subset v$ if $\mathcal{P}(u) \subset \mathcal{P}(v)$.[1]

THEOREM 11.4
The pair $(\Gamma_n, \subset)$ is a lattice which does not satisfy the Jordan-Dedekind condition.

---

[1]Do not confuse "correlation u is contained in correlation v" as defined here with "partial word u is contained in partial word v" as defined in Chapter 1.

11.3.1 Δ_n is a distributive lattice

For $u, v \in \Delta_n$, define $u \subset v$ if $\mathcal{P}(u) \subset \mathcal{P}(v)$, and $p \in u$ if $p \in \mathcal{P}(u)$.

THEOREM 11.5

The pair $(\Delta_n, \subset)$ is a lattice.

- *The meet of u and v, $u \cap v$, is the unique vector in Δ_n such that $\mathcal{P}(u \cap v) = \mathcal{P}(u) \cap \mathcal{P}(v)$.*

- *The join of u and v, $u \cup v$, is the unique vector in Δ_n such that $\mathcal{P}(u \cup v) = \mathcal{P}(u) \cup \mathcal{P}(v)$.*

- *The null element is 10^{n-1}.*

- *The universal element is 1^n.*

PROOF   We leave it to the reader to show that the pair $(\Delta_n, \subset)$ is a poset with null element 10^{n-1} and universal element 1^n. First, if $u, v \in \Delta_n$ then $(u \cap v) \in \Delta_n$. To see this, notice that if $p \in (u \cap v)$ then $p \in u$ and $p \in v$. Thus $\langle p \rangle_n \subset \mathcal{P}(u)$ and $\langle p \rangle_n \subset \mathcal{P}(v)$. So $\langle p \rangle_n \subset \mathcal{P}(u \cap v)$ and by Theorem 11.3 we have that $u \cap v$ is a valid binary correlation. Second, if $u, v \in \Delta_n$ then $(u \cup v) \in \Delta_n$. Indeed, if $p \in u$ then $\langle p \rangle_n \subset \mathcal{P}(u)$. Similarly if $p \in v$ then $\langle p \rangle_n \subset \mathcal{P}(v)$. Thus, if $p \in (u \cup v)$ then $\langle p \rangle_n \subset \mathcal{P}(u \cup v)$. Thus, by Theorem 11.3 we have that $u \cup v$ is a valid binary correlation. ⬚

Figure 11.2 depicts Δ_6, the set of partial word binary correlations of length 6, as a lattice and Figure 11.3 its associated nontrivial period sets.

Since the meet and the join of binary correlations are the *set intersection* and *set union* of the correlations, we have the following theorem.

THEOREM 11.6

The lattice $(\Delta_n, \subset)$ is distributive.

11.3.2 Δ'_n is a distributive lattice

We now expand our considerations to Δ'_n, the set of ternary correlations of partial words of length n, and show that Δ'_n is a lattice again with respect to inclusion, which we define suitably.

For $u, v \in \Delta'_n$, define $u \subset v$ if $\mathcal{P}(u) \subset \mathcal{P}(v)$ and $\mathcal{P}'(u) \subset \mathcal{P}'(v)$. Equivalently, $u \subset v$ provided that whenever $u_i > 0$ we have that $u_i \geq v_i > 0$. Or more explicitly, $u \subset v$ if the following two conditions hold:

- If $u_i = 1$, then $v_i = 1$.

- If $u_i = 2$, then $v_i = 1$ or $v_i = 2$.

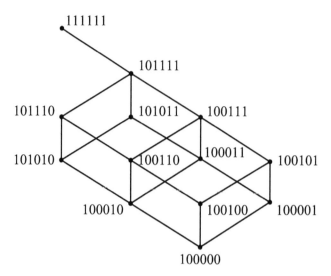

FIGURE 11.2:   A representation of the lattice Δ_6.

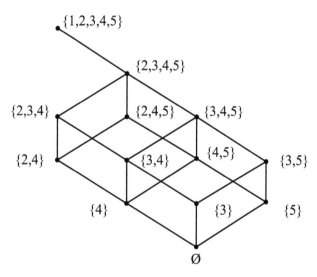

FIGURE 11.3:   The associated nontrivial period sets of Figure 11.2.

Under these definitions, we have the following lemma.

LEMMA 11.3

The pair $(\Delta'_n, \subset)$ is a poset. Its null element is 10^{n-1} and its universal element is 1^n.

PROOF If $u \in \Delta'_n$, then $u \subset u$ and so reflexivity holds.

For antisymmetry, consider $u, v \in \Delta'_n$ such that $u \subset v$ and $v \subset u$. Then whenever $u_i = 0$ we have that $v_i = 0$ since $v \subset u$. Moreover, whenever $u_i = 1$ we must have that $v_i = 1$ since $u \subset v$. Finally, whenever $u_i = 2$ we have that $v_i = 1$ or $v_i = 2$ since $u \subset v$ and that $v_i \neq 1$ since $v \subset u$. Thus, $v_i = 2$. Therefore, $u = v$.

For transitivity, let $u, v, w \in \Delta'_n$ satisfy $u \subset v$ and $v \subset w$. When $u_i = 1$ we have that $v_i = 1$, and so $w_i = 1$. And when $u_i = 2$ we have that $v_i = 1$ or $v_i = 2$. In the first case we have that $w_i = 1$ and in the second case we have that $w_i = 1$ or $w_i = 2$. Thus, in either case, $u_i \geq w_i > 0$. The inclusion $u \subset w$ follows. $\square$

Consider ternary correlations $u, v \in \Delta'_n$. We define the intersection of u and v as the ternary vector $u \cap v$ such that $\mathcal{P}(u \cap v) = \mathcal{P}(u) \cap \mathcal{P}(v)$ and $\mathcal{P}'(u \cap v) = \mathcal{P}'(u) \cap \mathcal{P}'(v)$. Equivalently,

$$(u \cap v)_i = \begin{cases} 0 & \text{if either } u_i = 0 \text{ or } v_i = 0 \\ 1 & \text{if } u_i = v_i = 1 \\ 2 & \text{otherwise} \end{cases} \qquad (11.7)$$

LEMMA 11.4

The set Δ'_n is closed under intersection.

PROOF Let $u, v \in \Delta'_n$. If $p \in \mathcal{P}(u \cap v)$ then $u_p = v_p = 1$, and so $u_{ip} = v_{ip} = 1$ and equivalently $(u \cap v)_{ip} = 1$ for all multiples ip of p. Moreover, if $p \in \mathcal{P}'(u \cap v) \setminus \mathcal{P}(u \cap v)$ then u_p or v_p is 2. Without loss of generality, assume that $u_p = 2$. Then Theorem 11.3 implies that for some multiple ip of p we have that $u_{ip} = 0$. But this means that $(u \cap v)_{ip} = 0$ and so Theorem 11.3 implies that $(u \cap v) \in \Delta'_n$. $\square$

We may define the union in the analogous way. Specifically, for $u, v \in \Delta'_n$, $\mathcal{P}(u \cup v) = \mathcal{P}(u) \cup \mathcal{P}(v)$ and $\mathcal{P}'(u \cup v) = \mathcal{P}'(u) \cup \mathcal{P}'(v)$. Equivalently, $u \cup v$ is the ternary vector satisfying

$$(u \cup v)_i = \begin{cases} 0 & \text{if } u_i = v_i = 0 \\ 1 & \text{if either } u_i = 1 \text{ or } v_i = 1 \\ 2 & \text{otherwise} \end{cases} \qquad (11.8)$$

However Δ'_n is not closed under union. Indeed, the union of the two cor-relations $u = 102000101$ and $v = 100010001$ is $(u \cup v) = 102010101$, which violates the second condition of Theorem 11.3. Indeed, there is no $i \geq 2$ such that $(u \cup v)_{i2} = 0$. On the other hand, we can modify the union slightly such that we obtain the join constructively. If we simply change $(u \cup v)_2$ from 2 to 1, then we have created the valid ternary correlation 101010101. Calling this vector $u \vee v$, we see that $u \subset (u \vee v)$ and that $v \subset (u \vee v)$.

THEOREM 11.7

The poset $(\Delta'_n, \subset)$ is a lattice.

- *The meet of u and v, $u \wedge v$, is the unique vector in Δ'_n defined by Equality 11.7.*

- *The join of u and v, $u \vee v$, is the unique vector in Δ'_n defined by*

$$\mathcal{P}'(u \vee v) = \mathcal{P}'(u) \cup \mathcal{P}'(v)$$

and

$$\mathcal{P}(u \vee v) = \mathcal{P}(u) \cup \mathcal{P}(v) \cup B(u \cup v)$$

where $B(u \cup v)$ is the set of all $0 < p < n$ such that $(u \cup v)_p = 2$ and there exists no $i \geq 2$ satisfying $(u \cup v)_{ip} = 0$.

PROOF   The proof is analogous to the proof of Theorem 11.5 except this time we do not have the union of the two correlations to explicitly define the join. One method of proving that the join exists is to notice that the join of $u, v \in \Delta'_n$ is the intersection of all elements of Δ'_n which contain u and v. This intersection is guaranteed to be nonempty since Δ'_n contains a universal element. Note that $B(u \cup v)$ is the set of positions in $u \cup v$ which do not satisfy the second condition of Theorem 11.3.

We claim that $u \vee v$ is the unique join of u and v (and thus justify the use of the traditional notation $\vee$ for the binary operation). Notice first that since $\mathcal{P}(u \cup v) = \mathcal{P}(u) \cup \mathcal{P}(v)$ and $\mathcal{P}'(u \cup v) = \mathcal{P}'(u) \cup \mathcal{P}'(v)$, we have that $(u \cup v) \subset (u \vee v)$. Thus we have that $u \subset (u \cup v) \subset (u \vee v)$ and that $v \subset (u \cup v) \subset (u \vee v)$. We also see that $(u \vee v) \in \Delta'_n$. This follows from the fact that if $p \in \mathcal{P}(u \vee v)$ then either $p \in \mathcal{P}(u) \cup \mathcal{P}(v)$ or for all $i \geq 1$ we have that $ip \in \mathcal{P}'(u) \cup \mathcal{P}'(v)$, and if $p \in \mathcal{P}'(u \vee v) \setminus \mathcal{P}(u \vee v)$ then $(u \cup v)_p = 2$ and $(u \cup v)_{ip} = 0$ for some $i \geq 2$, and so $(u \vee v)_{ip} = 0$ for some $i \geq 2$. In the first case where $p \in \mathcal{P}(u) \cup \mathcal{P}(v)$, we have that $\langle p \rangle_n \subset \mathcal{P}(u) \cup \mathcal{P}(v) \subset \mathcal{P}(u \vee v)$. In the second case where for all $i \geq 1$ we have that $ip \in \mathcal{P}'(u) \cup \mathcal{P}'(v)$, by the definition of $u \vee v$ and the fact that the multiples of all multiples of p are again multiples of p, we must have that $\langle p \rangle_n \subset \mathcal{P}(u \vee v)$. Thus, using the

$\vee$ operator instead of the $\cup$ operator resolves all conflicts with Theorem 11.3 and so $(u \vee v) \in \Delta'_n$. From here it suffices to show that $(u \vee v)$ is minimal.

Let $w \in \Delta'_n$ such that $u \subset w$ and $v \subset w$ and $w \subset (u \vee v)$. We must show that $w = (u \vee v)$. Note first that if $u_i = v_i = 0$ then $(u \vee v)_i = 0$, and so $w_i = 0$. Moreover, if $u_i = 1$ or $v_i = 1$ then $(u \vee v)_i = 1$ by construction, and also $w_i = 1$ by the definition of inclusion. Finally, we must consider the case when at least one of u_i and v_i is 2 while the other is either 0 or 2. In this case we have by the definition of inclusion that $w_i = 1$ or $w_i = 2$. If $w_i = 2$, then there must be some $k \geq 2$ such that $w_{ki} = 0$, and thus $u_{ki} = v_{ki} = 0$. Therefore, $(u \vee v)_{ki} = 0$ and $(u \vee v)_i = 2$. On the other hand, if $w_i = 1$, then $(u \vee v)_i = 1$ since $w \subset (u \vee v)$. Thus, $w = (u \vee v)$. ☐

Figure 11.4 depicts Δ'_5, the set of valid ternary correlations of length 5, as a lattice and Figure 11.5 its associated nontrivial period and weak period sets.

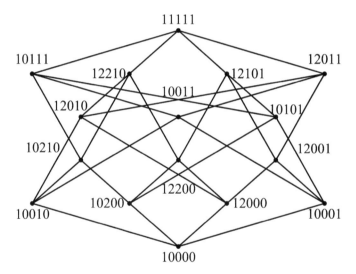

FIGURE 11.4: A representation of the lattice Δ'_5.

Strangely, even though the join operation of Δ'_n is more complicated than the join operation of Δ_n, we still have that Δ'_n is distributive and thus satisfies the Jordan-Dedekind condition. This is stated in the following theorem.

THEOREM 11.8

The lattice $(\Delta'_n, \subset)$ is distributive.

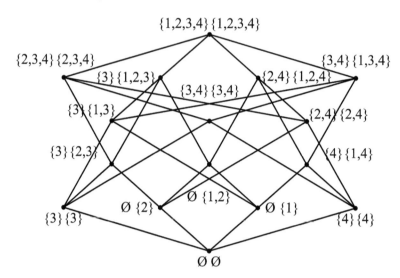

FIGURE 11.5:   The associated nontrivial period and weak period sets of
Figure 11.4.

PROOF   By definition, we must show the following two equalities:

$$u \wedge (v \vee w) = (u \wedge v) \vee (u \wedge w) \qquad (11.9)$$

$$u \vee (v \wedge w) = (u \vee v) \wedge (u \vee w) \qquad (11.10)$$

for all $u, v, w \in \Delta'_n$. We recall first that the archetypal distributive lattice
is a subset of a power set closed under set theoretic union and intersection.
Since the sets of weak periods of the meet and join of two ternary correlations
are defined as the intersection and union of the weak period sets of the two
correlations, we need not worry about showing the definition of equality for
the sets of weak periods. That is, the only difference in either equality between
the left and right hand sides could be in the sets of periods.

Consider first Equality 11.9. We must show that $p \in \mathcal{P}(u \wedge (v \vee w)) = \mathcal{P}(u) \cap$
$\mathcal{P}(v \vee w)$ if and only if $p \in \mathcal{P}((u \wedge v) \vee (u \wedge w))$. We note that $p \in \mathcal{P}(u) \cap \mathcal{P}(v \vee w)$
if and only if $p \in \mathcal{P}(u)$ and $p \in \mathcal{P}(v \vee w)$. But $p \in \mathcal{P}(v \vee w)$ if and only if
either $p \in \mathcal{P}(v) \cup \mathcal{P}(w)$ *or* for all $i \geq 1$ we have that $ip \in \mathcal{P}'(v) \cup \mathcal{P}'(w)$. In
the first case, p is in one of $\mathcal{P}(u \wedge v)$ and $\mathcal{P}(u \wedge w)$ and is thus in the union.
In the second case, we see that since $p \in \mathcal{P}(u)$ that $\langle p \rangle_n \subset \mathcal{P}(u) \subset \mathcal{P}'(u)$.
Therefore, for all $i \geq 1$ we have that $ip \in \mathcal{P}'(u) \cap \mathcal{P}'(v)$ or $ip \in \mathcal{P}'(u) \cap \mathcal{P}'(w)$,
and $ip \in \mathcal{P}'(u \wedge v) \cup \mathcal{P}'(u \wedge w)$. Thus, by the definition of $\vee$, we have that
$p \in \mathcal{P}((u \wedge v) \vee (u \wedge w))$. But all these assertions are bidirectional implications,
and therefore we have the equality we seek.

Next consider Equality 11.10. We must show that $p \in \mathcal{P}(u \vee (v \wedge w))$ if
and only if $p \in \mathcal{P}((u \vee v) \wedge (u \vee w)) = \mathcal{P}(u \vee v) \cap \mathcal{P}(u \vee w)$. Assume that

$p \in \mathcal{P}(u \vee (v \wedge w))$. If $p \in \mathcal{P}(u)$ or $p \in \mathcal{P}(v \wedge w) = \mathcal{P}(v) \cap \mathcal{P}(w)$ then we are done. Otherwise, for all $i \geq 1$ we have that $ip \in \mathcal{P}'(u) \cup \mathcal{P}'(v \wedge w) = \mathcal{P}'(u) \cup (\mathcal{P}'(v) \cap \mathcal{P}'(w)) = (\mathcal{P}'(u) \cup \mathcal{P}'(v)) \cap (\mathcal{P}'(u) \cup \mathcal{P}'(w))$. But then we have

$$ip \in \mathcal{P}'(u) \cup \mathcal{P}'(v) \text{ and } ip \in \mathcal{P}'(u) \cup \mathcal{P}'(w)$$

for all $i \geq 1$, thus $p \in \mathcal{P}(u \vee v) \cap \mathcal{P}(u \vee w)$. The proof in the other direction is the same and thus Equality 11.10 holds. □

So unlike the lattice of correlations of full words which does not even satisfy the Jordan-Dedekind condition, the lattices of both binary and ternary correlations of partial words are distributive.

11.4   Irreducible period sets

We notice that in the case of full words, some periods are implied by other periods because of the forward propagation rule (see Definition 11.3). For instance, if a twelve-letter word has periods 7 and 9, then it must also have period 11 because $11 = 7 + 2(9 - 7)$. The period set $\{7, 9, 11\}$ can then be reduced to the set $\{7, 9\}$, while the latter is *irreducible*.

We will denote by Λ_n the set of these irreducible period sets of full words of length n. We invite the reader to prove the following proposition (we will prove its partial word counterpart later in this section).

PROPOSITION 11.1
The pair $(\Lambda_n, \subset)$ is not a lattice but does satisfy the Jordan-Dedekind condition as a poset.

The forward propagation rule does not hold in the case of partial words. For example, $abbbbbb\diamond b\diamond bb$ has periods 7 and 9 but does *not* have period 11. Thus, $\{7, 9, 11\}$ is irreducible in the sense of partial words, but not in the sense of full words.

This leads us to the notion of *generating set*.

DEFINITION 11.10   *A set $P \subset \{1, \ldots, n-1\}$ **generates** the correlation $v \in \Delta_n$ provided that for each $0 < i < n$ we have that $v_i = 1$ if and only if there exists $p \in P$ and $0 < k < \frac{n}{p}$ such that $i = kp$.*

One such P is $\mathcal{P}(v) \setminus \{n\}$. But in general there are strictly smaller P which have this property. For example, if $v = 1001001101$ then

$$\{3, 6, 7, 9\}$$

$$\{3, 6, 7\}$$
$$\{3, 7, 9\}$$
$$\{3, 7\}$$

generate v. The set $\{3, 7\}$ is the *minimal generating set* of v.

For every $v \in \Delta_n$, there is a well defined minimal generating set for v as stated in the next lemma.

LEMMA 11.5

For every $v \in \Delta_n$, there exists a unique set P that generates v and is such that for all sets P' that generate v we have that $P \subset P'$. Namely, P is the set of all $p \in \mathcal{P}(v) \setminus \{n\}$ such that for all $q \in \mathcal{P}(v) \setminus \{n\}$ with $q \neq p$ we have that q does not divide p.

PROOF   If there exists q distinct from p such that q divides p, then $\langle p \rangle_n \subset \langle q \rangle_n$. Moreover, since there are no divisors of the elements of P in $\mathcal{P}(v) \setminus \{n\}$ the only $p \in \mathcal{P}(v) \setminus \{n\}$ which can generate $r \in P$ is r itself. Thus we have achieved minimality.   ▯

We call the unique minimal generating set P of Lemma 11.5 the *irreducible period set* of v and denote it by $R(v)$. In the example above where $v = 1001001101$, $R(v) = \{3, 7\}$ and $\{3, 7\}$ is the irreducible period set of v.

We will denote by Φ_n the set of irreducible period sets of partial words of length n. There is an obvious one-to-one correspondence (or bijection) between Δ_n and Φ_n given by

$$R : \Delta_n \rightarrow \Phi_n$$
$$v \mapsto R(v)$$

$$E : \Phi_n \rightarrow \Delta_n$$
$$P \mapsto \bigcup_{p \in P} \langle p \rangle_n$$

For instance, the correspondence between Δ_6 and Φ_6 is as depicted in Figure 11.6 (to make it easier to read, we have deleted the trivial period 6 from the period sets).

For $n \geq 3$, we see immediately that the poset $(\Phi_n, \subset)$ *is not* a join semilattice since the sets $\{1\}$ and $\{2\}$ do not have a join because $\{1\}$ is maximal. On the other hand, the following holds.

PROPOSITION 11.2

The pair $(\Phi_n, \subset)$ is a meet semilattice that satisfies the Jordan-Dedekind condition. Here the null element is $\emptyset$, and the meet of two elements is simply their intersection.

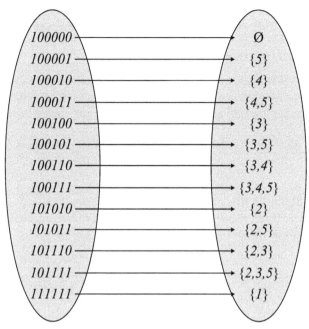

FIGURE 11.6: Bijective correspondence between Δ_6 and Φ_6.

PROOF The proof is left as an exercise for the reader. ☐

Figure 11.7 depicts Φ_6 as a meet semilattice.

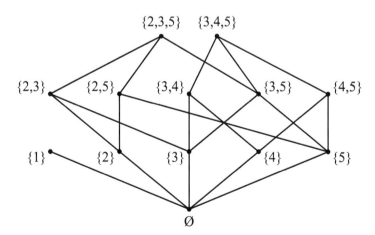

FIGURE 11.7: A representation of the meet semilattice Φ_6.

11.5   Counting correlations

In this section we look at the number of partial word correlations of a given length. In the case of binary correlations, we give bounds and link the problem to one in number theory, and in the case of ternary correlations we give an exact count.

To begin, we recall the definition of a primitive set of integers from number theory.

DEFINITION 11.11   *Let S be a subset of* $\mathbb{N} = \{1, 2, \ldots\}$. *We say that S is* **primitive** *if for any two distinct elements* $s, s' \in S$ *we have that neither s divides s' nor s' divides s.*

Example 11.11
The sets $\emptyset, \{1\}$ and $\{p \mid p$ is prime$\}$ are examples of primitive sets.   ⬚

The amazing thing is that the irreducible period sets of correlations $v \in \Delta_n$ are precisely the finite primitive subsets of $\{1, 2, \ldots, n-1\}$. So if we can count the number of finite primitive sets of integers less than n then we can count the number of partial word binary correlations of length n. We present some results on approximating this number.

THEOREM 11.9
Let S be a finite primitive set of size k with elements less than n. Then $k \le \lfloor \frac{n}{2} \rfloor$. *Moreover, this bound is sharp.*

PROOF   We show this by induction. First note that the basis when $n = 2$ is obvious since the only such (nonempty) set is $\{1\}$. So consider $n > 2$. Then the inductive hypothesis is that the maximum size of a finite primitive set with elements less than $n - 1$ is $\lfloor \frac{n-1}{2} \rfloor$. This tells us that for any primitive set with elements less than n, the subset of elements less than $n - 1$ can have a maximum size of $\lfloor \frac{n-1}{2} \rfloor$. Thus, the whole set can have maximum length $\lfloor \frac{n-1}{2} \rfloor + 1$. When n is even, $\lfloor \frac{n-1}{2} \rfloor + 1 = \lfloor \frac{n}{2} \rfloor$ and we are done.
So consider the case when n is odd. We show that given a primitive set of size $\lfloor \frac{n-1}{2} \rfloor$ with elements less than $n - 1$ that $n - 1$ cannot be added to this set. First notice that this statement is true for $n = 3$. So taking $n \ge 5$, we see that if S is a maximal primitive set with elements less than $n - 1$ then $1 \notin S$ since 1 divides all integers and the set $\{2, 3\}$ is primitive and of size greater than the set $\{1\}$. We claim that if $\frac{n-1}{2}$ is prime then $\frac{n-1}{2} \in S$. We show this by demonstrating that neither a divisor nor a multiple of $\frac{n-1}{2}$ can lie in S. Indeed, the only proper divisor of $\frac{n-1}{2}$ is 1 and we have shown that $1 \notin S$.

Moreover, the least proper multiple of $\frac{n-1}{2}$ is $n-1$ itself. Thus, if $\frac{n-1}{2} \notin S$ then we may add it to S and obtain a strictly larger set than S.

If $\frac{n-1}{2}$ is not prime and $\frac{n-1}{2} \in S$, then $n-1$ cannot be added to S. If $\frac{n-1}{2}$ is not prime and $\frac{n-1}{2} \notin S$, then some proper divisor of $\frac{n-1}{2}$ must be in S. For if not we could again increase the size of the set while maintaining primitivity simply by adding $\frac{n-1}{2}$ since the least multiple of $\frac{n-1}{2}$ is $n-1$. But every divisor of $\frac{n-1}{2}$ is again a divisor of $n-1$, so $n-1$ cannot be added to S. Thus the inductive step is proven and first statement of the lemma follows.

For the sharpness of the bound consider the set of integers which are greater than or equal to $\lfloor \frac{n+1}{2} \rfloor$ and which are less than n. All multiples of each element of this set are at least n. Therefore, this is a primitive set of the desired size.
$\square$

This bound shows that the number of partial word binary correlations of length n is *at most* the number of subsets of $\{1, 2, \ldots, n-1\}$ of size at most $\lfloor \frac{n}{2} \rfloor$. This number is

$$\sum_{i=0}^{\lfloor \frac{n}{2} \rfloor} \binom{n-1}{i} = \begin{cases} 2^{n-2} + \frac{1}{2}\binom{n-1}{\frac{n-1}{2}} & \text{if } n \text{ is odd} \\ 2^{n-2} + \binom{n-1}{\lfloor \frac{n-1}{2} \rfloor} & \text{if } n \text{ is even} \end{cases} \tag{11.11}$$

Moreover, the sharpness of the bound derived in Theorem 11.9 gives us that

$$\|\Delta_n\| \geq 2^{\lfloor \frac{n}{2} \rfloor}$$

Thus

$$\frac{\ln 2}{2} \leq \frac{\ln \|\Delta_n\|}{n} \leq \ln 2$$

The bounds we give show explicitly that $\ln \|\Delta_n\| \in \Theta(n)$. [2]

Several values of this sequence are listed in Table 11.1. The time and space needed to continue this sequence farther is very great and so the problem does not lend itself to much empirical observation.

We now show that the set of partial word ternary correlations is actually much more tractable to count than the set of partial word binary correlations. Specifically, we show that $\|\Delta'_n\| = 2^{n-1}$.

To this end we first note an interesting consequence of Theorem 11.3.

LEMMA 11.6
Let u be a partial word of length n and let $p \in \mathcal{P}'(u)$. Then $p \in \mathcal{P}(u)$ if and only if $ip \in \mathcal{P}'(u)$ for all $0 < i \leq \lfloor \frac{n}{p} \rfloor$. That is, a weak period is a strong period if and only if all of its multiples are also weak periods.

---

[2] For $f, g : \mathbb{Z}^+ \rightarrow \mathbb{R}$, we say that f is "big theta of g," and write $f \in \Theta(g)$, if there exist $k_1, k_2 \in \mathbb{R}^+$ and a positive integer N such that $k_1|g(n)| \leq |f(n)| \leq k_2|g(n)|$ for all integers $n \geq N$.

TABLE 11.1: Number of primitive sets of integers less than n.

n	Number	n	Number	n	Number
1	1	15	733	29	355729
2	2	16	1133	30	711457
3	3	17	1529	31	879937
4	5	18	3057	32	1759873
5	7	19	3897	33	2360641
6	13	20	7793	34	3908545
7	17	21	10241	35	5858113
8	33	22	16513	36	10534337
9	45	23	24593	37	12701537
10	73	24	49185	38	25403073
11	103	25	59265	39	38090337
12	205	26	109297	40	63299265
13	253	27	163369	41	81044097
14	505	28	262489	42	162088193

PROOF　If $p \in \mathcal{P}'(u)$ and all of its multiples are also in $\mathcal{P}'(u)$, then we have by Theorem 11.3 that $p \notin \mathcal{P}'(u) \setminus \mathcal{P}(u)$. Thus, $p \in \mathcal{P}(u)$. On the other hand, if $p \in \mathcal{P}(u)$ then we have again by Theorem 11.3 that all of its multiples are in $\mathcal{P}(u) \subset \mathcal{P}'(u)$. Therefore, the lemma follows. ☐

This lemma leads us to the following.

LEMMA 11.7

If $S \subset \{1, 2, \ldots, n-1\}$, then there is a unique ternary correlation $v \in \Delta'_n$ such that $\mathcal{P}'(v) \setminus \{n\} = S$.

PROOF　For each $p \in S$, let $v_p = 1$ provided that all of the multiples of p are in S and let $v_p = 2$ provided that there is some multiple of p which is not in S. For all other $0 < p < n$, let $v_p = 0$. Notice that v satisfies the conditions of Theorem 11.3 to belong to Δ'_n. Moreover, it is obvious that these are the conditions *forced* on the ternary vector by Theorem 11.3. Thus, this correlation is unique. ☐

We note that Lemma 11.7 agrees with the definition of the join forced upon us in Section 11.3. Considering all periods as *weak* periods and then determining which ones are actually strong periods is how we defined that operation.

So the cardinality of the set of partial word ternary correlations is the same as the cardinality of the power set of $\{1, 2, \ldots, n-1\}$.

PROPOSITION 11.3
The equality $\|\Delta'_n\| = 2^{n-1}$ *holds.*

Table 11.2 lists all the 32 partial word ternary correlations of Δ'_6.

TABLE 11.2: The partial word ternary correlations of Δ'_6.

100000	101010	111111	120111
100001	101011	120000	121010
100010	101110	120001	121011
100011	101111	120010	121110
100100	102000	120011	122000
100101	102001	120100	122001
100110	102100	120101	122100
100111	102101	120110	122101

Exercises

11.1 What is the binary correlation of the partial word $u = aaba\diamond bab\diamond a\diamond a$? What is the ternary correlation of u?

11.2 Does the vector $v = 100001001000$ satisfy the backward propagation rule?

11.3 Is the ternary vector $v = 1002000001$ a valid ternary correlation?

11.4 Give an example of a nonvalid binary correlation and an example of a nonvalid ternary correlation.

11.5 $\boxed{\text{s}}$ What is the *population size* of the correlation 102100 over the alphabet $\{a, b\}$, that is, what is the number of partial words over $\{a, b\}$ sharing the correlation 102100?

11.6 Prove Lemma 11.2.

11.7 Run Algorithm 11.1 on the valid ternary correlation $v = 122211011$.

11.8 Show that the pair $(\Delta_n, \subset)$ is a poset with null element 10^{n-1} and universal element 1^n.

11.9 Show that if (S, ρ) is a poset with a null element, then it is unique. Show a similar statement if (S, ρ) has a universal element.

11.10 Is the poset $(A_\diamond^n, \subset)$ a join semilattice with $u \vee v$ defined as the least upper bound of u and v?

11.11 $\boxed{\text{s}}$ Show that if $v \in \Delta_n'$, then $v_p \neq 2$ for all $p > \lfloor \frac{n-1}{2} \rfloor$.

11.12 $\boxed{\text{H}}$ List all the correlations in Δ_4' and their corresponding nontrivial period and weak period sets.

11.13 Draw Δ_4' as a lattice.

Challenging exercises

11.14 $\boxed{\text{H}}$ Represent the lattice Γ_9 in a Hasse diagram. Show two maximal chains of different lengths between 10^8 and 1^9.

11.15 Prove that if a poset violates the Jordan-Dedekind condition, then the poset is not distributive.

11.16 Prove Theorem 11.4.

11.17 Show that Equality 11.11 holds.

11.18 $\boxed{\text{s}}$ While there is a natural bijection between the lattice Δ_6 and the meet semilattice Φ_6 given by the maps R and E, show that these maps are not morphisms.

11.19 Define $\varphi : \Delta_n' \to A_\diamond^n$ by

$$v \mapsto \left(\bigwedge_{p \in \mathcal{P}(v) \setminus \{n\}} \omega_p \right) \wedge \left(\bigwedge_{p \in \mathcal{P}'(v) \setminus \mathcal{P}(v)} \psi_p \right)$$

The proof of Theorem 11.3 shows that φ is a lattice morphism from the join semilattice Δ_n' to the meet semilattice $A_\diamond^n$. Verify that for all $v, w \in \Delta_n'$, we have that $\varphi(v \vee w) = \varphi(v) \wedge \varphi(w)$.

11.20 Prove Proposition 11.1.

11.21 Show that for any $v \in \Gamma_n$, based on forward propagation an irreducible period set associated with v exists and is unique.

11.22 $\boxed{\text{s}}$ Prove Proposition 11.2.

11.23 $\boxed{\text{H}}$ Referring to Table 11.1, find a function f that approximates the number of primitive sets of integers less than n.

Programming exercises

11.24 Design an applet that provides an implementation of Algorithm 11.1, that is, given as input a valid ternary correlation v of length n, the applet outputs the partial word u in Equality 11.5 with correlation v.

11.25 Design an applet that when given as input a partial word u over an alphabet A, outputs the ternary correlation of u.

11.26 Write a program that counts the number of partial words over the alphabet $\{a, b\}$ sharing a given ternary correlation. Run your program on correlation 10200101.

11.27 Write a program that lists all the valid ternary correlations of a given length. Run your program on length 7.

11.28 In Chapter 5, we showed that given a partial word u with one hole, we can compute a partial word v over the binary alphabet such that $\mathcal{P}(v) = \mathcal{P}(u)$, $\mathcal{P}'(v) = \mathcal{P}'(u)$, and $H(v) \subset H(u)$. This last condition cannot be satisfied in the two-hole case. Write a program to check that the pword $abaca\diamond\diamond acaba$ has no such binary reduction.

Website

A World Wide Web server interface at

$$\texttt{http://www.uncg.edu/mat/research/correlations}$$

has been established for automated use of a program that when given a partial word u over an alphabet A, computes a binary partial word v of length $|u|$ such that $\mathcal{P}(v) = \mathcal{P}(u)$ and $\mathcal{P}'(v) = \mathcal{P}'(u)$. Another website related to correlations of partial words is

$$\texttt{http://www.uncg.edu/cmp/research/correlations2}$$

Bibliographic notes

In [82], Guibas and Odlyzko considered the period sets of full words of length n over a finite alphabet, and specific representations of them, *(auto) correlations*, which are bit vectors of length n indicating the periods. Among the possible 2^n binary vectors, only a small subset are valid correlations. There, they provided characterizations of correlations (Theorem 11.1), asymptotic bounds on their number, and a recurrence for the population size of a correlation, that is, the number of full words sharing a given correlation.

In [125], Rivals and Rahmann showed that there is redundancy in period sets and introduced the notion of an *irreducible* period set based on the forward propagation rule (Proposition 11.1). They proved that Γ_n, the set of all correlations of full words of length n, is a lattice under set inclusion and does not satisfy the Jordan-Dedekind condition (Theorem 11.4). They proposed the first efficient enumeration algorithm for Γ_n and improved upon the previously known asymptotic lower bounds on the cardinality of Γ_n. Finally, they provided a new recurrence to compute the number of full words sharing a given period set, and exhibited an algorithm to sample uniformly period sets through irreducible period sets.

In [31], Blanchet-Sadri, Gafni and Wilson introduced partial word binary and ternary correlations and all results on such correlations discussed in this chapter are from there. The bound on the size of primitive sets with elements less than n (Theorem 11.9) is due to Erdös [76].

Chapter 12

Unavoidable Sets of Partial Words

The notion of an unavoidable set of words appears frequently in the fields of mathematics and theoretical computer science, in particular with its connection to the study of combinatorics on words. The theory of unavoidable sets has seen extensive study over the past twenty years. An *unavoidable* set of words X over an alphabet A is a set for which any sufficiently long word over A will have a factor in X. It is clear from the definition that from each unavoidable set we can extract a finite unavoidable subset, so the study can be reduced to finite unavoidable sets.

In this chapter, we introduce unavoidable sets of partial words. In Section 12.1, we recall the definition of unavoidable sets of words and some useful elementary properties. There, we present a definition for unavoidable sets of partial words and introduce the problem of classifying such sets of small cardinality and in particular those with two elements. In Section 12.2, we show that the problem of classifying unavoidable sets of size two reduces to the problem of classifying unavoidable sets of the form

$$\{a\diamond^{m_1} a \dots a\diamond^{m_k} a, b\diamond^{n_1} b \dots b\diamond^{n_l} b\}$$

where $m_1, \dots, m_k, n_1, \dots, n_l$ are nonnegative integers and a, b are distinct letters. In Section 12.3, we give an elegant characterization of the particular case of this problem when $k = 1$ and $l = 1$. In Section 12.4, we propose a conjecture characterizing the case where $k = 1$ and $l = 2$. There, we prove one direction of the conjecture. We then give partial results towards the other direction and in particular prove that the conjecture is easy to verify in a large number of cases. Finally in Section 12.5, we prove that verifying this conjecture is sufficient for solving the problem for larger values of k and l.

12.1 Unavoidable sets

We begin this section with the following basic terms and definitions.

Let $\mathbb{Z}$ denote the set of integers. A *two-sided infinite word* w is a function $w : \mathbb{Z} \to A$. A finite word u is a factor of w if u is a finite subsequence of w, that is, there exists some integer i such that $u = w(i)w(i+1)\dots w(i+|u|-1)$. The empty word ε is trivially a factor of w.

For a positive integer p, we say that a two-sided infinite word w has period p, or that w is *p-periodic*, if $w(i) = w(i + p)$ for all integers i. If w has period p for some p, then we call w *periodic*.

If v is a finite word, then we denote by $v^{\mathbb{Z}}$ the unique two-sided infinite word w with period $|v|$ and such that $w(0) \ldots w(|v| - 1) = v$.

If X is a set of partial words, then we use $\hat{X}$ to denote the set of all full words compatible with a member of X. In other words,

$$\hat{X} = C(X) \cap A^*$$

For example, if $A = \{a, b\}$ and $X = \{\diamond a, b\diamond\}$, then $\hat{X} = \{aa, ba, bb\}$.

The concept relevant to this chapter is that of an *unavoidable set* of partial words. We start with the full word concept and some relevant properties.

DEFINITION 12.1   *Let $X \subset A^*$.*

- *A two-sided infinite word w avoids X if no factor of w is in X.*

- *The set X is* unavoidable *if no two-sided infinite word avoids X, that is, X is unavoidable if every two-sided infinite word has a factor in X.*

Example 12.1
Let $A = \{a, b\}$. Then

- The set $X_1 = \{\varepsilon\}$ is unavoidable since ε is a factor of every two-sided infinite word.

- The set $X_2 = \{a, bbb\}$ is unavoidable. Indeed, if a two-sided infinite word w does not have a as a factor, then $w = b^{\mathbb{Z}}$ and w has bbb as a factor.

                                                                        ⬜

Following are two useful lemmas giving alternative characterizations of unavoidable sets of full words.

LEMMA 12.1
Let $X \subset A^*$. Then X is unavoidable if and only if there are only finitely many words in A^* with no member of X as a factor.

LEMMA 12.2
Let $X \subset A^*$ be finite. Then X is unavoidable if and only if no periodic two-sided infinite word avoids X.

We now give our extension of the definition of unavoidable sets of words to unavoidable sets of partial words.

DEFINITION 12.2 *Let* $X \subset W(A)$.

- *A two-sided infinite word w avoids X if no factor of w is in $\hat{X}$.*

- *The set X is* unavoidable *if no two-sided infinite word avoids X, that is, X is unavoidable if every two-sided infinite word has a factor in $\hat{X}$.*

We first explore some trivial examples (some less trivial examples will come soon).

Example 12.2
Let $A = \{a, b\}$. Then

- For any nonnegative integer n, the set $Y_1 = \{\diamond^n\}$ is unavoidable as well as any set containing Y_1 as a subset. Let us call such sets the *trivial* unavoidable sets.

- The set $Y_2 = \{aa, b\diamond b\}$ is unavoidable. Clearly $b^{\mathbb{Z}}$ does not avoid Y_2. Thus if there were a two-sided infinite word w avoiding Y_2 it would have an a as a factor. Without loss of generality $w(0) = a$. Then since w avoids aa, $w(-1) = w(1) = b$. Then $bab \in \hat{Y_2}$ is a factor of w.

$\square$

Clearly if every member of X is full, then the concept of unavoidable set in Definition 12.2 is equivalent to the one in Definition 12.1. There is another simple connection between sets of partial words and sets of full words that is worth noting.

REMARK 12.1 By the definition of $\hat{X}$, a two-sided infinite word w has a factor in $\hat{X}$ if and only if that factor is compatible with a member of X. Thus the two-sided infinite words which avoid $X \subset W(A)$ are exactly those which avoid $\hat{X} \subset A^*$, and

$$X \subset W(A) \text{ is unavoidable if and only if } \hat{X} \subset A^* \text{ is unavoidable}$$

$\square$

Thus with regards to unavoidability, a set of partial words serves as a representation of a set of full words. The set

$$X = \{a\diamond\diamond\diamond a, b\diamond b\}$$

represents the set of full words over $\{a, b\}$ with two a's separated by three letters and two b's separated by one letter, or the set

$$\hat{X} = \{aaaaa, aaaba, aabaa, aabba, abaaa, ababa, abbaa, abbba, bab, bbb\}$$

It is natural to begin investigating the unavoidable sets of partial words with small cardinality. Of course, every two-sided infinite word avoids the empty set and thus, there are no unavoidable sets of size 0.

It is clear that unless the alphabet is unary, the only unavoidable sets of size 1 are trivial. If the alphabet is unary, then every nonempty set is unavoidable and in that case there is only one two-sided infinite word. Thus the unary alphabet is not interesting and we will not consider it further. Classifying the unavoidable sets of size 2 is the focus of the next section.

12.2   Classifying unavoidable sets of size two

In this section, we restrict ourselves to two-element sets. If X is an unavoidable set, then every two-sided infinite unary word has a factor compatible with a member of X. In particular, X cannot have fewer elements than the alphabet. Thus since X has size 2, the alphabet is unary or binary. We hence assume that the alphabet A is binary say with distinct letters a and b. So one element of X is compatible with a factor of $a^{\mathbb{Z}}$ and the other element is compatible with a factor of $b^{\mathbb{Z}}$, since this is the only way to guarantee that both $a^{\mathbb{Z}}$ and $b^{\mathbb{Z}}$ will not avoid X. Thus we may restrict ourselves to nontrivial unavoidable sets of size 2 of the form

$$X_{m_1,\ldots,m_k \mid n_1,\ldots,n_l} = \{a\diamond^{m_1}a\ldots a\diamond^{m_k}a, b\diamond^{n_1}b\ldots b\diamond^{n_l}b\}$$

for some nonnegative integers $m_1,\ldots,m_k$ and $n_1,\ldots,n_l$. The question we ask is:

> For which $m_1,\ldots,m_k$ and $n_1,\ldots,n_l$ is the set $X_{m_1,\ldots,m_k \mid n_1,\ldots,n_l}$ unavoidable?

The following lemma shows that we need only answer the question for cases where $m_1+1,\ldots,m_k+1,n_1+1,\ldots,n_l+1$ are relatively prime, or

$$\gcd(m_1+1,\ldots,m_k+1,n_1+1,\ldots,n_l+1) = 1$$

LEMMA 12.3
If p is a nonnegative integer, then set

$$X = X_{m_1,\ldots,m_k \mid n_1,\ldots,n_l}$$

and $Y = X_{p(m_1+1)-1,\ldots,p(m_k+1)-1 \mid p(n_1+1)-1,\ldots,p(n_l+1)-1}$. Then X is unavoidable if and only if Y is unavoidable.

PROOF   In terms of notation, it will be helpful to define

$$M_j = \sum_{i=1}^{j}(m_i + 1)$$

Suppose that a two-sided infinite word w avoids X, and set

$$v = \ldots (w(-1))^p (w(0))^p (w(1))^p \ldots$$

We claim that v avoids Y. Suppose otherwise. Then v has a factor compatible with some $x \in Y$. Without loss of generality say that

$$x = a\diamond^{p(m_1+1)-1}a \ldots a\diamond^{p(m_k+1)-1}a$$

Then to say that v has a factor compatible with x is equivalent to saying that there exists some integer i for which

$$v(i) = v(i + pM_1) = \cdots = v(i + pM_k) = a$$

But if we set $h = \lfloor \frac{i}{p} \rfloor$, then this implies that

$$w(h) = w(h + M_1) = \cdots = w(h + M_k) = a$$

contradicting the fact that w avoids X.

We prove the other direction analogously. Suppose that a two-sided infinite word w avoids Y, and set

$$v = \ldots w(-p)w(0)w(p) \ldots$$

We claim that v avoids X. Otherwise v has a factor compatible with some $x \in X$ which we may suppose without loss of generality is $a\diamond^{m_1}a \ldots a\diamond^{m_k}a$. Then there exists some integer i for which

$$v(i) = v(i + M_1) = \cdots = v(i + M_k) = a$$

but this implies that

$$w(pi) = w(pi + pM_1) = \cdots = w(pi + pM_k) = a$$

which contradicts the fact that w avoids Y. ⬚

Two simple facts of symmetry are worth noting.

REMARK 12.2   Say that w avoids $X = X_{m_1,\ldots,m_k|n_1,\ldots,n_l}$. The reverse word $\ldots w(1)w(0)w(-1) \ldots$ avoids $Y = X_{m_k,\ldots,m_1|n_l,\ldots,n_1}$, and the word obtained from w by swapping the a's and b's avoids $Z = X_{n_1,\ldots,n_l|m_1,\ldots,m_k}$. Hence one of the sets X, Y and Z is unavoidable precisely when all three of them are. ⬚

In order to solve the problem of identifying when $X_{m_1,\ldots,m_k|n_1,\ldots,n_l}$ is unavoidable, we start with small values of k and l. Of course, the set

$$\{a, b\diamond^{n_1}b \ldots b\diamond^{n_l}b\}$$

is unavoidable for if w is a two-sided infinite word which does not have a as a factor, then $w = b^{\mathbb{Z}}$. This handles the case where $k = 0$ (and symmetrically the case where $l = 0$).

12.3   The case where $k = 1$ and $l = 1$

We now consider the case where $k = 1$ and $l = 1$, that is, we consider the set

$$X_{m|n} = \{a\diamond^m a, b\diamond^n b\}$$

In this case, we can give an elegant characterization of which integers m, n make this set avoidable: $X_{m|n}$ is avoidable if and only if the greatest powers of 2 dividing $m + 1$ and $n + 1$ are equal.

THEOREM 12.1
Write $m + 1 = 2^s r_0$ and $n + 1 = 2^t r_1$ where r_0, r_1 are odd. Then $X_{m|n}$ is avoidable if and only if $s = t$.

PROOF   Let w be a two-sided infinite word avoiding $X_{m|n}$. Then w also avoids $b\diamond^m b$. Otherwise for some integer i, $w(i) = b$ and $w(i+m+1) = b$. Since w avoids $b\diamond^n b$ we must have that $w(i+n+1) = a$ and $w(i+m+1+n+1) = a$, which contradicts the fact that w avoids $a\diamond^m a$. A symmetrical argument shows that w avoids $a\diamond^n a$.

For ease of notation, define $\bar{a} = b$ and $\bar{b} = a$. If p is a nonnegative integer, then we call a two-sided infinite word w p-alternating if for all integers i, $w(i) = \overline{w(i + p)}$. By our previous observation, it is easy to see that w avoids $X_{m|n}$ if and only if w is $m + 1$- and $n + 1$-alternating. Notice that if w is p-alternating, then it has period $2p$: for every integer i,

$$w(i) = \overline{w(i + p)} = \overline{\overline{w(i + 2p)}} = w(i + 2p)$$

Set $p = m + 1$ and $q = n + 1$. Thus to prove the theorem it is sufficient to show that a two-sided infinite word exists which is both p- and q-alternating if and only if the greatest power of 2 dividing p is equal to the greatest power of 2 dividing q. Write $p = 2^s r_0$ and $q = 2^t r_1$ with r_0 and r_1 odd.

Suppose that $s \neq t$. Without loss of generality say $s < t$. Then $s + 1 \leq t$. Let l be the least common multiple of p and q. The prime factorization of l must have no greater power of 2 than the prime factorization of q. Thus there exists an odd number k such that kq is a multiple of $2p$. If there were a two-sided infinite word w which was p-alternating and q-alternating, then we would have $w(0) = w(2p) = w(kq)$ since w is $2p$-periodic. But since k is odd and w is q-alternating, we also have $w(0) = \overline{w(kq)}$. This is a contradiction. We have half of the necessary implication.

We now prove the other half. Suppose that $s = t$, and so $p = 2^s r_0$ and $q = 2^s r_1$. We only need to prove that there exists some w which is p-alternating and q-alternating and we do this by induction on s.

If $s = 0$, then p and q are odd, and the word $(ab)^{\mathbb{Z}}$ is p-alternating and q-alternating. This handles our basis. Now say w is $2^s r_0$- and $2^s r_1$-alternating. Then $v = \ldots w(-1)w(-1)w(0)w(0)w(1)w(1)\ldots$ is $2^{s+1}r_0$- and $2^{s+1}r_1$-alternating. This finishes the induction and the result follows. $\square$

12.4   The case where $k = 1$ and $l = 2$

We next consider the case where $k = 1$ and $l = 2$, that is, sets of the form

$$X_{m|n_1,n_2} = \{a\diamond^m a, b\diamond^{n_1} b\diamond^{n_2} b\}$$

On the one hand, we have identified a large number of avoidable sets of the form $\{a\diamond^m a, b\diamond^n b\}$. For $X_{m|n_1,n_2}$ to be avoidable it is sufficient that one of the sets

$$\{a\diamond^m a, b\diamond^{n_1} b\} \text{ or } \{a\diamond^m a, b\diamond^{n_2} b\} \text{ or } \{a\diamond^m a, b\diamond^{n_1+n_2+1} b\}$$

be avoidable. Thus by first identifying the avoidable sets for smaller values of k and l, our job has gotten a little easier. On the other hand, the structure of words avoiding $\{a\diamond^m a, b\diamond^{n_1} b\diamond^{n_2} b\}$ is not nearly as nice as those avoiding $\{a\diamond^m a, b\diamond^n b\}$. Thus a simple characterization seems unlikely, unless perhaps there are no unavoidable sets of this form at all.

But there are. We check that the set

$$\{a\diamond^7 a, b\diamond b\diamond^3 b\}$$

is unavoidable. Seeing that it is provides a nice example of the techniques we use.

Example 12.3
The set $\{a\diamond^7 a, b\diamond b\diamond^3 b\}$ is unavoidable. Suppose instead that there exists a two-sided infinite word w which avoids it. We know from Theorem 12.1 that $\{a\diamond^7 a, b\diamond b\}$ is unavoidable, thus w must have a factor compatible with $b\diamond b$. Say without loss of generality that $w(0) = w(2) = b$. This implies that $w(6) = a$, which in turn implies that $w(-2) = b$. Then we have that $w(-2) = w(0) = b$, forcing $w(4) = a$. This propagation continues: $w(-4) = w(-2) = b$ and so $w(2) = a$, a contradiction. $\square$

This example is part of a more general phenomenon. Notice how in this example as the patterns reoccur, we have a sequence of a's traveling to the left toward the b at $w(0)$. There is a symmetric situation in which the b's travel to the right towards the a at $w(n_1 + 1)$.

In proving that a set of the form $X_{m|n_1,n_2}$ is unavoidable our strategy is to derive a contradiction using structural properties that any potential two-sided infinite word w avoiding X would have. These properties take the form of certain rules involving the occurrences of letters in w. For example, whenever $w(i) = w(i+n_1+1) = b$ in w, we must have that $w(i+n_1+n_2+2) = a$. The presence of an a also has implications: if $w(i) = a$ then $w(i-m-1) = b$ and $w(i+m+1) = b$. Often particular values of m, n_1 and n_2 have a relationship that cause these patterns to reoccur and perpetuate themselves, making a contradiction easy to find. In order for this to happen we also need a starting point for the perpetuation. For this Theorem 12.1 is a very handy tool.

Both scenarios are covered by the following proposition.

PROPOSITION 12.1

Suppose either $m = 2n_1 + n_2 + 2$ *or* $m = n_2 - n_1 - 1$, *and* $n_1 + 1$ *divides* $n_2 + 1$. *Then* $X_{m|n_1,n_2}$ *is unavoidable if and only if* $X_{m|n_1}$ *is unavoidable.*

PROOF   If a two-sided infinite word w avoids $\{a\diamond^m a, b\diamond^{n_1} b\}$, then it also avoids $X_{m|n_1,n_2}$.

Now we suppose instead that $\{a\diamond^m a, b\diamond^{n_1} b\}$ is unavoidable. We will just consider the case $m = 2n_1 + n_2 + 2$ (the case where $m = n_2 - n_1 - 1$ is similar and is left as an exercise). Suppose for contradiction that the two-sided infinite word w avoids $X_{m|n_1,n_2}$. Since $\{a\diamond^m a, b\diamond^{n_1} b\}$ is unavoidable and w avoids $a\diamond^m a$, w must have a factor compatible with $b\diamond^{n_1} b$. Suppose without loss of generality that $w(0) = w(n_1 + 1) = b$. We must have that $w(n_1 + n_2 + 2) = a$ which immediately gives us

$$w(n_1 + n_2 + 2 - m - 1) = w(n_1 + n_2 + 1 - 2n_1 - n_2 - 2) = w(-n_1 - 1) = b$$

Since $w(-n_1 - 1) = w(0) = b$, we must have $w(n_2 + 1) = a$. By induction we can verify that this process continues, and we ultimately find that

$$a = w(n_2 + 1) = w(n_2 + 1 - (n_1 + 1)) = w(n_2 + 1 - 2(n_1 + 1)) = \dots$$

Since $n_1 + 1$ divides $n_2 + 1$ we find that $w(0) = a$, a contradiction. $\square$

One notable consequence of Proposition 12.1 is that if m is odd, then both $\{a\diamond^m a, bb\diamond^{m+1} b\}$ and $\{a\diamond^m a, bb\diamond^{m-2} b\}$ are unavoidable.

The next theorem takes advantage of the perpetuating pattern phenomenon in a more complicated context. Proposition 12.1 held because each a forced a b into the next position of an occurence of $w(i) = w(i+n_1+1) = b$, which in turn forced a new a in w. This created a single traveling sequence of a's and b's, causing an a to overlap with the b at $w(0)$, yielding a contradiction. In the next argument, we take notice of the fact that each a occurring in w may contribute to two occurrences of $w(i) = w(i+n_1+1) = b$ simultaneously so that a contradiction will occur after many traveling sequences of letters appear and overlap.

THEOREM 12.2

Say that $m = n_2 - n_1 - 1$ or $m = 2n_1 + n_2 + 2$, and that the highest power of 2 dividing $n_1 + 1$ is less than the highest power of 2 dividing $m + 1$. Then $X_{m|n_1,n_2}$ is unavoidable.

PROOF Since the highest power of 2 dividing $n_1 + 1$ is different than the highest power of 2 dividing $m + 1$, we have that the set $Y = \{a \diamond^m a, b \diamond^{n_1} b\}$ is unavoidable. Consider the case where $m = n_2 - n_1 - 1$ and suppose for contradiction that there exists a two-sided infinite word w that avoids $X = X_{m|n_1,n_2}$. Then w has no factor compatible with $a \diamond^m a$, and so since Y is unavoidable it must have a factor compatible with $b \diamond^{n_1} b$. Assume without loss of generality that $w(0) = b$ and $w(n_1 + 1) = b$.

We now generate an infinite table of facts about w. Two horizontally adjacent entries in the table will represent positions in w which are $n_1 + 1$ letters apart. Two vertically adjacent entries in the table will represent positions in w which are $m + 1 = n_2 - n_1$ letters apart. The two upper left entries of our table are $w(0) = b$ and $w(n_1 + 1) = b$, two facts we have already assumed. Since w avoids X we have more information relevant to the table: two horizontally adjacent b entries force an a entry diagonally down and to the right from them as seen in Figure 12.1. And an a entry forces a b entry in the

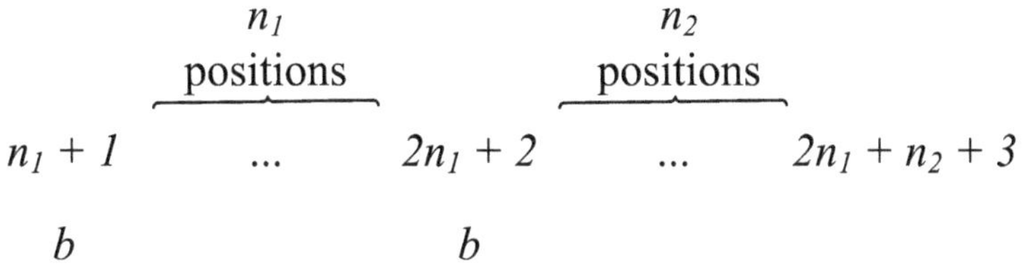

FIGURE 12.1: Horizontal arrows.

vertically adjacent positions as seen in Figure 12.2.

From these rules we can build the table of Figure 12.3, labeling the columns $C_0, C_1, \ldots$.

For a nonnegative integer i, we shall define v_i to be the factor of w represented by C_i. If i is odd then C_i has i entries, and if i is even then C_i has

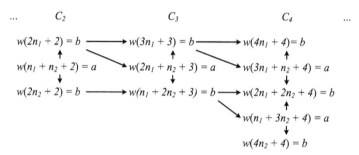

$$
\underbrace{}_{\substack{n_2 - n_1 \text{ - } 1 \\ \text{positions}}}
$$

$$
2n_1 + 2 \quad \dots \quad n_1 + n_2 + 2 \quad \dots \quad 2n_2 + 2
$$

a

FIGURE 12.2: Vertical arrows.

FIGURE 12.3: The table.

$i + 1$ entries. Thus we define

$$
v_i = \begin{cases} w(in_1 + i)w(in_1 + i + 1)\ldots w(in_2 + i) & \text{if } i \text{ even} \\ w(in_1 + i)w(in_1 + i + 1)\ldots w(n_1 + (i - 1)n_2 + i) & \text{if } i \text{ odd} \end{cases}
$$

Two adjacent entries in C_i represent a distance of $m + 1$ positions between letters in v_i. Thus for i even we have that $|v_i| = im + 1$ and for i odd we have that $|v_i| = (i - 1)m + 1$. We can also use the table to get some partial information about the positions of a's and b's in v_i. For a nonnegative integer j, $v_i(j) = b$ if $j \equiv 0 \bmod 2m + 2$, and $v_i(j) = a$ if $j \equiv m + 1 \bmod 2m + 2$.

Because the highest power of 2 dividing $n_1 + 1$ is no greater than the highest power of 2 dividing $m + 1$, there exists some k for which $k(n_1 + 1) \equiv m + 1 \bmod 2m + 2$. Take i sufficiently large so that Columns C_i and C_{i+k} overlap, or in other words $|v_i| > kn_1 + k$. Because of how k was chosen, we have that $v_i(kn_1 + k) = a$. However examining the table we see that

$$
w((i + k)n_1 + i + k) = v_i(kn_1 + k) = v_{i+k}(0) = b
$$

a contradiction. This handles the situation where $m = n_2 - n_1 - 1$. The reader can check that the case where $m = 2n_1 + n_2 + 2$ is similar, the only difference being that the table will represent increasingly negative positions of w, rather than increasingly positive ones. $\quad\square$

REMARK 12.3   Take $m = 1$ in Theorem 12.2. Let us see for which nonnegative integers n_1 the hypotheses of the theorem hold to make $X_{m|n_1,n_2}$

unavoidable. The highest power of 2 dividing $n_1 + 1$ should be less than the highest power of 2 dividing $m + 1 = 2$. Thus $n_1 + 1$ must be odd, n_1 is even. Since $m = 1$ we cannot have $m = 2n_1 + n_2 + 2$. Say we have $m = n_2 - n_1 - 1$. Then $n_2 = n_1 + 2$. So we have that for any even n_1, the set $\{a \diamond a, b \diamond^n b \diamond^{n+2} b\}$ is unavoidable. We will prove that this is a complete characterization of unavoidability of $X_{m|n_1,n_2}$ for $m = 1$. ⬚

Propositions 12.2 and 12.3 are other results for $k = 1$ and $l = 2$.

The next proposition identifies another large class of unavoidable sets using a modification of the strategies discussed so far.

PROPOSITION 12.2
If $n_1 < n_2$, $2m = n_1 + n_2$ and $m - n_1$ divides $m + 1$, then $X_{m|n_1,n_2}$ is unavoidable.

By taking $n_1 = m - 1$ and $n_2 = m + 1$, Proposition 12.2 yields a nice fact: the set $\{a \diamond^m a, b \diamond^{m-1} b \diamond^{m+1} b\}$ is unavoidable for all $m > 0$.

We believe that together Proposition 12.1, Proposition 12.2, and Theorem 12.2 nearly give a complete characterization of when $X_{m|n_1,n_2}$ is unavoidable. The case $m = 6, n_1 = 1$ and $n_2 = 3$ is what we believe to be the only exception.

PROPOSITION 12.3
The set $X_{6|1,3} = \{a \diamond^6 a, b \diamond b \diamond^3 b\}$ is unavoidable.

Extensive experimentation suggests that these results, and their symmetric equivalents, give a complete characterization of when $X_{m|n_1,n_2}$ is unavoidable. Using Lemma 12.3, we may assume without loss of generality that $m+1, n_1+1, n_2+1$ are relatively prime.

Conjecture 1 *Let m, n_1, n_2 be nonnegative integers satisfying $n_1 \leq n_2$ and $\gcd(m+1, n_1+1, n_2+1) = 1$. The set $X_{m|n_1,n_2}$ is unavoidable precisely when the hypotheses of at least one of Proposition 12.1, Proposition 12.2, Proposition 12.3 or Theorem 12.2 hold. In other words, $X_{m|n_1,n_2}$ is unavoidable if and only if one of the following cases (or symmetric equivalents) holds:*

- *The case where $X_{m|n_1}$ is unavoidable, $m = 2n_1 + n_2 + 2$ or $m = n_2 - n_1 - 1$, and $n_1 + 1$ divides $n_2 + 1$.*

- *The case where $m = n_2 - n_1 - 1$ or $m = 2n_1 + n_2 + 2$, and the highest power of 2 dividing $n_1 + 1$ is less than the highest power of 2 dividing $m + 1$.*

- *The case where $n_1 < n_2$, $2m = n_1 + n_2$ and $m - n_1$ divides $m + 1$.*

- *The case where $m = 6$, $n_1 = 1$ and $n_2 = 3$.*

The reader may verify that for any fixed m the only one of the above cases that contributes infinitely many unavoidable sets to $X_{m|n_1,n_2}$ is Theorem 12.2, and that this theorem never applies to even m. Thus the conjecture states that there are only finitely many values of m, n_1, n_2 with m fixed and even and $X_{m|n_1,n_2}$ unavoidable. We will prove that this is indeed the case.

An important consequence of the conjecture is that in order for $X_{m|n_1,n_2}$ to be unavoidable it is necessary that either $m = 6$ and $\{n_1, n_2\} = \{1, 3\}$, or that one of the following equations holds:

$$m = 2n_1 + n_2 + 2 \tag{12.1}$$

$$m = 2n_2 + n_1 + 2 \tag{12.2}$$

$$m = n_1 - n_2 - 1 \tag{12.3}$$

$$m = n_2 - n_1 - 1 \tag{12.4}$$

$$2m = n_1 + n_2 \tag{12.5}$$

In order to prove Conjecture 1, only one direction remains. We must show that if none of the aforementioned cases hold, then $X_{m|n_1,n_2}$ is avoidable. We now give partial results towards this goal.

We have found that in general identifying sets of the form $X_{m|n_1,n_2}$ as avoidable tends to be a more difficult task than identifying them as unavoidable. In the case of unavoidability we needed only consider a single word then derive a contradiction from its necessary structural properties. To find a class of avoidable sets we must invent some general procedure for producing a two-sided infinite word which avoids each such set. This is precisely what we move towards in the following propositions in which we verify that the conjecture holds for certain values of m and n_1.

It is easy to see that none of Equations 12.1, 12.2, 12.3, 12.4 or 12.5 are satisfied when $\max(n_1, n_2) < m \leq n_1 + n_2 + 2$. Thus the conjecture for such values is that $X_{m|n_1,n_2}$ is avoidable. The following fact verifies that this is indeed the case.

PROPOSITION 12.4
If $\max(n_1, n_2) < m < n_1 + n_2 + 2$, then $X_{m|n_1,n_2}$ is avoidable.

The next proposition gives an easy way of verifying the conjecture for even values of m.

PROPOSITION 12.5
Assume that m is even and $2m \leq \min(n_1, n_2)$. Then $X_{m|n_1,n_2}$ is avoidable.

PROOF If either n_1 or n_2 is even, then either

$$\{a \diamond^m a, b \diamond^{n_1} b\} \text{ or } \{a \diamond^m a, b \diamond^{n_2} b\}$$

is avoidable by Theorem 12.1. Both situations imply that $X_{m|n_1,n_2}$ is avoidable. Thus we only need to prove that $X_{m|n_1,n_2}$ is avoidable for n_1, n_2 odd. Say without loss of generality that $n_1 \leq n_2$.

Let $v = b^m$, and let $u = baba \ldots ab$ with $|u| = n_2 + 2$. We claim that $w = (uv)^{\mathbb{Z}}$ avoids $X_{m|n_1,n_2}$. Clearly w avoids $\{a \diamond^m a\}$. Because of periodicity, it is enough to prove that for any $i \in \{0, \ldots, n_2+2+m-1\}$ if $w(i-n_1-1) = b$ and $w(i) = b$ then $w(i+n_2+1) = a$. We claim that such an i must be greater than m. Suppose for contradiction that $i \leq m$. Then $-|uv| = -n_2 - 2 - m < i - n_1 - 1 < -m = -|v|$. Thus $w(i - n_1 - 1)$ occurs in the repetition of u at $w(-n_2 - 1 - m) \ldots w(-m - 1)$, and so since $i - n_1 - 1$ is an even number, $w(i - n_1 - 1) = a$ which is a contradiction.

Since $i > m$ we have that

$$|uv| = n_2+2+m \leq i+n_2+1 < n_2+2+m-1+n_2+2 = 2n_2+m+3 = |uv|+|u|-1$$

Thus $w(i + n_2 + 1)$ occurs in the second repetition of u at $w(n_2 + 2 + m) \ldots w(2n_2+m+3)$, and so since $i+n_2+1$ is an even number, $w(i+n_2+1) = a$. $\square$

Thus for any fixed even m we only need to verify the conjecture for finitely many values of n_1 and n_2, which is generally easy. The reader may verify that this is consistent with the conjecture. Similarly the conjecture for $m = 2$ is that $X_{2|n_1,n_2}$ is avoidable except for $n_1 = 1, n_2 = 3$ or $n_2 = 3, n_1 = 1$. It is easy to find avoiding two-sided infinite words for other values of n_1 and n_2 less than 5 when $m = 2$. By Proposition 12.5 this is all that is necessary to confirm the conjecture for $m = 2$. In this way we have been able to verify the conjecture for all even m up to very large values.

The following proposition shows that the conjecture is true for $m = 1$.

PROPOSITION 12.6
The conjecture holds for $m = 1$, that is, $X_{1|n_1,n_2}$ is unavoidable if and only if n_1 and n_2 are even numbers with $|n_1 - n_2| = 2$.

PROOF That $X_{1|n_1,n_2}$ is unavoidable for n_1 and n_2 even with $|n_1-n_2| = 2$ is a direct consequence of Theorem 12.2 and was explained in Section 12.3. Thus we only need to prove that the set is avoidable for other values of n_1 and n_2. We divide these values of n_1 and n_2 into cases and prove that $X_{1|n_1,n_2}$ is avoidable in each case. By symmetry we may assume that $n_1 \leq n_2$ throughout.

Claim 1. For n_1 or n_2 odd, $X_{1|n_1,n_2}$ is avoidable. If both n_1 and n_2 are odd then $m+1, n_1+1, n_2+1$ are all divisible by 2. Thus by applying Lemma 12.3 with Proposition 12.5, we find that $X_{1|n_1,n_2}$ is avoidable. If either n_1 or n_2 is equivalent to 1 mod 4, then $X_{1|n_1,n_2}$ is avoidable by Theorem 12.1. The last case to consider is when n_1 or n_2 is equivalent to 3 mod 4. By symmetry we may suppose that $n_1 \equiv 3 \bmod 4$ and n_2 is even. Thus we divide into two further cases.

First say $n_2 \equiv 0 \bmod 4$ and write $n_2 = 4k$. Let $v = (aabb)^k abb$. We prove that $w = v^{\mathbb{Z}}$ avoids $X_{1|n_1,n_2}$. Certainly it avoids $a \diamond a$. By periodicity we may assume without loss of generality that $i \in \{0, \ldots, |v| - 1\}$ and $w(i + n_1 + 1) = w(i + n_1 + n_2 + 2) = b$. We then only need to prove that $w(i) = a$. Since w is $|v| = 4k + 3$-periodic, we have that $w(i + n_1 + n_2 + 2) = w(i + n_1 + n_2 + 2 - n_2 - 3) = w(i + n_1 - 1)$. Examining v we see that $w(i + n_1 - 1) = w(i + n_1 + 1) = b$ can only occur if $i + n_1 + 1 = 4k + 1$. It is easy to see that since $n_1 + 1 \equiv 0 \bmod 4$ and $n_1 \leq n_2$ that $w(i + n_1 + 1 - n_1 - 1) = a$, and so $w(i) = a$.

For the second case say $n_2 \equiv 2 \bmod 4$ and write $n_2 = 4k + 2$. The reader may verify using a similar argument that $((aabb)^k aabbb)^{\mathbb{Z}}$ avoids $X_{1|n_1,n_2}$ in this case and the claim is proved.

Claim 2. If $n_1 < n_2 - 2$ and either $n_1 \equiv 0 \bmod 4$ and $n_2 \equiv 2 \bmod 4$, or $n_1 \equiv 2 \bmod 4$ and $n_2 \equiv 0 \bmod 4$, then $X_{1|n_1,n_2}$ is avoidable. Take the first case, $n_1 \equiv 0 \bmod 4$ and $n_2 \equiv 2 \bmod 4$. Write $n_2 = 4k + 2$, with $k \geq 1$ which is valid since we have assumed $n_1 < n_2 - 2$. Let $v = (aabb)^{k-1} aabbb$. Our argument is similar to those used for the last claim. In particular we show that $w = v^{\mathbb{Z}}$ avoids $X_{1|n_1,n_2}$. Certainly it avoids $a \diamond a$. By periodicity we may assume without loss of generality that $i \in \{0, \ldots, |v| - 1\}$ and $w(i + n_1 + 1) = w(i + n_1 + n_2 + 2) = b$. We then only need to prove that $w(i) = a$. Since w is $|v| = 4k + 1$-periodic, we have that $w(i + n_1 + n_2 + 2) = w(i + n_1 + n_2 + 2 - n_2 + 1) = w(i + n_1 + 3)$. Examining v we see that $w(i + n_1 + 1) = w(i + n_1 + 3) = b$ can only occur for $w(i + n_1 + 1) = 4k - 2$. It is easy to see that since $n_1 + 1 \equiv 1 \bmod 4$ and $n_1 < n_2 - 2$ that $w(i + n_1 + 1 - n_1 - 1) = a$, and so $w(i) = a$.

For the second case, where $n_1 \equiv 2 \bmod 4$ and $n_2 \equiv 0 \bmod 4$, write $n_2 = 4k$. Then $((aabb)^{k-1} abb)$ avoids $X_{1|n_1,n_2}$ and the claim is proved.

The only possible values of n_1 and n_2 left to consider are where $n_1, n_2 \equiv 0 \bmod 4$ or $n_1, n_2 \equiv 2 \bmod 4$.

Claim 3. If $n_1 < n_2 - 2$ and either $n_1, n_2 \equiv 0 \bmod 4$ or $n_1, n_2 \equiv 2 \bmod 4$ then $X_{1|n_1,n_2}$ is avoidable. First say $n_1, n_2 \equiv 0 \bmod 4$ and write $n_2 = 4k$. In this case $(aabb)^k abb$ avoids $X_{1|n_1,n_2}$. Second suppose $n_1, n_2 \equiv 2 \bmod 4$ and write $n_2 = 4k + 2$. In this case $(aabb)^k aabbb$ avoids $X_{1|n_1,n_2}$. ∎

The other odd values of m seem to be much more difficult and will most likely require more sophisticated techniques.

The following proposition intuitively says that if m and n_1 are close enough in value, then $X_{m|n_1,n_2}$ is avoidable for large enough n_2.

PROPOSITION 12.7

Let s be a nonnegative integer satisfying $s < m - 2$. Then for $n > 2(m + 1)^2 + m - 1$, $X_{m|m+s,n} = \{a \diamond^m a, b \diamond^{m+s} b \diamond^n b\}$ is avoidable.

12.5   Larger values of k and l

We end this chapter with the following proposition which implies that if Conjecture 1 is true, then $X_{m_1,\ldots,m_k|n_1,\ldots,n_l}$ is avoidable for all $k = 1, l \geq 3$, and for $k \geq 2, l \geq 2$: avoidable sets of the form $X_{m_1,\ldots,m_k|n_1,\ldots,n_l}$ for small values of k and l translate directly to avoidable sets for larger values of k and l. If indeed Conjecture 1 is true, then we have completely classified the unavoidable sets of size two.

PROPOSITION 12.8

If Conjecture 1 is true, then $X_{m_1,\ldots,m_k|n_1,\ldots,n_l}$ is avoidable for all $k = 1, l \geq 3$, and for $k \geq 2, l \geq 2$.

PROOF   Assume Conjecture 1 holds. To prove the proposition it is sufficient to prove that both $X_{m_1,m_2|n_1,n_2}$ and $X_{m|n_1,n_2,n_3}$ are avoidable for all m_1, m_2, n_1, n_2.

First let us consider $X_{m_1,m_2|n_1,n_2}$. Assume without loss of generality that m_1, m_2, n_1, n_2 are relatively prime. In order for this set to be unavoidable, it is necessary that the sets $\{a\diamond^{m_1}a, b\diamond^{n_1}b\diamond^{n_2}b\}$, $\{a\diamond^{m_2}a, b\diamond^{n_2}b\diamond^{n_2}b\}$, $\{a\diamond^{m_1}a\diamond^{m_2}a, b\diamond^{n_1}b\}$ and $\{a\diamond^{m_1}a\diamond^{m_2}a, b\diamond^{n_2}b\}$ be unavoidable as well. For each of these sets, Conjecture 1 gives a necessary condition: either $m = 6$ and $n_1 = 1, n_2 = 3$ (or symmetrically $n_1 = 3, n_2 = 1$) or one of Equations 12.1, 12.2, 12.3, 12.4 or 12.5 must hold. Consider the following tables:

$m_1 = 2n_1 + n_2 + 2$	$m_2 = 2n_1 + n_2 + 2$
$m_1 = 2n_2 + n_1 + 2$	$m_2 = 2n_2 + n_1 + 2$
$m_1 = n_1 - n_2 - 1$	$m_2 = n_1 - n_2 - 1$
$m_1 = n_2 - n_1 - 1$	$m_2 = n_2 - n_1 - 1$
$m_1 = 6, n_1 = 1, n_2 = 3$	$m_2 = 6, n_1 = 1, n_2 = 3$
$m_1 = 6, n_2 = 1, n_1 = 3$	$m_2 = 6, n_2 = 1, n_1 = 3$
$2m_1 = n_1 + n_2$	$2m_2 = n_1 + n_2$

$n_1 = 2m_1 + m_2 + 2$	$n_2 = 2m_1 + m_2 + 2$
$n_1 = 2m_2 + m_1 + 2$	$n_2 = 2m_2 + m_1 + 2$
$n_1 = m_1 - m_2 - 1$	$n_2 = m_1 - m_2 - 1$
$n_1 = m_2 - m_1 - 1$	$n_2 = m_2 - m_1 - 1$
$n_1 = 6, m_1 = 1, m_2 = 3$	$n_2 = 6, m_1 = 1, m_2 = 3$
$n_1 = 6, m_2 = 1, m_1 = 3$	$n_2 = 6, m_2 = 1, m_1 = 3$
$2n_1 = m_1 + m_2$	$2n_2 = m_1 + m_2$

In order for $X_{m_1,m_2|n_1,n_2}$ to be unavoidable it is necessary that at least one equation from each column be satisfied. It is easy to verify using a computer algebra system that this is impossible except in the case where the last equation in each column is satisfied. However in this case $m_1 = m_2 = n_1 = n_2$ and so by Theorem 12.1, the set is avoidable.

Now let us consider $X_{m|n_1,n_2,n_3}$. In order for this set to be unavoidable, it is necessary that $\{a\diamond^m a, b\diamond^{n_1} b\diamond^{n_2} b\}$, $\{a\diamond^m a, b\diamond^{n_2} b\diamond^{n_3} b\}$, $\{a\diamond^m a, b\diamond^{n_1+n_2+1} b\diamond^{n_3} b\}$ and $\{a\diamond^m a, b\diamond^{n_1} b\diamond^{n_2+n_3+1} b\}$ be unavoidable as well. Again, for each of these sets Conjecture 1 gives a necessary condition: either $m = 6$ and $n_1 = 1, n_2 = 3$ (or $n_1 = 3, n_2 = 1$) or one of Equations 12.1, 12.2, 12.3, 12.4 or 12.5 must hold. Consider now the following tables:

$m = 2n_1 + n_2 + 2$	$m = 2n_2 + n_3 + 2$
$m = 2n_2 + n_1 + 2$	$m = 2n_3 + n_2 + 2$
$m = n_1 - n_2 - 1$	$m = n_2 - n_3 - 1$
$m = n_2 - n_1 - 1$	$m = n_3 - n_2 - 1$
$2m = n_1 + n_2$	$2m = n_2 + n_3$
$m = 6, n_1 = 1, n_2 = 3$	$m = 6, n_2 = 1, n_3 = 3$
$m = 6, n_2 = 1, n_1 = 3$	$m = 6, n_3 = 1, n_2 = 3$

$m = 2(n_1 + n_2 + 1) + n_3 + 2$	$m = 2n_1 + n_2 + n_3 + 3$
$m = 2n_3 + (n_1 + n_2 + 1) + 2$	$m = 2(n_2 + n_3 + 1) + n_1 + 2$
$m = (n_1 + n_2 + 1) - n_3 - 1$	$m = n_1 - (n_2 + n_3 + 1) - 1$
$m = n_3 - (n_1 + n + 2 + 1) - 1$	$m = (n_2 + n_3 + 1) - n_1 - 1$
$2m = (n_1 + n_2 + 1) + n_3$	$2m = n_1 + (n_2 + n_3 + 1)$
$m = 6, n_1 + n_2 + 1 = 1, n_3 = 3$	$m = 6, n_1 = 1, n_2 + n_3 + 1 = 3$
$m = 6, n_3 = 1, n_1 + n_2 = 3$	$m = 6, n_2 + n_3 = 1, n_1 = 3$

Again unavoidability of $X_{m|n_1,n_2,n_3}$ requires that one equation from each column be satisfied. It is easy to verify that no such system of equations has a nonnegative solution. ⬚

Conjecture 1 has been tested in numerous cases via computer, and verified for $m = 1$ and a large number of even values of m.

Exercises

12.1 Let n be a nonnegative integer. Is the set A^n, or the set of all words of length n, avoidable?

12.2 If $A = \{a, b\}$, then show that $X = \{a\diamond, \diamond b\}$ is unavoidable.

12.3 Show that the set $X = \{a\diamond\diamond\diamond a, b\diamond b\}$ is unavoidable.

12.4 [s] Let $A = \{a, b\}$. Characterize the two-sided infinite words that avoid $X = \{a\diamond^n b, b\diamond^n a\}$ where n is a nonnegative integer.

12.5 Setting $A = \{a, b\}$, is the set $\{a\diamond^6 a, b\diamond b\diamond^3 b\}$ unavoidable?

12.6 No nontrivial unavoidable set can have fewer elements than the alphabet. True or false?

12.7 Show that the set $X_{4|2,3}$ is avoidable by giving a word v such that $v^{\mathbb{Z}}$ avoids it.

12.8 [s] Repeat Exercise 12.7 for the set $X_{5|1,3}$.

12.9 Show that the set $\{a\diamond^m a, bbb\}$ is avoidable.

12.10 Describe C_5 of Figure 12.3.

12.11 [s] Classify the sets $X_{4|3,4}$ and $X_{2|3,2}$ as avoidable or unavoidable.

12.12 [s] Prove Proposition 12.2.

12.13 Verify that the conjecture for $m = 0$ is that $X_{0|n_1,n_2}$ is always avoidable, which is given by Proposition 12.5.

Challenging exercises

12.14 Prove Lemma 12.1.

12.15 [H] Prove Lemma 12.2.

12.16 Prove the case where $m = n_2 - n_1 - 1$ of Proposition 12.1.

12.17 Check the case where $m = 2n_1 + n_2 + 2$ of Theorem 12.2.

12.18 [s] Prove Proposition 12.3.

12.19 [s] Prove Proposition 12.4.

12.20 [s] Prove Proposition 12.7.

Programming exercises

12.21 Referring to the first two tables in the proof of Proposition 12.8, verify using a computer algebra system that it is impossible that at least one equation from each column be satisfied except in the case where the last equation in each column is satisfied.

12.22 Referring to the last two tables in the proof of Proposition 12.8, verify for each column that no such system of equations has a nonnegative solution.

Website

A World Wide Web server interface at

`http://www.uncg.edu/mat/research/unavoidablesets`

has been established for automated use of a program that classifies a set of partial words $X_{m_1,\ldots,m_k|n_1,\ldots,n_l}$ of size two as avoidable or unavoidable. If the set is avoidable, then the program gives a word v such that $v^{\mathbb{Z}}$ avoids the set. Another related website is

`http://www.uncg.edu/cmp/research/unavoidablesets2`

for classifying sets of size three.

Bibliographic notes

The concept of an avoidable set of full words was explicitly introduced in 1983 in connection with an attempt to characterize the rational languages among the context-free ones [73]. Since then it has been consistently studied by researchers in both mathematics and theoretical computer science. Testing the unavoidability of a finite set X can be done in different ways [51]: Check whether there is a loop in the finite automaton of Aho and Corasick [1] recognizing $A^* \setminus A^* X A^*$, or simplify X as much as possible. These same algorithms can be used to decide if a finite set of partial words X is unavoidable by determining the unavoidability of $\hat{X}$. However this incurs a dramatic loss in efficiency, as each pword u in X can contribute as many as $\|A\|^{\|H(u)\|}$ elements to $\hat{X}$. We refer the reader to [50, 126] for more information on unavoidable sets.

Unavoidable sets of partial words were introduced by Blanchet-Sadri, Brownstein and Palumbo [22]. The results in this chapter are from there. In terms of unavoidability, sets of partial words serve as efficient representations of sets of full words. This is strongly analogous to the study of unavoidable patterns, in which sets of patterns are used to represent infinite sets of full words [107].

Solutions to Selected Exercises

CHAPTER 1

1.2

1. $\{9, 10\}$
2. $\{3, 9, 10\}$
3. $0010\diamond10110$

1.4 If $\|\alpha(u)\| \leq 1$, then there exists a letter $a \in A$ such that $u \subset a^p$ with $p \geq 2$. Conversely, if u is not primitive, then there exists a word v such that $u \subset v^n$ with $n \geq 2$. But then n divides $|u| = p$, and since p is prime we get $n = p$. We conclude that $|v| = 1$ and so $\|\alpha(u)\| \leq 1$.

1.10 We prove the first statement (the second one is similar). We use Figure 1 to illustrate our ideas. If $|u| \geq |v|$, then set $u = wz$ with $|v| = |w|$. Then

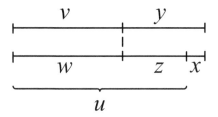

FIGURE 1: Picture for Lemma 1.2.

$wzx = ux \uparrow vy$ and the simplification rule gives the result.

1.11 First, assume that u is unbordered. Suppose to the contrary that $p(u) < |u|$. Then $u \subset v^n w$ for some word v satisfying $|v| = p(u)$, some prefix w of v distinct from v, and some positive integer n. If $w = \varepsilon$, then $n \geq 2$ and $u \subset vv^{n-1}$ and $u \subset v^{n-1}v$. If $w \neq \varepsilon$, then put $v = wy$ for some nonempty word y. In this case, $u \subset wyv^{n-1}w$ and $u \subset v^n w$. In either case, we get a contradiction with the fact that u is unbordered.

Second, let u be an unbordered partial word and assume that u is not primitive. Then $u \subset x^k$ for some word x and integer $k \geq 2$. But then $|x|$ is a period of u smaller than $|u|$.

1.13 Conjugacy on full words is reflexive ($u = u\varepsilon$ and $u = \varepsilon u$) and trivially symmetric. It is also transitive. To see this, if u and v are conjugate and v and w are conjugate, then there exist words x_1, y_1, x_2, y_2 such that $u = x_1 y_1$, $v = y_1 x_1 = x_2 y_2$ and $w = y_2 x_2$. We first assume that

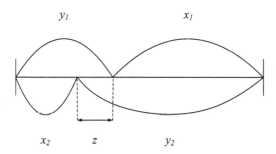

FIGURE 2: Conjugacy on full words is transitive.

$|y_1| \geq |x_2|$ (the case where $|y_1| < |x_2|$ is handled similarly). There exists z such that $y_1 = x_2 z$ and $y_2 = z x_1$, and $u = x_1 y_1 = x_1 x_2 z$ and $w = y_2 x_2 = z x_1 x_2$. Therefore, u and w are conjugate (see Figure 2).

Conjugacy on partial words is reflexive ($u \subset u\varepsilon$ and $u \subset \varepsilon u$) and trivially symmetric. However, conjugacy on partial words is not transitive as the following example shows. Consider, $u = a \diamond babb \diamond a$, $v = \diamond b \diamond \diamond aa \diamond \diamond$, and $w = ba \diamond bbbaa$. By setting $x = a \diamond b$ and $y = abb \diamond a$, we get $u \subset xy$ and $v \subset yx$ showing that u and v are conjugate. Similarly, by setting $x' = \diamond bbbaa$ and $y' = ba$, we get $v \subset x'y'$ and $w \subset y'x'$ showing that v and w are conjugate. But we can see that u and w are not conjugate.

1.16 The partial word $x \diamond x$ where $x \in \{0, 1\}^*$ is as desired.

1.17 Write u as $v_1 v_2 \ldots v_k r$ where $|v_1| = |v_2| = \cdots = |v_k| = p$ and $0 \leq |r| < p$, and v_k as st where $|s| = |r|$. Set $x_1 = v_1 \ldots v_{k-1} s$ and $x_2 = v_2 \ldots v_k r$.

1.18 The conclusion is immediate for the base case $|u| = 1$. Now suppose the statement is true for partial words whose length is smaller than $|u|$. If u is primitive, then let v be any word such that $u \subset v$. Then v is primitive as well and the result follows in this case. If u is not primitive, then $u \subset v^n$ for some word v and integer $n \geq 2$. Since $|v| < |u|$, by the inductive hypothesis, there exists a primitive word w and a positive integer m such that $v \subset w^m$. We have then $u \subset w^{mn}$.

Uniqueness does not hold for partial words. The partial word $u = \diamond a$ serves as a counterexample ($u \subset a^2$ and $u \subset ba$ for distinct letters a, b).

1.19 If $x \uparrow y$, then $x \subset (x \vee y)$ and $y \subset (x \vee y)$. Thus $xy \subset (x \vee y)^2$ and $yx \subset (x \vee y)^2$. Therefore $(xy \vee yx) \subset (x \vee y)^2$. The reverse containment is true.

1.21 Assume that $p(u) = |u|$. If $|v| = |x|$, then put $u = u_1 u_2$ where $|u_1| = |x|$. So $u \subset xx$ and $|x|$ is a period of u smaller than $p(u)$, a contradiction. If $|v| > |x|$, then put $v = v_1 v_2$ where $|v_2| = |x|$. So $u \subset xv_1x$ and $|x|$ is a period of u smaller than $p(u)$, a contradiction.

The statement does not necessarily hold when $|v| < |x|$ as the partial word $u = aba\diamond babb$ shows. Here u is bordered since $u \subset (ababb)(abb)$ and $u \subset (aba)(ababb)$ but $p(u) = |u|$.

CHAPTER 2

2.5 If $k = 4$ and $l = 10$, then $u = a\diamond baab\diamond aabaa\diamond\diamond$ is $(4,10)$-special since $\mathrm{seq}_{4,10}(0)$ contains the positions 6, 12 which are in $H(u) = \{1,6,12,13\}$ while

$$u(0)u(4)u(8)u(12)u(2)u(6)u(10)u(0) = aaa\diamond b\diamond aa$$

is not 1-periodic. However, the partial word $v = \diamond babab\diamond bababab\diamond b$ is not $(4,10)$-special.

2.10 If $k = 3$ and $l = 6$, then the partial word $w = ab\diamond\diamond bc\diamond bc$ is $\{3,6\}$-special since $\mathrm{seq}_{3,6}(0) = (0,3,6,0)$ contains the consecutive positions 3 and 6 which are in $H(w) = \{2,3,6\}$ (but w is not $(3,6)$-special).

2.11 If $k = 3$ and $l = 6$, then the partial word $w = ab\diamond\diamond bc\diamond bc$ is $\{3,6\}$-special since $\mathrm{seq}_{3,6}(0) = (0,3,6,0)$ contains the consecutive positions 3 and 6 which are in $H(w) = \{2,3,6\}$ (but w is not $(3,6)$-special). Here, by letting $u = abc$ and $v = abcbbc$, we have $w \subset uv$ and $w \subset vu$ and $uv \neq vu$. Our answer does not contradict Lemma 2.5.

2.17 Assume that $uz \uparrow zv$ with $\|H(z)\| = 1$ (the case where z is full comes from Corollary 2.1). Let m be such that $m|u| > |z| \geq (m-1)|u|$. Put $u = x_1 y_1$ and $v = y_2 x_2$ where $|x_1| = |x_2| = |z| - (m-1)|u|$ and $|y_1| = |y_2|$ (here $|u| = |v|$). Put $z = x_1' y_1' x_2' y_2' \ldots x_{m-1}' y_{m-1}' x_m'$ where $|x_1'| = \cdots = |x_{m-1}'| = |x_m'| = |x_1| = |x_2|$ and $|y_1'| = \cdots = |y_{m-1}'| = |y_1| = |y_2|$. Since $uz \uparrow zv$, we get

$$x_1 \; y_1 \; x_1' \; y_1' \; x_2' \; y_2' \ldots x_{m-2}' \; y_{m-2}' \; x_{m-1}' \; y_{m-1}' \; x_m'$$
$$\uparrow$$
$$x_1' \; y_1' \; x_2' \; y_2' \; x_3' \; y_3' \ldots x_{m-1}' \; y_{m-1}' \; x_m' \quad y_2 \quad x_2$$

If the hole is in x'_m, then $y_1 = y'_1 = y'_2 = \cdots = y'_{m-1} = y_2$, $x'_m \subset x_2$, and $x'_m \subset x'_{m-1} = \cdots = x'_1 = x_1$. Here, $u = x_1 y_1$, $v = y_1 x_2$, $z = (x_1 y_1)^{m-1} x'_m$. Now, if the hole is in x'_i for some $1 \le i < m$, then $y_1 = y'_1 = y'_2 = \cdots = y'_{m-1} = y_2$, $x'_i \subset x'_{i+1} = \cdots = x'_m = x_2$, and $x'_i \subset x'_{i-1} = \cdots = x'_1 = x_1$. Here, $u = x_1 y_1$, $v = y_1 x_2$, $z = (x_1 y_1)^{i-1} x'_i y_1 (x_2 y_1)^{m-i-1} x_2$ and Statement 1 holds.

If the hole is in y'_i for some $1 \le i < m$, then $x_1 = x'_1 = x'_2 = \cdots = x'_m = x_2$, $y'_i \subset y'_{i+1} = \cdots = y'_{m-1} = y_2$, and $y'_i \subset y'_{i-1} = \cdots = y'_1 = y_1$. Here, $u = x_1 y_1$, $v = y_2 x_1$, $z = (x_1 y_1)^{i-1} x_1 y'_i (x_1 y_2)^{m-i-1} x_1$ and Statement 2 holds.

2.18 By weakening, $uz \uparrow zv$. If Statement 1 of Exercise 2.17 holds, then there exist partial words x, y, x_1, x_2 such that $u = x_1 y$, $v = y x_2$, $x \subset x_1$, $x \subset x_2$, and $z = (x_1 y)^m x (y x_2)^n$ for some integers $m, n \ge 0$. Since u, v are full, we have y, x_1, x_2 full and thus, $\|H(x)\| = 1$. Since $z \uparrow z'$, there exists a word x' such that $x \sqsubset x'$ and $z' = (x_1 y)^m x' (y x_2)^n$. Now, $uz \uparrow z'v$ implies $(x_1 y)^{m+1} x (y x_2)^n \uparrow (x_1 y)^m x' (y x_2)^{n+1}$ and by simplification, $x_1 y x \uparrow x' y x_2$. Thus, $x_1 \uparrow x'$. The latter along with the fact that both x_1 and x' are full lead to $x' = x_1$, and Statement 1 holds in this case. If Statement 2 of Exercise 2.17 holds, then Statement 2 follows.

2.21 True. Indeed, a word u is primitive if and only if u is not a proper factor of uu, that is, $uu = xuy$ implies $x = \varepsilon$ or $y = \varepsilon$. To see this, assume that u is primitive and that $uu = xuy$ for some nonempty partial words x, y. Since $|x| < |u|$, by Lemma 1.2, there exist nonempty partial words z, v such that $u = zv$, $z = x$, and $vu = uy$. Then $zvzv = xzvy$ yields $vz = zv$ by simplification. By Theorem 2.5, v and z are powers of a common word, a contradiction with the fact that u is primitive.

Now, assume that $uu = xuy$ for some partial words x, y implies $x = \varepsilon$ or $y = \varepsilon$. Suppose to the contrary that u is not primitive. Then there exists a nonempty word v and an integer $n \ge 2$ such that $u = v^n$. But then $uu = v^{n-1} uv$, and using our assumption we get $v^{n-1} = \varepsilon$ or $v = \varepsilon$, a contradiction.

2.22 Put $i = l + j$ where $0 \le j < k$. Since $xy \subset u$ and $yx \subset u$, we have

$x(j) \subset u(j)$ and $y(j) \subset u(j)$,

$y(j) \subset u(j + k)$ and $y(j + k) \subset u(j + k)$,

$y(j + k) \subset u(j + 2k)$ and $y(j + 2k) \subset u(j + 2k)$,

$y(j + 2k) \subset u(j + 3k)$ and $y(j + 3k) \subset u(j + 3k)$,

$\vdots$

$y(j + (m-2)k) \subset u(j + (m-1)k)$ and $y(j + (m-1)k) \subset u(j + (m-1)k)$,

$y(j + (m-1)k) \subset u(j + mk)$ and $x(j) \subset u(j + mk)$.

Put $x(j)y(j)y(j+k)\ldots y(j+(m-1)k)x(j) = v_j$. As in Case 1, the partial word v_j is 1-periodic, say with letter a_j in $A \cup \{\diamond\}$. By letting $z = a_0 a_1 \ldots a_{k-1}$, we get $x \subset z$ and $y \subset z^m$ as desired.

CHAPTER 3

3.3 By Theorem 3.1(2), $\gcd(p'(u), q)$ is a period of u since $|u| \geq p'(u) + q$. Since $p(u)$ is the minimal period of u and $p'(u)$ is the minimal weak period of u, we get $p'(u) \leq p(u) \leq \gcd(p'(u), q)$. We conclude that $p'(u) = \gcd(p'(u), q)$ and so $p'(u)$ divides q.

3.4 The bound is optimal here as can be seen with $abaaba\diamond$ of length 7 which is 3-periodic and 5-periodic but not 1-periodic.

3.7

1. Using Definition 3.2, $H(u) = \{5, 6, 7, 11, 14\}$ 1-isolates $S = \{0, 2, 4, 9\}$.
   Left If $i \in S$ and $i \geq q$, then $i - q \in S$ or $i - q \in H(u)$.
   For $i = 9$, we have $i - q = 9 - 5 = 4 \in S$.
   Right If $i \in S$, then $i + q \in S$ or $i + q \in H(u)$.
   For $i = 0$, we have $i + q = 0 + 5 = 5 \in H(u)$;
   for $i = 2$, $i + q = 2 + 5 = 7 \in H(u)$;
   for $i = 4$, $i + q = 4 + 5 = 9 \in S$;
   for $i = 9$, $i + q = 9 + 5 = 14 \in H(u)$.
   Above If $i \in S$ and $i \geq p$, then $i - p \in S$ or $i - p \in H(u)$.
   For $i = 2$, we have $i - p = 2 - 2 = 0 \in S$;
   for $i = 4$, $i - p = 4 - 2 = 2 \in S$;
   for $i = 9$, $i - p = 9 - 2 = 7 \in H(u)$.
   Below If $i \in S$, then $i + p \in S$ or $i + p \in H(u)$.
   For $i = 0$, we have $i + p = 0 + 2 = 2 \in S$;
   for $i = 2$, $i + p = 2 + 2 = 4 \in S$;
   for $i = 4$, $i + p = 4 + 2 = 6 \in H(u)$;
   for $i = 9$, $i + p = 9 + 2 = 11 \in H(u)$.

2. Using Definition 3.3, $H(v) = \{7, 9, 10, 16, 17, 19\}$ 2-isolates $S = \{12, 14\}$.

3. Using Definition 3.4, $H(w) = \{14, 17, 20, 21\}$ 3-isolates $S = \{19, 22\}$.

3.9 Although $G_{(4,7)}(u)$ is disconnected, the partial word u is not $(2, 4, 7)$-special by using Definition 3.1. The undirected graph $G_{(4,7)}(u)$ is shown in Figure 3.

3.13 The proof is divided into two cases.

First, if $p = 1$ and $q > 1$, then by Definition 3.1(2)(d), $i - p, i + p, i + q \in H(v_n)$ with $i = 1$.

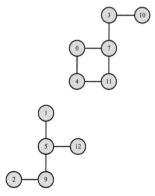

FIGURE 3: The disconnected graph $G_{(4,7)}(u)$.

Second, if $p > 1$, then by Definition 3.1(1)(b), $i + p, i + q \in H(v_n)$ with $i = 1$. The weakly p- and weakly q-periodicity can be seen in Figure 4.

$$\overbrace{\Diamond\, a\, b ... b}^{p \text{ columns}} \qquad \overbrace{\Diamond\, a\, b ... b \Diamond b ... b}^{q \text{ columns}}$$

$$b \Diamond b ... \qquad\qquad b \Diamond b ...$$

FIGURE 4: A $(3, p, q)$-special binary partial word.

3.16 The sequence $(ab^{p-1} \diamond b^{q-p-1} \diamond^{H-1} b^n)_{n>0}$ satisfies the desired properties.

3.20 See Reference [14].

CHAPTER 4

4.3 Let x, y, s be nonempty partial words satisfying $y \subset x$, $u = rx$ and $y = vs$ for some pword r. Here $w = uv = rxv$, and since v is the maximal suffix with respect to $\preceq_r$, we get $xv \preceq_r v$. Since $y \subset x$, we get $yv \preceq_r v$. Replacing y by vs in the latter inequality yields $vsv \preceq_r v$, leading to a contradiction.

4.9 The minimal local periods are: 3, 3, 1, 1, 3 and 3. The maximum among all minimal local periods is 3. Since $p'(w) = 3$, w has four critical factorizations.

4.10 The statement is true. The partial word $w = a \diamond bc$ serves as an example. Note that if $a \prec_l b \prec_l c$, then w is special according to Definition 4.4(1). Here $(a \diamond b)(c) = uv = w = u'v' = (\varepsilon)(a \diamond bc)$ and $|v| \leq |v'|$. We have $p(w, |u| - 1) = 2 < 3 = |u|$ and $r = a \notin C(S(u))$.

4.11 Here $v = ccb \diamond ab \diamond ba$ and $v' = ab \diamond ba$ are the maximal suffixes of w with respect to $\preceq_l$ and $\preceq_r$ respectively. We have $|v'| < |v|$ and $w = u'v' = (ccb \diamond)(ab \diamond ba)$. Since $p(w, |u'| - 1) = p(w, 3) = 1 < 4 = |u'|$ and $r = ccb \notin C(S(u'))$, w is special.

4.12 The result being trivial for $v \in A^+$, assume that $\|H(v)\| = 1$. If $u \in P(v)$, then both $u \preceq_l v$ and $u \preceq_r v$. Conversely, if both $u \preceq_l v$ and $u \preceq_r v$, then either u is a prefix of v, or $u = \mathrm{pre}(u, v)ax, v = \mathrm{pre}(u, v)by$ with $a, b \in A \cup \{\diamond\}$ satisfying $a \prec_l b$ and $a \prec_r b$. The latter possibility leads to $a = \diamond$, contradicting the fact that u is full.

4.19 Below are tables for the nonempty suffixes of the partial word $w = a \diamond cbba$ and its reversal $\mathrm{rev}(w) = abbc \diamond a$. These suffixes are ordered in two different ways: The first ordering is on the left and is an $\prec_l$-ordering according to the order $\diamond \prec a \prec b \prec c$, and the second is on the right and is an $\prec_r$-ordering where $\diamond \prec c \prec b \prec a$. The tables also contain the indices used by the algorithm, k_0, l_0, k_1, l_1, and the local periods that needed to be calculated in order to compute the critical factorization $(a \diamond c, bba)$. The minimal weak period of w turns out to be equal to 5.

k_0	p_{0,k_0}	v_{0,k_0}	v'_{0,l_0}	p_{0,l_0}	l_0
5		$\diamond cbba$	$\diamond cbba$		5
4		a	$cbba$		4
3		$a \diamond cbba$	bba		3
2		ba	ba		2
1		bba	a		1
0	1	$cbba$	$a \diamond cbba$		0

k_1	p_{1,k_1}	v_{1,k_1}	v'_{1,l_1}	p_{1,l_1}	l_1
5		$\diamond a$	$\diamond a$		5
4		a	$c \diamond a$		4
3		$abbc \diamond a$	$bc \diamond a$		3
2		$bbc \diamond a$	$bbc \diamond a$		2
1		$bc \diamond a$	a		1
0	5	$c \diamond a$	$abbc \diamond a$		0

Algorithm 4.3 starts with

$$(v_{0,0}, v'_{0,0}) = (cbba, a\diamond cbba) \text{ and } (v_{1,0}, v'_{1,0}) = (c\diamond a, abbc\diamond a)$$

and selects the shortest component of each pair, that is, $v_{0,0}$ and $v_{1,0}$. In Step 2, mwp is set to 0. In Step 7, $p_{0,0}$ and $p_{1,0}$ are calculated to be 1 and 5 respectively. In Step 10, the value of mwp is updated to be $p_{1,0} = 5$. In Part 3 of Step 10, the critical factorization is output as $(a\diamond c, bba)$.

4.21

Case 2. $p_{0,k_0} < |u_{0,k_0}|$ and $p_{0,k_0} > |v_{0,k_0}|$
Here Definition 4.2(3) is satisfied and there exist partial words x, y, r, s, γ such that $|x| = p_{0,k_0}$, $\gamma \uparrow v_{0,k_0}$, $u_{0,k_0} = rx = r\gamma s$, and $y = v_{0,k_0}s$. Note that if $k_0 = 0$ and $v_{0,k_0} \subset \gamma$, then $y \subset x$ and we get a contradiction with Lemma 4.3. If $r \notin C(S(u_{0,k_0}))$, then w is $((k_0, l_0))$-special by Definition 4.4(1). If $r \in C(S(u_{0,k_0}))$, then there exists x' such that $x'r \uparrow rx$. The result follows as in Case 2.

Case 3. $p_{0,k_0} < |u_{0,k_0}|$ and $p_{0,k_0} \leq |v_{0,k_0}|$
Here Definition 4.2(1) is satisfied and there exist partial words x, y, r, s such that $|x| = p_{0,k_0}$, $x \uparrow y$, $u_{0,k_0} = rx$, and $v_{0,k_0} = ys$. Note that if $k_0 = 0$ and $y \subset x$, then we get a contradiction with Lemma 4.2. Here w is $((k_0, l_0))$-special by Definition 4.4 unless $r \in C(S(u_{0,k_0}))$ and $s \in C(P(v_{0,k_0}))$. If the two conditions hold, then $x'r \uparrow rx$ and $ys \uparrow sy'$ for some x', y'. The result follows as in Case 3.

Case 4. $p_{0,k_0} \geq |u_{0,k_0}|$ and $p_{0,k_0} < |v_{0,k_0}|$
Here Definition 4.2(2) is satisfied and there exist partial words x, y, r, s such that $|x| = p_{0,k_0}$, $x \uparrow y$, $x = ru_{0,k_0}$ and $v_{0,k_0} = ys$. Note that if $k_0 = 0$ and $r = \varepsilon$ and $y \subset x$, then we get a contradiction with Lemma 4.2. Here w is $((k_0, l_0))$-special by Definition 4.4(2) unless $s \in C(P(v_{0,k_0}))$. If $s \in C(P(v_{0,k_0}))$, then $ys \uparrow sy'$ for some y' and the result follows as in Case 4.

CHAPTER 5

5.1

- Find the minimal period $p(u)$ of u.
- Find integers m and r such that $|u| = mp(u) + r$ where $0 \leq r < p(u)$.
  - If $r = 0$, then $v = \varepsilon$, $w = u(0) \ldots u(p(u) - 1)$ and $k = m$.
  - If $r \neq 0$, then $v = u(0) \ldots u(r - 1)$, $w = u(r) \ldots u(p(u) - 1)$ and $k = m$.

- Output $u = (vw)^k v$.

5.3 The partial word $0\diamond00$ has a period of 1 but u does not; $0\diamond01$ has a period of 2 but u does not; $0\diamond10$ has a period of 3 but u does not; and $0\diamond11$ has a weak period of 1 but u does not.

5.10 If $u = (adabc)(d)(\diamond dabc)(d)(adabc) = v_1wv_2wv_3$, then

$$\text{Bin}(v_1wv_1) = 01111101111$$

and $T(u) = [011111\diamond1111101111, a, a]$. Both u and $\text{Bin}'(u)$ have only the periods 6, 12, 17 and the weak periods 6, 12, 17. This example illustrates Item 2(b)(iii).

5.12 On input $u = (abcdabedabcd)(ab\diamond dabedabcd)(abedabedabcd)$, the algorithm proceeds as follows:

- The partial words found satisfy Lemma 5.7 with $1 < i < k$ and $a \neq b$. Indeed,

  $$u = (ab\underline{c}dabedabcd)(ab\diamond dabedabcd)(ab\underline{e}dabedabcd) = w_1w_2w_3,$$

  where $v = \varepsilon$, $k = 3$, $i = 2$, $x = ab$, $y = dabedabcd$ and $c \neq e$.
- And $T(vw_iv) = [\text{Bin}'(vw_iv), \alpha, \beta]$ is such that ($\alpha = \square$ and $\beta = \square$) or ($\beta \neq \square$ and $x \neq \varepsilon$). Indeed, $T(w_2) = [011111110111, \square, c]$ is such that $\beta = c \neq \square$ and $x \neq \varepsilon$.

In this case,

1. Compute $\text{Bin}'(vw_iv) = v'w'v'$ where $|v'| = |v|$ and $|w'| = |w_i|$. Here $\text{Bin}'(w_2) = (0111)(1111)(0111)$.
2. Compute $h' = h - (i-1)p'(u)$ or $h' = h - (2-1)p'(u) = 14 - (2-1)12 = 2$, and compute $d \in \{0,1\}$ as follows: Since $\alpha = \square$, $\beta \neq \square$, $x \neq \varepsilon$, the "a" value is equal to β and $|x| < |y|$, we have $d = \text{Bin}'(vw_iv)(h' + p'(vw_iv)) = \text{Bin}'(w_2)(2 + p'(w_2)) = (011111110111)(10) = 1$.
3. Output $T(u) =$
   $[((v'w')(h',d))^{i-1}(v'w')(h',\diamond)((v'w')(h',\bar{d}))^{k-i}v', a, b] =$
   $= [(01\underline{1}111110111)(01\diamond111110111)(01\underline{0}111110111), c, e]$
   Both u and $\text{Bin}'(u)$ have only the periods 32, 36 and the weak periods 12, 32, 36.

5.13 Statement 3 is impossible.

5.17 For any $0 \leq j < |u| - q = p(u) + |v| - r$, we have $u(j) = (vwv)(j)$ and $u(j+q) = (vwv)(j+r)$. Hence $u(j) = u(j+q)$ if and only if $(vwv)(j) = (vwv)(j+r)$. The latter implies that $q \in \mathcal{P}(u)$ if and only if $r \in \mathcal{P}(vwv)$, as claimed.

5.19 Let us first consider finding the minimal period of a word. A linear pattern matching algorithm can be easily adapted to compute the minimal period of a given word u. Given words v and w, the algorithm finds the leftmost occurrence, if any, of v as a factor of w. The comparisons done by the algorithm are of the type $a \overset{?}{=} b$, for letters a and b. We consider a new letter (wild card) $\#$ which passes the test $\# \overset{?}{=} a$ for any letter a. Put $u = au'$, where a is a letter. Then we run the algorithm on the inputs $u, u'\#^{|u|}$. Clearly, an integer p, $1 \leq p \leq |u|$, is a period of u if and only if u is a factor starting at position $p - 1$ of $u'\#^{|u|}$. Therefore, the leftmost occurrence of u as a factor of $u'\#^{|u|}$ (which always exists) gives the minimal period of u. Consequently, the computing of $p(u)$ can be performed in linear time. Finding a positive integer k and words v, w satisfying Lemma 5.3 is performed in linear time, since we know that $p(u) = |vw|$ from computing the minimal period as described above. Step 2(2) is obviously performed in linear time. At Step 2(1), we have to test which of the words $\text{Bin}(v)1^{|w|-1}0$ or $\text{Bin}(v)1^{|w|-1}1$ is primitive. Primitivity can be tested in linear time for full words as will be shown in Chapter 6. Indeed, a word u is primitive if and only if $u^2 = xuy$ implies that either $x = \varepsilon$ or $y = \varepsilon$.

The algorithm is recursive, so let us compute the complexity of a single call of the procedure Bin, say $f(n)$, where n is the length of the current word for this call, say u. Consequently, we have shown so far that a single call of Bin requires $f(n) = O(n)$ time. More precisely, there is a constant c such that $f(n) \leq cn$, for any $n \geq 0$.

To calculate the time required for the whole algorithm on an input u of length n, we first determine how fast the length of the current word decreases from a call to the next call. Consider u_1 and u_2 the current words for two consecutive calls of Bin on u, respectively. We have that either $u_1 = (vw)^k v$ and $u_2 = vwv$ with $k \geq 2$ (if $\text{Bin}(u_2)$ is called at Step 2(2) in $\text{Bin}(u_1)$), or $u_1 = vwv$ and $u_2 = v$ (if $\text{Bin}(u_2)$ is called at Step 2(1) in $\text{Bin}(u_1)$). In either case, $|u_2| \leq 2/3|u_1|$. Therefore, the time required by the algorithm to compute $\text{Bin}(u)$ is at most

$$\Sigma_{i \geq 0} f((2/3)^i n) \leq \Sigma_{i \geq 0} c(2/3)^i n \leq 3cn$$

hence it is linear, as claimed. Finally, it is clear that the algorithm is optimal, as the problem requires at least linear time.

5.20 Let u be a nonempty partial word over A with minimal weak period $p'(u)$. Then $|u| = kp'(u) + r$ where $0 \leq r < p'(u)$. Put $u = v_1 w_1 v_2 w_2 \ldots v_k w_k v_{k+1}$ where $|v_1 w_1| = |v_2 w_2| = \cdots = |v_k w_k| = p'(u)$ and $|v_1| = |v_2| = \cdots = |v_k| = |v_{k+1}| = r$. If w_i is empty, then $r = |v_{k+1}| = |v_k| = p'(u)$, a contradiction. If $k = 0$, then $u = v_{k+1}$ and u has weak period $|v_{k+1}| < p'(u)$ contradicting the fact that $p'(u)$

is the minimal weak period of u. Since $p'(u)$ is the minimal weak period of u, we get $v_i w_i \uparrow v_{i+1} w_{i+1}$ for all $1 \le i < k$ and $v_k \uparrow v_{k+1}$. The result follows.

5.24 There exists c such that $\mathrm{Bin}(v_2)1^{|w|-1}c$ is primitive by Lemma 5.1. The equality $\mathcal{P}'(u) = \mathcal{P}(u)$ holds since every weak period of u is greater than or equal to $p'(u)$, and the equality $\mathcal{P}'(u') = \mathcal{P}(u')$ holds trivially.

To see that $\mathcal{P}(u) \subset \mathcal{P}(u')$, first note that $\mathcal{P}(\mathrm{Bin}(v_2)) = \mathcal{P}(v_2)$, and all periods q of u satisfy $q \ge p'(u)$. If $q = p'(u)$, then q is a period of u'. If $q > p'(u)$, put $q = p'(u) + r$ where $r > 0$. Then r is a weak period of v_1. Since $\beta = \square$, $h + r \ge h + p'(v_1) \ge |v_1|$ where $H(v_1) = \{h\}$. In this case, r is a period of v_2 and hence of $\mathrm{Bin}(v_2)$, and so $q \in \mathcal{P}(u')$.

Assume then that there exists $q \in \mathcal{P}(u') \backslash \mathcal{P}(u)$ and also that q is minimal with this property. Either $q < |\mathrm{Bin}(v_2)|$ or $|\mathrm{Bin}(v_2)| + |w| - 1 \le q < |u|$, since $\mathrm{Bin}(v_2)$ does not begin with 1. If $q < |\mathrm{Bin}(v_2)|$, then, by the minimality of q, q is the minimal period of u', and Lemma 5.2 implies that $p'(u)$ is a multiple of q, and so $\mathrm{Bin}(v_2)1^{|w|-1}c$ is not primitive, a contradiction. If $q = |\mathrm{Bin}(v_2)| + |w| - 1$, then $c = 0$. In this case, if $|w| > 1$, we get $\mathrm{Bin}(v_2)1 = 0\mathrm{Bin}(v_2)$, which is impossible, and if $|w| = 1$, we get that $\mathrm{Bin}(v_2)$ consists of 0's only and therefore $\mathrm{Bin}(v_2)1^{|w|-1}c = \mathrm{Bin}(v_2)0$ is not primitive. Hence $q > |\mathrm{Bin}(v_2)| + |w| - 1$, and $q > p'(u)$ since $p'(u) \notin \mathcal{P}(u') \backslash \mathcal{P}(u)$. By putting $q = p'(u) + r$ where $r > 0$, we get that r is a period of $\mathrm{Bin}(v_2)$ and hence of v_2. Therefore $q \in \mathcal{P}(u)$.

5.25 See Reference [23].

CHAPTER 6

6.1 Here $u = abca\diamond\diamond\diamond bc$ where $D(u) = \{0, 1, 2, 3, 7, 8\}$ and $H(u) = \{4, 5, 6\}$. The algorithm proceeds as follows:

$k = 1, l = 8$: Compatibility of u with $U[1..10]$ is nonsuccessful.

$k = 2, l = 7$: Compatibility of u with $U[2..11]$ is nonsuccessful.

$k = 3, l = 6$: Compatibility of u with $U[3..12]$ is successful.

$$a\, b\, c\, a\, \diamond\, \diamond\, \diamond\, b\, c\, a\, b\, c\, a\, \diamond\, \diamond\, \diamond\, b\, c$$
$$a\, b\, c\, a\, \diamond\, \diamond\, \diamond\, b\, c$$

The partial word u is not $(3, 6)$-special and is thus nonprimitive ($u \subset (abc)^3$).

6.6 The values are 408 and 513.

6.9 The result follows from the following list of equalities:

$$\begin{aligned}
P_{1,k}(n) &= T_{1,k}(n) - N_{1,k}(n) \\
&= T_{1,k}(n) - nN_{0,k}(n) \\
&= T_{1,k}(n) - n(T_{0,k}(n) - P_{0,k}(n)) \\
&= nk^{n-1} - nk^n + nP_{0,k}(n) \\
&= n(P_{0,k}(n) + k^{n-1} - k^n)
\end{aligned}$$

6.12 Consider for example the partial word $u = b \diamond b \diamond b$. Neither ua nor ub is primitive since $ua \subset (ba)^3$ and $ub \subset (bb)^3$.

6.18 Let w' be the prefix of length $|u| + |v|$ of w. Both $|u|$ and $|v|$ are periods of w'. By Theorem 3.1, $\gcd(|u|, |v|)$ is also a period of w', and hence there exists a word x of length $\gcd(|u|, |v|)$ such that w' is contained in a power of x. If $H(w') = \emptyset$, then the result clearly follows. Otherwise, put $H(w') = \{i\}$ where $0 \leq i < |w'|$. Let r, $0 \leq r < |x|$, be the remainder of the division of i by $|x|$. If $i < |x|$, then $i = r$ and $w'(i + |x|) = x(r)$, and if $i \geq |x|$, then $w'(i - |x|) = x(r)$. Hence for all $0 \leq j < |x|$ and $j \neq r$, we have $x(j) = w'(j)$, and we have $x(r) = w'(i + |x|)$ or $x(r) = w'(i - |x|)$. Since $|x|$ divides both $|u|$ and $|v|$, we conclude that $u = x^k$ and $v = x^l$ for some integers k, l.

6.19 First, assume that $n = 1$. Let x be a primitive word such that $uv = x$. By Proposition 6.10, since uv is primitive, vu is also primitive. The result follows with $y = vu$.

Now, assume that $n > 1$. Since $uv = x^n$, there exist words x_1, x_2 such that $x = x_1 x_2$, $u = (x_1 x_2)^k x_1$ and $v = x_2 (x_1 x_2)^l$ with $k + l = n - 1$. Since $x = x_1 x_2$ is primitive, $x_2 x_1$ is also primitive by Proposition 6.10. The result follows since $vu = (x_2 x_1)^n$.

Now, suppose that uv is a primitive partial word. If vu is not primitive, then there exists a word y such that $vu = y^m$ for some $m \geq 2$. So there exist words y_1, y_2 such that $y = y_1 y_2$, $v = (y_1 y_2)^k y_1$ and $u = y_2 (y_1 y_2)^l$ with $k + l = m - 1$. Hence $uv = (y_2 y_1)^m$ and uv is not primitive, a contradiction. Therefore, if uv is primitive, then vu is primitive.

6.20 Put $u = u_1 \diamond u_2 \diamond u_3 \diamond u_4$ where the u_j's do not contain any holes.

$m = 2$: There exist a word x and integers $0 = i_0 < i_1 \leq 3$ such that

$$u_{i_0+1} \diamond \ldots \diamond u_{i_1} \subset x$$
$$u_{i_1+1} \diamond \ldots \diamond u_4 \subset x$$

$m = 3$: There exist a word x and integers $0 = i_0 < i_1 < i_2 \leq 3$ such that

$$u_{i_0+1}\diamond\ldots\diamond u_{i_1} \subset x$$
$$u_{i_1+1}\diamond\ldots\diamond u_{i_2} \subset x$$
$$u_{i_2+1}\diamond\ldots\diamond u_4 \subset x$$

$m = 4$: There exist a word x and integers $0 = i_0 < i_1 < i_2 < i_3 \leq 3$ such that

$$u_{i_0+1}\diamond\ldots\diamond u_{i_1} \subset x$$
$$u_{i_1+1}\diamond\ldots\diamond u_{i_2} \subset x$$
$$u_{i_2+1}\diamond\ldots\diamond u_{i_3} \subset x$$
$$u_{i_3+1}\diamond\ldots\diamond u_4 \subset x$$

Consequently, the set S_3 consists of the partial words of the form

$$x_1 a x_2 b x_3 \diamond x_1 \diamond x_2 \diamond x_3 \text{ or } x_1 \diamond x_2 \diamond x_3 \diamond x_1 a x_2 b x_3$$

for words x_1, x_2, x_3 and letters a, b; or $x_1 \diamond x_2 \diamond x_3 \diamond x_4$ for words x_1, x_2, x_3, x_4 and letters a, b satisfying $x_1 a x_2 = x_3 b x_4$; or

$$x_1 \diamond x_2 \diamond x_1 a x_2 \diamond x_1 a x_2 \text{ or } x_1 a x_2 \diamond x_1 \diamond x_2 \diamond x_1 a x_2 \text{ or } x_1 a x_2 \diamond x_1 a x_2 \diamond x_1 \diamond x_2$$

for words x_1, x_2 and letter a; or $x \diamond x \diamond x \diamond x$ for a word x.

CHAPTER 7

7.5 Note that Proposition 7.9 implies that if u is a full bordered word, then $x_1 = x$ is unbordered. In this case, $u = x u' x$ where x is the minimal border of u. Hence a bordered full word is always simply bordered.

7.8 Yes. Here $u = (abaa)(aba)(abaaac)(a)$ where $abaa, aba, abaaac$, and a are prefixes of $v = abaaacc$.

7.9 The following table depicts the information submitted:

partial word v	$ab\diamond\diamond ba\diamond\diamond abba$
prefix sequence	$(ab\diamond\diamond, ab, ab\diamond\diamond ba\diamond, a, ab\diamond\diamond ba\diamond\diamond a)$

The set S contains all nonempty prefixes of v, while the set S' contains all nonempty unbordered prefixes of v. The set S consists of the elements

$$a, ab, ab\diamond, ab\diamond\diamond, ab\diamond\diamond b, ab\diamond\diamond ba, ab\diamond\diamond ba\diamond, ab\diamond\diamond ba\diamond\diamond,$$
$$ab\diamond\diamond ba\diamond\diamond a, ab\diamond\diamond ba\diamond\diamond ab, ab\diamond\diamond ba\diamond\diamond abb, ab\diamond\diamond ba\diamond\diamond abba$$

and the set S' of

$$a, ab$$

In the first iteration, the multiset T' contains the input sequence. During subsequent iterations, it is determined whether each object in T' is well ordered, badly bordered, or unbordered. If the object is well bordered, it is split into two smaller objects, and T' is updated. Otherwise T' is updated and the algorithm continues until either a badly bordered object is found or $T' \subset S'$.

Iteration	T'
1	$\{ab\diamond\diamond, ab, ab\diamond\diamond ba\diamond, a, ab\diamond\diamond ba\diamond\diamond a\}$
2	$\{ab\diamond, a, ab, ab\diamond\diamond ba, a, a, ab\diamond\diamond ba\diamond\diamond, a\}$
3	$\{ab, a, a, ab, ab\diamond\diamond b, a, a, a, ab\diamond\diamond ba\diamond, a, a\}$
4	$\{ab, a, a, ab, ab\diamond, ab, a, a, a, ab\diamond\diamond ba, a, a, a\}$
5	$\{ab, a, a, ab, ab, a, ab, a, a, a, ab\diamond\diamond b, a, a, a, a\}$
6	$\{ab, a, a, ab, ab, a, ab, a, a, a, ab\diamond, ab, a, a, a, a\}$
7	$\{ab, a, a, ab, ab, a, ab, a, a, a, ab, a, ab, a, a, a, a\}$

Since $T' \subset S'$, a sequence of unbordered prefixes of v does exist that is compatible with the original sequence:

$$ab, a, a, ab, ab, a, ab, a, a, a, ab, a, ab, a, a, a, a$$

7.12 The factorization $(u, v) = (aa, bc\diamond bc)$ of w is critical and $w' = vu = bc\diamond bcaa$ is unbordered. The position $|v| - 1 = 4$ is a critical point of w'.

7.15 These equalities can be seen from the fact that if a word has odd length $2n+1$ then it is unbordered if and only if it is unbordered after removing the middle letter. If a word has even length $2n$ then it is unbordered if and only if it is obtained from an unbordered word of length $2n - 1$ by adding a letter next to the middle position unless doing so creates a word that is a perfect square.

7.16 Let $B_k(j, n)$ be the number of full words of length n over a k-letter alphabet that have a minimal border of length j:

$$B_k(j, n) = U_k(j)k^{n-2j}$$

If we let $B_k(n)$ be the number of full words of length n over a k-letter alphabet with a border of any length, then we have that

$$B_k(n) = \sum_{j=1}^{\lfloor \frac{n}{2} \rfloor} B_k(j, n)$$

7.23 Let x be a minimal border of w. Because w is well bordered, we can write $w = x_1 w' x_2$ with $x_1 \subset x$ and $x_2 \subset x$ and x_1 unbordered. Suppose first that $au' = au$. If $|x| = |au| = |au'|$, then we have that $w = au'w_1 \subset xw_1$ and we also have that $w = w_2bu' \subset w_2x$. But because $au' \subset x$ and $bu' \subset x$ this leads us to conclude that $a = b$, contradicting (2). If $|x| < |au|$, then x_1 is a prefix of $au = au'$ and x_2 is a suffix of u' so au' is bordered by x, which contradicts (3). So we must have that $|x| > |au|$, and the conclusion follows. If au' is a proper prefix of au, then we have three cases similar to the above:

    Case 1. $|x| < |au'|$

Then we conclude, as above, that au' is bordered. This contradicts (3).

    Case 2. $|x| = |au'|$

Then, as above, we conclude that $au' \subset x$ and $bu' \subset x$ and $a = b$ which contradicts (2).

    Case 3. $|au'| < |x| \leq |au|$

Here x_1 is a longer unbordered prefix of au than au'. This contradicts (3) and so we must have that $|x| > |au|$, and au must be contained in a proper prefix of x.

7.24 Suppose there exists w a factor of u with $w = hv'h$ such that h is not compatible with any factor of v'. The equality $h = \text{unb}(hv'h)$ holds. If v is a full word, then v' is full and the conditions of Proposition 7.5 are met for $hv'h$ and we have that $v'h$ is unbordered. So $|v'h| \leq \mu(u)$ but this only holds if $v' = \varepsilon$. So we must have $u = h(h')^{k-2}h$ for some integer $k \geq 2$ and word h' satisfying $h \subset h'$. So u is $|h|$-periodic and thus weakly $|h|$-periodic and we have that $p'(u) \leq |h| = \mu(u)$. Proposition 7.4 states that $p'(u) \geq \mu(u)$ so we must have that $p'(u) = \mu(u)$.

The equality $u = h(h')^{k-2}h$ cannot be replaced by $u = h^k$ as is seen by considering $u = a \diamond bcabbcabbca \diamond bc$. We have $u = hvh$ where $h = a \diamond bc$ is unbordered and $v = abbcabbc$ is full.

CHAPTER 8

8.1 The inclusion $\subset$ in Statement 1 follows from the fact that $\mathcal{P}(u) \subset \mathcal{P}(u^k)$ for all u, k. To see that the inclusion $\sqsubset$ holds in case $(i, j) \neq (1, 1)$, we argue as follows: if $i > 1$, then we consider $u = \diamond$ and $v = a^{i-1}b$ which satisfy $(u, v) \notin \delta$ and $u\delta_{i,j}v$; and if $j > 1$, then we consider $u = a^{j-1}b$ and $v = \diamond$ which satisfy $(u, v) \notin \delta$ and $u\delta_{i,j}v$. Statement 2 follows from the fact that $\mathcal{P}(u) = A$ for all u.

8.8 No since $bbb \in W(A) \setminus F(C(X^*))$.

8.10 Let $u = a\diamond b$ and $v = aababbaababb$ so that $|v| = 4|u|$ and $\|H(v)\| = 0$. A nontrivial compatibility relation does exist, $u^2 v \uparrow vu^2$,

$$a\diamond ba\diamond baababbaababb \uparrow aababbaababba\diamond ba\diamond b$$

thus this set is not a pcode. Note that this set yields a nontrivial compatibility relation prior to the upper bound of $k = 4$.

8.12 If $\{u, v\}$ is a pcode, then clearly $uv \not\Uparrow vu$. Conversely, assume that $\{u, v\}$ is not a pcode and $uv \not\Uparrow vu$. Then there exist an integer $n \geq 1$ and partial words $u_1, u_2, \ldots, u_n, v_1, v_2, \ldots, v_n \in \{u, v\}$ such that

$$u_1 u_2 \ldots u_n \uparrow v_1 v_2 \ldots v_n$$

and with $|u_1 u_2 \ldots u_n|$ as small as possible contradicting Proposition 8.4. We hence have $u_1 \neq v_1$ and $u_n \neq v_n$, and we may assume that $n \geq 2$. There are four possibilities: $u_1 = u_n = u, v_1 = v_n = v$; $u_1 = v_n = u, v_1 = u_n = v$; $u_1 = v_n = v, v_1 = u_n = u$; and $u_1 = u_n = v, v_1 = v_n = u$. In all cases, put $u_2 \ldots u_{n-1} = x$ and $v_2 \ldots v_{n-1} = y$. These possibilities can be rewritten as

$$(1)\ uxu \uparrow vyv \quad (2)\ uxv \uparrow vyu \quad (3)\ vxu \uparrow uyv \quad (4)\ vxv \uparrow uyu$$

Since $|u| > |v|$, for any of the possibilities (1)-(4), there exist nonempty pwords w, w', z, z' such that $u = wz = z'w', w \uparrow v$, and $w' \uparrow v$. The latter two relations give $w \subset v$ and $w' \subset v$ since v is full. Since $|w| = |z|$, we get $u = ww' \subset v^2, uv \subset v^3, vu \subset v^3$, and thus $uv \uparrow vu$, a contradiction.

First, consider the set $\{aba, a\diamond\diamond bab\}$. Let $u = a\diamond\diamond bab$ and $v = aba$ so that $|u| = 2|v|$ and $\|H(v)\| = 0$. In this case, $uv \not\Uparrow vu$,

$$a\diamond\diamond b\underline{ababa} \not\Uparrow aba\underline{a}\diamond\diamond\underline{bab}$$

and thus this set is a pcode.

Second, consider the set $\{b\diamond abb\diamond, bba\}$. Let $u = b\diamond abb\diamond$ and $v = bba$ so that $|u| = 2|v|$ and $\|H(v)\| = 0$. A nontrivial compatibility relation does exist, $uv \uparrow vu$,

$$b\diamond abb\diamond bba \uparrow bbab\diamond abb\diamond$$

and thus this set is not a pcode.

8.14 Suppose that there exist two distinct conjugate partial words u and v in X, and let x, y be partial words such that $u \subset xy, v \subset yx$. If $x = \varepsilon$ or $y = \varepsilon$, then $u \uparrow v$, contradicting the fact that X is a pcode. So we may assume that $x \neq \varepsilon$ and $y \neq \varepsilon$. Since X is a circular pcode, the two conditions $yux \uparrow vv$ and $u \subset xy$ imply $x = \varepsilon$, a contradiction.

8.16 To see this, suppose the contrary. Then there exists a partial word $u \notin X$ such that $Y = X \cup \{u\}$ is a pcode. Since $|u^n|$ is a multiple of n, the partial word u^n can be written as $u_1 u_2 \ldots u_{|u|}$ where $|u_i| = n$ for all $i = 1, \ldots, |u|$. Thus u^n belongs to Y^*, and there exist $v_1, v_2, \ldots, v_{|u|} \in X$ such that $u_1 \uparrow v_1, \ldots, u_{|u|} \uparrow v_{|u|}$ showing that u^n also belongs to $C(X^*)$. We get the nontrivial compatibility relation

$$u_1 u_2 \ldots u_{|u|} \uparrow v_1 v_2 \ldots v_{|u|}$$

and so Y is not a pcode and X is maximal.

8.17 Let $\varphi : B^* \to W(A)$ be a morphism such that φ is a bijection of B onto X. Let $u, v \in B^*$ be words such that $\varphi(u) \uparrow \varphi(v)$. If $u = \varepsilon$, then $v = \varepsilon$. To see this, $\varphi(b) \neq \varepsilon$ for each letter $b \in B$ since $\varphi(b) \in X$ and X does not contain ε. If $u \neq \varepsilon$ and $v \neq \varepsilon$, put $u = b_1 \ldots b_m$ and $v = b'_1 \ldots b'_n$ with positive integers m, n and $b_1, \ldots, b_m, b'_1, \ldots, b'_n \in B$. Since φ is a morphism, we have

$$\varphi(b_1) \ldots \varphi(b_m) \uparrow \varphi(b'_1) \ldots \varphi(b'_n)$$

But X is a pcode and $\varphi(b_i), \varphi(b'_j) \in X$. Thus $m = n$ and $\varphi(b_i) = \varphi(b'_i)$ for $i = 1, \ldots, m$. Now φ is injective on B. Thus $b_i = b'_i$ for $i = 1, \ldots, m$, and $u = v$. This shows that φ is pinjective.

Conversely, let $\varphi : B^* \to W(A)$ be a pinjective morphism such that $X = \varphi(B)$. If

$$u_1 u_2 \ldots u_m \uparrow v_1 v_2 \ldots v_n$$

for some positive integers m, n, and $u_1, \ldots, u_m, v_1, \ldots, v_n \in X$, then consider the elements $b_i, b'_j \in B$ such that $\varphi(b_i) = u_i$, $\varphi(b'_j) = v_j$ for $i = 1, \ldots, m, j = 1, \ldots, n$. Since φ is pinjective, the above compatibility relation implies that $b_1 \ldots b_m = b'_1 \ldots b'_n$. Thus $m = n$ and $b_i = b'_i$ for $i = 1, \ldots, m$. Whence $u_i = v_i$ for $i = 1, \ldots, m$.

For Corollary 1, let $\psi : B^* \to W(A)$ be a pcoding morphism for X. Then $\varphi(\psi(B)) = \varphi(X)$, and since $\varphi \circ \psi : B^* \to W(C)$ is a pinjective morphism, Proposition 8.13 shows that $\varphi(X)$ is a pcode over C.

For Corollary 2, let $\varphi : B^* \to W(A)$ be a pcoding morphism for X. Then $X^n = \varphi(B^n)$. But B^n is a code over B. Thus the conclusion follows from Corollary 1.

8.18 Put $N = M \setminus \{\varepsilon\}$. First, we prove that X generates M. Since $X \subset M$, we have $X^* \subset M$. To show the other inclusion, we use induction on the length of partial words. Clearly, $\varepsilon \in X^*$. If $m \in N \setminus N^2$, then $m \in X$. If $m \in N^2$, then put $m = m_1 m_2$ where m_1 and m_2 are elements of N shorter than m. Therefore m_1, m_2 belong to X^* and $m \in X^*$.

Now, we prove that X is contained in any set $Y \subset W(A)$ generating M. We may assume that $\varepsilon \notin Y$. Then each $x \in X$ is in Y^* and therefore can be written as $x = y_1 y_2 \ldots y_n$ where $y_1, y_2, \ldots, y_n \in Y$ and $n \geq 0$. The facts that $x \neq \varepsilon$ and $x \notin N^2$ imply $n = 1$ and $x \in Y$. This shows that $X \subset Y$ and thus X is a minimal generating set. The uniqueness of such a minimal set follows.

8.20 Assume first that M is stable. Put $X = (M \setminus \{\varepsilon\}) \setminus (M \setminus \{\varepsilon\})^2$. To prove that X is a pcode, suppose the contrary. Then there exist positive integers m, n and partial words $u_1, \ldots, u_m, v_1, \ldots, v_n \in X$ such that

$$u_1 u_2 \ldots u_m \uparrow v_1 v_2 \ldots v_n$$

and with $|u_1 u_2 \ldots u_m|$ as small as possible contradicting the definition of a pcode. We hence have $u_1 \neq v_1$. We may suppose $|u_1| \leq |v_1|$. If $|u_1| = |v_1|$, then $u_1 \uparrow v_1$. Since M is stable, we deduce that $u_1 = v_1$, a contradiction. If $|u_1| < |v_1|$, then $v_1 = u_1' w$ for some partial words u_1', w satisfying $u_1 \uparrow u_1'$ and $w \neq \varepsilon$. It follows that $u_1, u_1' w, v_2 \ldots v_n$ are all in M, and $w v_2 \ldots v_n \in C(M)$. Since M is stable, $u_1 = u_1'$ and $w \in M$. Consequently, $v_1 = u_1' w \notin X$, which yields a contradiction. Thus X is a pcode.

Conversely, assume that M is pfree and let X be its base. Let u, u', v, w be partial words with $u \uparrow u'$, $u, u'w, v \in M$ and $wv \in C(M)$. Put $u = u_1 \ldots u_k$, $wv \uparrow u_{k+1} \ldots u_m$, $u'w = v_1 \ldots v_l$, and $v = v_{l+1} \ldots v_n$ where $u_1, \ldots, u_m, v_1, \ldots, v_n \in X$. The compatibility relation $u u_{k+1} \ldots u_m \uparrow u'wv$ implies

$$u_1 \ldots u_k u_{k+1} \ldots u_m \uparrow v_1 \ldots v_l v_{l+1} \ldots v_n$$

Thus $m = n$ and $u_i = v_i$ for $i = 1, \ldots, m$ since X is a pcode. Moreover, $l \geq k$ because $|u'w| \geq |u|$, showing that

$$u'w = u_1 \ldots u_k u_{k+1} \ldots u_l = u u_{k+1} \ldots u_l$$

Hence $u = u'$ and $w = u_{k+1} \ldots u_l \in M$. Thus M is stable.

8.24 If $\{u, v\}$ is a pcode, then clearly $u^2 v \not\uparrow v u^2$. Conversely, assume that $\{u, v\}$ is not a pcode and $u^2 v \not\uparrow v u^2$. Then there exist an integer $n \geq 1$ and partial words $u_1, u_2, \ldots, u_n, v_1, v_2, \ldots, v_n \in \{u, v\}$ such that

$$u_1 u_2 \ldots u_n \uparrow v_1 v_2 \ldots v_n$$

and with $|u_1 u_2 \ldots u_n|$ as small as possible contradicting Proposition 8.11. We hence have $u_1 \neq v_1$ and $u_n \neq v_n$, and we may assume that $n \geq 2$. There are four possibilities: $u_1 = u_n = u, v_1 = v_n = v$; $u_1 = v_n = u, v_1 = u_n = v$; $u_1 = v_n = v, v_1 = u_n = u$; and $u_1 = u_n = v, v_1 = v_n = u$. In all cases, put $u_2 \ldots u_{n-1} = x$ and $v_2 \ldots v_{n-1} = y$. These possibilities can be rewritten as

$$(1)\ uxu \uparrow vyv\ (2)\ uxv \uparrow vyu\ (3)\ vxu \uparrow uyv\ (4)\ vxv \uparrow uyu$$

Since $|u| < |v|$, for any of the possibilities (1)-(4), there exist nonempty pwords w, w', z, z' such that $v = wz = z'w', w \uparrow u$, and $w' \uparrow u$. Considering $|v| = 2|u|$, it is clear that $|w| = |z|$ and so $v = ww'$. Since $u \uparrow w$ and $u \uparrow w'$, by multiplication $ww' \uparrow uu$. Therefore $v \uparrow u^2$ and consequently $u^2v \uparrow vu^2$, a contradiction.

It should be noted that in the case where $w \uparrow w'$, it is apparent that the power of u^2 is not necessary in determining the potential of a nontrivial compatibility relation. In this case, $uv \not\uparrow vu$ if and only if $\{u, v\}$ is a pcode.

First, consider the set $\{bab\diamond, baa\diamond\diamond bab\}$. Let $u = bab\diamond$ and $v = baa\diamond\diamond bab$ so that $|v| = 2|u|$. In this case, $u^2v \not\uparrow vu^2$,

$$ba\underline{b}\diamond b\underline{ab}\diamond b\underline{aa}\diamond\diamond bab \not\uparrow ba\underline{a}\diamond\diamond ba\underline{b}ba\underline{b}\diamond bab\diamond$$

and thus this set is a pcode.

Second, consider the set $\{a\diamond b, aa\diamond abb\}$. Let $u = a\diamond b$ and $v = aa\diamond abb$ so that $|v| = 2|u|$. In this case, a nontrivial compatibility relation does exist, $u^2v \uparrow vu^2$,

$$a\diamond ba\diamond baa\diamond abb \uparrow aa\diamond abba\diamond ba\diamond b$$

and thus this set is not a pcode.

Third, consider the set $\{\diamond b\diamond abb, ab\diamond\}$. Let $u = ab\diamond$ and $v = \diamond b\diamond abb$ so that $|v| = 2|u|$. Factor the partial word v such that $v = ww'$ where $w = \diamond b\diamond$ and $w' = abb$. In this case, $w \uparrow w'$, $\diamond b\diamond \uparrow abb$, so the compatibility of $uv \uparrow vu$ will suffice to determine if this set is a pcode. A nontrivial compatibility relation does exist, $uv \uparrow vu$,

$$ab\diamond\diamond b\diamond abb \uparrow \diamond b\diamond abbab\diamond$$

and thus this set is not a pcode.

CHAPTER 9

9.2 The set X is pairwise noncompatible. Here

$$U_1 = \{a\}$$
$$U_2 = \{\diamond b, baaa, bba\}$$
$$U_3 = \{aa, aaa, b, ba\}$$
$$U_4 = \{b, baaa\}$$

$U_5 = \{baaa\}$

$U_6 = \emptyset$

and X is a pcode because $\varepsilon \notin U_i$ for any $i \geq 1$.

- The set U_1 is obtained by the following. Consider $u = a \diamond b$. In this case, $a \diamond b \uparrow ab\underline{ba}$ and therefore $x = a$. No other choices of u are successful and thus $U_1 = \{a\}$.

- In obtaining U_2, the first set is empty since every $u \in X$ is greater in length than every word in U_1. However, comparing U_1 with X produces a nonempty set:

  1. $a \uparrow a\diamond\underline{b}$ and thus $x = \diamond b$
  2. $a \uparrow \diamond b\underline{aaa}$ and thus $x = baaa$
  3. $a \uparrow a\underline{bba}$ and thus $x = bba$

- For U_3, comparing X with U_2 produces the empty set since each of the elements of X is either greater in length or equal in length and not compatible with the elements of U_2. However, comparing U_2 with X produces the following:

  1. $\diamond b \uparrow a\diamond\underline{b}$ and thus $x = b$
  2. $\diamond b \uparrow \diamond b\underline{aaa}$ and thus $x = aaa$
  3. $\diamond b \uparrow a\underline{bba}$ and thus $x = ba$
  4. $bba \uparrow \diamond b\underline{aaa}$ and thus $x = aa$

- Similarly, the set U_4 is computed:

  1. $aa \uparrow a\diamond\underline{b}$ and thus $x = b$
  2. $b \uparrow \diamond b\underline{aaa}$ and thus $x = baaa$

- The set U_5 is generated with the single comparison of $b \uparrow \diamond b\underline{aaa}$. The set U_6 is equal to the empty set since no comparisons between U_5 and X produce any results. Therefore, it is evident that $\varepsilon \notin U_i$ for any $i \geq 1$ and thus X is a pcode.

9.6 This is because $U_1 = \emptyset$ for such sets.

9.10 By definition, E_1-edges only originate at the *open* node and do not terminate there. Hence, an E_1-edge cannot be bidirectional. A similar statement holds for E_2-edges.

Let e be an E_3-edge. Then e is of the form $\left(\binom{u}{\varepsilon}, \binom{uv}{\varepsilon}\right)$ (or symmetrically, $\left(\binom{\varepsilon}{u}, \binom{\varepsilon}{uv}\right)$) for some $u \in C(P(X)) \setminus \{\varepsilon\}$ and $v \in X$. By definition, v is nonempty, and so $|u| < |uv|$. If e were bidirectional, then $\left(\binom{uv}{\varepsilon}, \binom{u}{\varepsilon}\right)$ would be an edge, implying $u = uvv'$ for some nonempty pword $v' \in X$. Thus, $|u| < |uv| < |uvv'| = |u|$, which is impossible. Thus an E_3-edge cannot be bidirectional.

9.12 Here X_1, X_2, X_3 and X_5 are pcodes since $U_1 = U_2 = U_3 = \cdots = \emptyset$ in each case and so none of the U_i-sets contain ε. But X_4 is not a pcode since by setting $u_1 = \diamond b, u_2 = a \diamond b$ and $u_3 = aa \diamond bba$ we get $u_2 u_1 u_2 \uparrow u_3 u_1$.

9.18 See Reference [16].

CHAPTER 10

10.2 Here $x^2 \uparrow y^4$ and so $m = 2n = 4$ and $u = \varepsilon$. We have

$$(\varepsilon)(\diamond bb)(\varepsilon)(ab\diamond)(\varepsilon) = u_0 v_0 u_1 v_1 u_2 = x = v_2 u_3 v_3 = (\diamond bb)(\varepsilon)(ab\diamond)$$

and $y = uv = (\varepsilon)(a\diamond b)$.

10.8 Set $x = (u_1)^{k_1}$, $y = (u_2)^{k_2}$ and $z = (u_3)^{k_3}$ for some primitive words u_1, u_2, u_3 and some positive integers k_1, k_2, k_3.

10.11 Integers for the first triple are $m = 2, n = 1$ and $p = 4$.

10.12 Note that if the conditions hold, then trivially $x^2 \uparrow y^m$ for some positive integer m. If $x^2 \uparrow y^m$ for some positive integer m, then we consider the cases where m is even or odd. If $m = 2n + 1$ for some integer n, then there exist partial words u, v such that $y = uv, x \uparrow (uv)^n u$ and $x \uparrow v(uv)^n = (vu)^n v$. From this, we deduce that $|u| = |v|$. Now note that x may be factored as $x = (u_0 v_0) \ldots (u_{n-1} v_{n-1}) u_n = v_n(u_{n+1} v_{n+1}) \ldots (u_{m-1} v_{m-1})$ where $u_i \uparrow u$ and $v_i \uparrow v$ for all $0 \le i < m$, If $m = 2n$ for some n, then $x \uparrow y^n$ and set $u = \varepsilon$ in the above.

10.17 We show that $z = \varepsilon$ and $m = 2$ (the result will then follow by simplification). Suppose to the contrary that $z \ne \varepsilon$ or $m > 2$. In either case, we have $|x| > |y| > 0$. By Corollary 10.2, u and v are contained in powers of a common word, say $u \subset t^k$ and $v \subset t^l$ for some word t and nonnegative integers k, l. Indeed, this is trivially true when either $u = \varepsilon$ or $v = \varepsilon$. When both $u \ne \varepsilon$ and $v \ne \varepsilon$, Condition 1 of Corollary 10.2 is satisfied. Since $y = uv$ and y is primitive, we have ($k = 0$ and $l = 1$) or ($k = 1$ and $l = 0$). In the former case, $u = \varepsilon$ and in the latter case, $v = \varepsilon$. By Theorem 10.2, $z = \varepsilon$ or $z = y$. If $z = \varepsilon$, then $m > 2$. If m is even, then by Proposition 10.1, $m = 2n$ and $u = \varepsilon$. Therefore, $x = v_0 \ldots v_{n-1}$ with $n > 1$, and $x \subset v^n$ leading to a contradiction with the fact that x is primitive. If m is odd, then $m = 2n + 1$ by Proposition 10.1 and $|u| = |v| = 0$ leading to a contradiction with the fact that $|y| = |uv| > 0$. Now, if $z = y$, then $x^2 \uparrow y^{m+1}$. If $m + 1 = 2n$, then $u = \varepsilon$ and $n > 1$, and if $m + 1 = 2n + 1$, then $|u| = |v| = 0$. In either case, we get a contradiction as above.

10.19 Set $x = abaabb$, $y = ab\diamond$ and $z = \varepsilon$.

10.20 A nontrivial solution is $x = a\diamond b$, $y = b\diamond a$ and $z = abba$.

CHAPTER 11

11.5 The population size is 8. The strings are as follows (up to a renaming of letters): $aa\diamond\diamond ab$, $ab\diamond\diamond bb$, $ba\diamond\diamond aa$ and $bb\diamond\diamond ba$.

11.11 If $v_p = 2$, then there must exist some $i > 1$ such that $v_{ip} = 0$. But if $p > \lfloor \frac{n-1}{2} \rfloor$ then $p \geq \lfloor \frac{n+1}{2} \rfloor$ and thus $2p \geq n$. So for all $i \geq 2$ we have that $ip \geq n$ and thus p violates the second condition of Theorem 11.3.

11.12 There are 8 of them.

11.14 See Reference [125].

11.18 Consider the nontrivial period sets $\{1, 2, 3, 4, 5\}$ and $\{2, 4\}$. Then $R(\{1, 2, 3, 4, 5\}) = \{1\}$ and $R(\{2, 4\}) = \{2\}$, and so $R(\{1, 2, 3, 4, 5\}) \cap R(\{2, 4\}) = \{1\} \cap \{2\} = \emptyset \neq \{2\} = R(\{1, 2, 3, 4, 5\} \cap \{2, 4\})$.

11.22 Let $S = \{p_1, p_2, \ldots, p_k\}$ and $T = S \cup \{q_1, q_2, \ldots, q_l\}$ be elements of Φ_n. Moreover, let $S \sqsubset C_1 \sqsubset C_2 \sqsubset \cdots \sqsubset C_m \sqsubset T$ be a maximal chain from S to T. We claim that for all $1 \leq i < m$, $\|C_i\| = \|C_{i+1}\| - 1$. For if $C_{i+1} \setminus C_i \supset \{q_{i_1}, q_{i_2}\}$ were of order at least 2, then both the sets $C_i \cup \{q_{i_1}\}$ and $C_i \cup \{q_{i_2}\}$ would both lie in Φ_n since $C_{i+1} \subset T$ and no element of T divides any other so all subsets of T lie in Φ_n. Moreover, we see that $C_i \cup \{q_{i_1}\}$ and $C_i \cup \{q_{i_2}\}$ both lie *strictly* between C_i and C_{i+1} in the poset Φ_n. Thus the chain is not maximal and we have produced a contradiction. This shows that $m = l$ and therefore, Φ_n must satisfy the Jordan-Dedekind condition and for any two distinct $S, T \in \Phi_n$ we have that the maximal chain length is $\|T \setminus S\| + 1$.

11.23 The number of primitive sets of integers less than n seems to be between 1.55^{n-1} and 1.60^{n-1}.

CHAPTER 12

12.4 Let w be a two-sided infinite word that avoids X. Whenever $w(i + n + 1) = b$, because w avoids $a\diamond^n b$ we have that $w(i) = b$. Similarly whenever $w(i + n + 1) = a$, we have that $w(i) = a$. Thus we can characterize the words avoiding X as exactly those with period $n + 1$.

12.8 The word $v = aabb$ is such that $v^{\mathbb{Z}}$ avoids $X_{5|1,3}$.

12.11 Both are avoidable by $(ab)^{\mathbb{Z}}$.

12.12 Say that the two-sided infinite word w avoids $X_{m|n_1,n_2}$. An a occurs in w, say without loss of generality that $w(0) = a$. Then $w(-m-1) = w(m+1) = b$. We have $n_1 + 1 + n_2 + 1 = 2m + 2$, and so since w avoids $b\diamond^{n_1}b\diamond^{n_2}b$ we necessarily have $w(-m-1+n_1+1) = w(n_1-m) = a$. Repeating this argument,

$$w(2(n_1 - m)) = w(3(n_1 - m)) = \cdots = a$$

But since $n_1 - m$ divides $m + 1$, $w(-m-1) = a$, a contradiction.

12.15 See Reference [107].

12.18 We first claim that any two-sided infinite word which avoids $X_{6|1,3}$ must also avoid $\{b\diamond b\diamond b\}$. Suppose otherwise. Then w avoids $X_{6|1,3}$ but has a factor compatible with $b\diamond b\diamond b$. Without loss of generality say that $w(0) = w(2) = w(4) = b$. Then we have $w(6) = w(8) = a$, which in turn implies that $w(-1) = w(1) = b$. This implies that $w(5) = a$, which tells us that $w(-2) = b$. Since $w(-2) = b$, $w(4) = a$, a contradiction.

Now suppose for contradiction that the two-sided infinite word w avoids $X_{6|1,3}$. It must avoid $a\diamond^6 a$, and since $\{a\diamond^6 a, b\diamond b\}$ is unavoidable it has a factor compatible with $b\diamond b$. Say without loss of generality that $w(0) = w(2) = b$. The reader may verify that this ultimately leads to a contradiction, using the fact that w avoids $X_{6|1,3}$ and $\{b\diamond b\diamond b\}$.

12.19 Let $v = a^m b^{m+1}$ and $w = v^{\mathbb{Z}}$. We claim that w avoids $X_{m|n_1,n_2}$. Clearly it avoids $a\diamond^m a$. Let i be an integer. If $w(i) = w(i + n_1 + 1) = b$, then the gap in $b\diamond^{n_1}b$ cannot straddle a block of a's, since $n_1 < m$ and these blocks come in sequences m letters long. Thus we must have $w(i) \ldots w(i + n_1 + 1) = b^{n_1+2}$. Similarly if $w(i) = w(i + n_2 + 1) = b$ since $n_2 < m$ we have $w(i) \ldots w(i + n_2 + 1) = b^{n_2+2}$. Hence if there were an integer i with $w(i) = w(i + n_1 + 1) = w(i + n_1 + n_2 + 2) = b$, we would have $w(i) \ldots w(i + n_1 + n_2 + 2) = b^{n_1+n_2+3}$ which is impossible since $m + 1 < n_1 + n_2 + 3$.

12.20 For any nonnegative integer p, all integers greater than p^2 can be written as $pq + (p+1)r$ for some nonnegative integer q, r. This is because

$$pp, (p-1)p + p + 1, (p-2)p + 2(p+1), \ldots, p + (p-1)(p+1)$$

is a sequence of consecutive integers with p members.

Now let $C = \{b^{m+1}a^{m+1}, b^{m+2}a^{m+1}\}$. There exists $u \in C^*$ with $|u| = n - m - 2$. We claim that $w = u^{\mathbb{Z}}$ avoids $X_{m|m+s,n}$. It certainly avoids $a\diamond^m a$. We need to verify that whenever $w(i - m - s - 1) = b$ and $w(i) = b$ that $w(i + n + 1) = a$. Examining C we see that the only i's for which this is possible are those for which $w(i)$ is part of an initial segment of s b's in a sequence of b's. Say without loss of generality that

$$w(0)w(1)\ldots w(s)\ldots w(m) = b^m \text{ and } w(m+1) = a$$

Since w is $n-m-2$-periodic, $w(s+n+1) = w(s+m+3)$, $s+m+3 < m-2+m+3 = 2m+1$ so $w(s+n+1) = a$. Similarly $w(0+n+1) = w(m+3) = a$. Thus w has no factor compatible with $b\diamond^{m+s}b\diamond^{n}b$ and w avoids $X_{m|m+s,n}$.

References

[1] A.V. Aho and M.J. Corasick, Efficient string machines, an aid to bibliographic research, *Communications of the ACM* **18** (1975) 333–340.

[2] J.P. Allouche and J. Shallit, *Automatic Sequences: Theory, Applications, Generalizations* (Cambridge University Press, Cambridge, 2003).

[3] A. Amir and G.E. Benson, Two-dimensional periodicity and its applications, *3rd ACM-SIAM Symposium on Discrete Algorithms* (1992) 440–452.

[4] A. Amir and G.E. Benson, Alphabet independent two-dimensional pattern matching, *24th ACM Symposium on Theory of Computing* (1992) 59–68.

[5] A. Amir and G.E. Benson, Two-dimensional periodicity in rectangular arrays, *SIAM Journal of Computing* **27** (1998) 90–106.

[6] A. Amir, G.E. Benson and M. Farach, An alphabet independent approach to two dimensional pattern matching, *SIAM Journal of Computing* **23** (1994) 313–323.

[7] A. Amir and M. Farach, Efficient matching of nonrectangular shapes, *Annals of Mathematics and Artificial Intelligence* **4** (1991) 211–224.

[8] A. Amir and M. Farach, Two dimensional dictionary matching, *Information Processing Letters* **44** (1992) 233–239.

[9] D.G. Arqués and C.J. Michel, A possible code in the genetic code, Lecture Notes in Computer Science, Vol. 900 (Springer-Verlag, Berlin, 1995) 640–651.

[10] J. Berstel and L. Boasson, Partial words and a theorem of Fine and Wilf, *Theoretical Computer Science* **218** (1999) 135–141.

[11] J. Berstel and A. de Luca, Sturmian words, Lyndon words and trees, *Theoretical Computer Science* **178** (1997) 171–203.

[12] J. Berstel and D. Perrin, *Theory of Codes* (Academic Press, New York, 1985).

[13] F. Blanchet-Sadri, On unique, multiset, and set decipherability of three-word codes, *IEEE Transactions on Information Theory* **47** (2001) 1745–1757.

[14] F. Blanchet-Sadri, A periodicity result of partial words with one hole, *Computers and Mathematics with Applications* **46** (2003) 813–820.

[15] F. Blanchet-Sadri, Periodicity on partial words, *Computers and Mathematics with Applications* **47** (2004) 71–82.

[16] F. Blanchet-Sadri, Codes, orderings and partial words, *Theoretical Computer Science* **329** (2004) 177–202.

[17] F. Blanchet-Sadri, Primitive partial words, *Discrete Applied Mathematics* **148** (2005) 195–213.

[18] F. Blanchet-Sadri and A.R. Anavekar, Testing primitivity on partial words, *Discrete Applied Mathematics* **155** (2007) 279–287 (http://www.uncg.edu/mat/primitive).

[19] F. Blanchet-Sadri, D. Bal and G. Sisodia, Graph connectivity, partial words, and a theorem of Fine and Wilf, Preprint (http://www.uncg.edu/mat/research/finewilf3).

[20] F. Blanchet-Sadri, D. Blair and R.V. Lewis, Equations on partial words, in R. Královic and P. Urzyczyn (Eds.), *MFCS 2006, 31st International Symposium on Mathematical Foundations of Computer Science, August 28–September 1, 2006, Stará Lesná, Slovakia*, Lecture Notes in Computer Science, Vol. 4162 (Springer-Verlag, Berlin, 2006) 167–178 (http://www.uncg.edu/mat/research/equations).

[21] F. Blanchet-Sadri, L. Bromberg and K. Zipple, Tilings and quasiperiods of words, Preprint (http://www.uncg.edu/cmp/research/tilingperiodicity).

[22] F. Blanchet-Sadri, N.C. Brownstein and J. Palumbo, Two element unavoidable sets of partial words, in T. Harju, J. Karhumäki and A. Lepistö (Eds.), *DLT 2007, 11th International Conference on Developments in Language Theory, July 3–6, 2007, Turku, Finland*, Lecture Notes in Computer Science, Vol. 4588 (Springer-Verlag, Berlin, 2007) 96–107 (http://www.uncg.edu/mat/research/unavoidablesets).

[23] F. Blanchet-Sadri and A. Chriscoe, Local periods and binary partial words: an algorithm, *Theoretical Computer Science* **314** (2004) 189–216 (http://www.uncg.edu/mat/AlgBin).

[24] F. Blanchet-Sadri, E. Clader and O. Simpson, Border correlations of partial words, Preprint (http://www.uncg.edu/cmp/research/bordercorrelation).

[25] F. Blanchet-Sadri, K. Corcoran and J. Nyberg, Fine and Wilf's periodicity result on partial words and consequences, *LATA 2007, 1st International Conference on Language and Automata Theory and Applications, March 29–April 4, 2007, Tarragona, Spain* (http://www.uncg.edu/mat/research/finewilf).

[26] F. Blanchet-Sadri and M. Cucuringu, Counting primitive partial words, Preprint.

[27] F. Blanchet-Sadri, M. Cucuringu and J. Dodge, Counting unbordered partial words, Preprint.

[28] F. Blanchet-Sadri, C.D. Davis, J. Dodge, R. Mercaş and M. Moorefield, Unbordered partial words, Preprint
(http://www.uncg.edu/mat/border).

[29] F. Blanchet-Sadri and S. Duncan, Partial words and the critical factorization theorem, *Journal of Combinatorial Theory, Series A* **109** (2005) 221–245 (http://www.uncg.edu/mat/cft).

[30] F. Blanchet-Sadri, J. Fowler, J.D. Gafni and K.H. Wilson, Combinatorics on partial word correlations, Preprint
(http://www.uncg.edu/cmp/research/correlations2).

[31] F. Blanchet-Sadri, J.D. Gafni and K.H. Wilson, Correlations of partial words, in W. Thomas and P. Weil (Eds.), *STACS 2007, 24th International Symposium on Theoretical aspects of Computer Science, February 22-24, 2007, Aachen, Germany*, Lecture Notes in Computer Science, Vol. 4393 (Springer-Verlag, Berlin, 2007) 97–108 (http://www.uncg.edu/mat/research/correlations).

[32] F. Blanchet-Sadri and R.A. Hegstrom, Partial words and a theorem of Fine and Wilf revisited, *Theoretical Computer Science* **270** (2002) 401–419.

[33] F. Blanchet-Sadri, R. Jungers and J. Palumbo, Testing avoidability of sets of partial words is hard, Preprint.

[34] F. Blanchet-Sadri, A. Kalcic and T. Weyand, Unavoidable sets of partial words of size three, Preprint
(http://www.uncg.edu/cmp/research/unavoidablesets2).

[35] F. Blanchet-Sadri and D.K. Luhmann, Conjugacy on partial words, *Theoretical Computer Science* **289** (2002) 297–312.

[36] F. Blanchet-Sadri, T. Mandel and G. Sisodia, Connectivity in graphs associated with partial words, Preprint
(http://www.uncg.edu/cmp/research/finewilf4).

[37] F. Blanchet-Sadri, R. Mercaş and G. Scott, A generalization of Thue freeness for partial words, Preprint
(http://www.uncg.edu/cmp/research/freeness).

[38] F. Blanchet-Sadri, R. Mercaş and G. Scott, Counting distinct squares in partial words, Preprint
(http://www.uncg.edu/cmp/research/freeness).

[39] F. Blanchet-Sadri and M. Moorefield, Pcodes of partial words, (http://www.uncg.edu/mat/pcode).

[40] F. Blanchet-Sadri, T. Oey and T. Rankin, Partial words and a generalization of Fine and Wilf's theorem for arbitrarily many weak periods, (http://www.uncg.edu/mat/research/finewilf2).

[41] F. Blanchet-Sadri and B. Shirey, Periods, partial words, and a result of Guibas and Odlyzko, (http://www.uncg.edu/mat/bintwo).

[42] F. Blanchet-Sadri and N.D. Wetzler, Partial words and the critical factorization theorem revisited, *Theoretical Computer Science* **385** (2007) 179–192 (http://www.uncg.edu/mat/research/cft2).

[43] F. Blanchet-Sadri and J. Zhang, On the critical factorization theorem, Preprint.

[44] R.S. Boyer and J.S. Moore, A fast string searching algorithm, *Communications of the ACM* **20** (1977) 762–772.

[45] D. Breslauer, T. Jiang and Z. Jiang, Rotations of periodic strings and short superstrings, *Journal of Algorithms* **24** (1997) 340–353.

[46] P. Bylanski and D.G.W. Ingram, Digital transmission systems, IEE (1980).

[47] M.G. Castelli, F. Mignosi and A. Restivo, Fine and Wilf's theorem for three periods and a generalization of Sturmian words, *Theoretical Computer Science* **218** (1999) 83–94.

[48] S. Cautis, F. Mignosi, J. Shallit, M.W. Wang and S. Yasdani, Periodicity, morphisms, and matrices, *Theoretical Computer Science* **295** (2003) 107–121.

[49] Y. Césari and M. Vincent, Une caractérisation des mots périodiques, *Comptes Rendus de l'Académie des Sciences de Paris* **268** (1978) 1175–1177.

[50] C. Choffrut and K. Culik II, On extendibility of unavoidable sets, *Discrete Applied Mathematics* **9** (1984) 125–137.

[51] C. Choffrut and J. Karhumäki, Combinatorics of Words, in G. Rozenberg and A. Salomaa (Eds.), *Handbook of Formal Languages, Vol. 1, Ch. 6* (Springer-Verlag, Berlin, 1997) 329–438.

[52] D.D. Chu and H.S. Town, Another proof on a theorem of Lyndon and Schützenberger in a free monoid, *Soochow Journal of Mathematics* **4** (1978) 143–146.

[53] A. Colosimo and A. de Luca, Special factors in biological strings, Preprint 97/42, Dipartimento di Matematica, Università di Roma "La Sapienza."

[54] S. Constantinescu and L. Ilie, Generalised Fine and Wilf's theorem for arbitrary number of periods, *Theoretical Computer Science* **339** (2005) 49–60.

[55] T.M. Cover and J.A. Thomas, *Elements of Information Theory* (John Wiley & Sons, New York, NY, 1991).

[56] M. Crochemore, F. Mignosi, A. Restivo and S. Salemi, Text compression using antidictionaries, Lecture Notes in Computer Science, Vol. 1644 (Springer-Verlag, Berlin, 1999) 261–270.

[57] M. Crochemore and D. Perrin, Two-way string matching, *Journal of the ACM* **38** (1991) 651–675.

[58] M. Crochemore and W. Rytter, *Text Algorithms* (Oxford University Press, 1994).

[59] M. Crochemore and W. Rytter, Squares, cubes, and time-space efficient string searching, *Algorithmica* **13** (1995) 405–425.

[60] M. Crochemore and W. Rytter, *Jewels of Stringology* (World Scientific, NJ, 2003).

[61] K. Culik II and T. Harju, Splicing semigroups of dominoes and DNA, *Discrete Applied Mathematics* **31** (1991) 261–277.

[62] A. de Luca, Sturmian words: Structure, combinatorics and their arithmetics, *Theoretical Computer Science* **183** (1997) 45–82.

[63] A. de Luca, On the combinatorics of finite words, *Theoretical Computer Science* **218** (1999) 13–39.

[64] A. de Luca and F. Mignosi, Some combinatorial properties of Sturmian words, *Theoretical Computer Science* **136** (1994) 361–385.

[65] A. de Luca and S. Varricchio, *Finiteness and Regularity in Semigroups and Formal Languages* (Springer-Verlag, Berlin, 1999).

[66] P. Dömösi, Some results and problems on primitive words, *11th International Conference on Automata and Formal Languages* (2005).

[67] P. Dömösi, S. Horváth and M. Ito, *Primitive Words and Context-Free Languages* (Masami Ito (Eds.), 2006).

[68] J.P. Duval, Périodes et répétitions des mots du monoïde libre, *Theoretical Computer Science* **9** (1979) 17–26.

[69] J.P. Duval, Relationship between the period of a finite word and the length of its unbordered segments, *Discrete Mathematics* **40** (1982) 31–44.

[70] J.P. Duval, Périodes locales et propagation de périodes dans un mot, *Theoretical Computer Science* **204** (1998) 87–98.

[71] J.P. Duval, R. Kolpakov, G. Kucherov, T. Lecroq and A. Lefebvre, Linear time computation of local periods, *Theoretical Computer Science* **326** (2004) 229–240.

[72] J.P. Duval, F. Mignosi and A. Restivo, Recurrence and periodicity in infinite words from local periods, *Theoretical Computer Science* **262** (2001) 269–284.

[73] A. Ehrenfeucht, D. Haussler and G. Rozenberg, On regularity of context-free languages, *Theoretical Computer Science* **27** (1983) 311–322.

[74] A. Ehrenfeucht and D.M. Silberger, Periodicity and unbordered segments of words, *Discrete Mathematics* **26** (1979) 101–109.

[75] C. Epifanio, M. Koskas and F. Mignosi, On a conjecture on bidimensional words, *Theoretical Computer Science* **299** (2003) 123–150.

[76] P. Erdös, Note on sequences of integers no one of which is divisible by another, *Journal of the London Mathematical Society* **10** (1935) 126–128.

[77] N.J. Fine and H.S. Wilf, Uniqueness theorems for periodic functions, *Proceedings of the American Mathematical Society* **16** (1965) 109–114.

[78] Z. Galil and K. Park, Truly alphabet independent two-dimensional pattern matching, *FOCS 1992, 33rd Annual IEEE Symposium on Foundations of Computer Science* (1992) 247–256.

[79] Z. Galil and J. Seiferas, Time-space optimal string matching, *Journal of Computer and System Sciences* **26** (1983) 280–294.

[80] D. Giammarresi, S. Mantaci, F. Mignosi and A. Restivo, Periodicities on trees, *Theoretical Computer Science* **205** (1998) 145–181.

[81] R. Giancarlo and F. Mignosi, Generalizations of the periodicity theorem of Fine and Wilf, *Trees in algebra and programming - CAAP 1994, 1994, Edinburgh.* Lecture Notes in Computer Science, Vol. 787 (Springer-Varlag, Berlin, 1994) 130–141.

[82] L.J. Guibas and A.M. Odlyzko, Periods in strings, *Journal of Combinatorial Theory, Series A* **30** (1981) 19–42.

[83] L.J. Guibas and A. Odlyzko, String overlaps, pattern matching, and nontransitive games, *Journal of Combinatorial Theory, Series A* **30** (1981) 183–208.

[84] D. Gusfield, *Algorithms on Strings, Trees, and Sequences* (Cambridge University Press, Cambridge, 1997).

[85] F. Guzmán, Decipherability of codes, *Journal of Pure and Applied Algebra* **141** (1999) 13–35.

[86] L.H. Haines, On free monoids partially ordered by embedding, *Journal of Combinatorial Theory* **6** (1969) 94–98.

[87] V. Halava, T. Harju and L. Ilie, Periods and binary words, *Journal of Combinatorial Theory, Series A* **89** (2000) 298–303.

[88] T. Harju and D. Nowotka, Density of critical factorizations, *Theoretical Informatics and Applications* **36** (2002) 315–327.

[89] T. Harju and D. Nowotka, The equation $x^i = y^j z^k$ in a free semigroup, *Semigroup Forum* **68** (2004) 488–490.

[90] T. Head, Formal language theory and DNA: an analysis of the generative capacity of specific recombinant behaviors, *Bulletin of Mathematical Biology* **49** (1987) 737–759.

[91] T. Head, G. Păun and D. Pixton, Language Theory and Molecular Genetics, in G. Rozenberg and A. Salomaa (Eds.), *Handbook of Formal Languages, Vol. 2, Ch. 7* (Springer-Verlag, Berlin, 1997) 295–360.

[92] T. Head and A. Weber, Deciding multiset decipherability, *IEEE Transactions on Information Theory* **41** (1995) 291–297.

[93] G. Higman, Ordering with divisibility in abstract algebras, *Proceedings of the London Mathematical Society* **3** (1952) 326–336.

[94] M. Ito, H. Jürgensen, H.J. Shyr and G. Thierrin, Anti-commutative languages and n-codes, *Discrete Applied Mathematics* **24** (1989) 187–196.

[95] M.A. Jimenez-Montano, On the syntactic structure of protein sequences and the concept of grammar complexity, *Bulletin of Mathematical Biology* **46** (1984) 641–659.

[96] H. Jürgensen and S. Konstantinidis, Codes, in G. Rozenberg and A. Salomaa (Eds.), *Handbook of Formal Languages, Vol. 1, Ch. 8* (Springer-Verlag, Berlin, 1997) 511–607.

[97] J. Justin, On a paper by Castelli, Mignosi, Restivo, *Theoretical Informatics with Applications* **34** (2000) 373–377.

[98] D.E. Knuth, J.H. Morris and V.R. Pratt, Fast pattern matching in strings, *SIAM Journal of Computing* **6** (1977) 323–350.

[99] R. Kolpakov and G. Kucherov, Finding approximate repetitions under hamming distance, Lecture Notes in Computer Science, Vol. 2161 (Springer-Verlag, Berlin, 2001) 170–181.

[100] R. Kolpakov and G. Kucherov, Finding approximate repetitions under hamming distance, *Theoretical Computer Science* **33** (2003) 135–156.

[101] G. Landau and J. Schmidt, An algorithm for approximate tandem repeats, Lecture Notes in Computer Science, Vol. 684 (Springer-Verlag, Berlin, 1993) 120–133.

[102] G.M. Landau, J.P. Schmidt and D. Sokol, An algorithm for approximate tandem repeats, *Journal of Computational Biology* **8** (2001) 1–18.

[103] P. Leupold, Partial words: Results and Perspectives, Preprint.

[104] P. Leupold, Partial words for DNA coding, *DNA 10, Tenth International Meeting on DNA Computing*, Lecture Notes in Computer Science, Vol. 3384 (Springer-Verlag, Berlin, 2005) 224–234.

[105] G. Lischke, Restorations of punctured languages and similarity of languages, *Mathematical Logic Quarterly* **52** (2006) 20–28.

[106] M. Lothaire, *Combinatorics on Words* (Addison-Wesley, Reading, MA, 1983; Cambridge University Press, Cambridge, 1997).

[107] M. Lothaire, *Algebraic Combinatorics on Words* (Cambridge University Press, Cambridge, 2002).

[108] M. Lothaire, *Applied Combinatorics on Words* (Cambridge University Press, Cambridge, 2005).

[109] R.C. Lyndon and M.P. Schützenberger, The equation $a^m = b^n c^p$ in a free group, *Michigan Mathematical Journal* **9** (1962) 289–298.

[110] G.S. Makanin, The problem of solvability of equations in a free semigroup, *Sbornik Mathematics* **32** (1977) 129–198.

[111] F. Manea and R. Mercaş, Freeness of partial words, Preprint.

[112] D. Margaritisn and S. Skiena, Reconstructing strings from substrings in rounds, *FOCS 1995, 36th Annual IEEE Symposium on Foundations of Computer Science* (1995) 613–620.

[113] A.A. Markov, The theory of algorithms, *Proceedings of the Steklov Institute of Mathematics* **42** (1954).

[114] E.M. McCreight, A space-economical suffix tree construction algorithm, *Journal of the ACM* **23** (1976) 262–272.

[115] F. Mignosi, A. Restivo and S. Salemi, A periodicity theorem on words and applications, Lectures Notes in Computer Science, Vol. 969 (Springer-Verlag, Berlin, 1995) 337–348.

[116] F. Mignosi, A. Restivo and P.V. Silva, On Fine and Wilf's theorem for bidimensional words, *Theoretical Computer Science* **292** (2003) 245–262.

[117] F. Mignosi, J. Shallit and M.W. Wang, Variations on a theorem of Fine and Wilf, Lecture Notes in Computer Science, Vol. 2136 (Springer-Verlag, Berlin, 2001) 512–523.

[118] G. Pǎun, N. Santean, G. Thierrin and S. Yu, On the robustness of primitive words, *Discrete Applied Mathematics* **117** (2002) 239–252.

[119] H. Petersen, On the language of primitive words, *Theoretical Computer Science* **161** (1996) 141–156.

[120] W. Plandowski, Satisfiability of word equations with constants is in NEXPTIME, *STOC 1999, Annual ACM Symposium on Theory of Computing* (1999) 721–725.

[121] W. Plandowski, Satisfiability of word equations with constants is in PSPACE, *FOCS 1999, 40th Annual IEEE Symposium on Foundations of Computer Science* (1999) 495–500.

[122] M. Regnier and L. Rostami, A unifying look at d-dimensional periodicities and space coverings, *4th Symposium on Combinatorial pattern matching*, Lecture Notes in Computer Science, Vol. 684 (Springer-Verlag, Berlin, 1993) 215–227.

[123] A. Restivo and P.V. Silva, Periodicity vectors for labelled trees, *Discrete Applied Mathematics* **126** (2003) 241–260.

[124] G. Richomme, Sudo-Lyndon, *Bulletin of the European Association for Theoretical Computer Science* **92** (2007) 143–149.

[125] E. Rivals and S. Rahmann, Combinatorics of periods in strings, *Journal of Combinatorial Theory, Series A* **104** (2003) 95–113.

[126] L. Rosaz, Inventories of unavoidable languages and the word-extension conjecture, *Theoretical Computer Science* **201** (1998) 151–170.

[127] A.A. Sardinas and G.W. Patterson, A necessary and sufficient condition for the unique decomposition of coded messages, *IRE International Convention Record* **8** (1953) 104–108.

[128] J.P. Schmidt, All highest scoring paths in weighted grid graphs and their application to finding all approximate repeats in strings, *SIAM Journal of Computing* **27** (1998) 972–992.

[129] J. Setubal and J. Meidanis, *Introduction to Computational Molecular Biology* (PWS Publishing Company, Boston, MA, 1997).

[130] A.M. Shur and Y.V. Gamzova, Partial words and the periods' interaction property, *Izvestiya RAN* **68** (2004) 199–222.

[131] A.M. Shur and Y.V. Konovalova, On the periods of partial words, Lecture Notes in Computer Science, Vol. 2136 (Springer-Verlag, Berlin, 2001) 657–665.

[132] H.J. Shyr, *Free Monoids and Languages* (Hon Min Book Company, Taichung, Taiwan, 1991).

[133] H.J. Shyr and G. Thierrin, Codes and binary relations, Lecture Notes of Springer-Verlag, Vol. 586 (Séminaire d'Algèbre, Paul Dubreil, Paris, 1975–1976) 180–188.

[134] H.J. Shyr and G. Thierrin, Disjunctive languages and codes, Lecture Notes in Computer Science, Vol. 56 (Springer-Verlag, Berlin, 1977) 171–176.

[135] R.J. Simpson and R. Tijdeman, Multi-dimensional versions of a theorem of Fine and Wilf and a formula of Sylvester, *Proceedings of the American Mathematical Society* **131** (2003) 1661–1671.

[136] N.J.A. Sloane, *On-Line Encyclopedia of Integer Sequences*, (http://www.research.att.com/~njas/sequences).

[137] J.A. Storer, *Data Compression: Methods and Theory*, (Computer Science Press, Rockville, MD, 1988).

[138] N.D. Wetzler, Unbordered partial words and the critical factorization theorem, Personal communication.

[139] R. Tijdeman and L. Zamboni, Fine and Wilf words for any periods, *Indagationes Mathematicae* **14** (2003) 135–147.

[140] A. Thue, Über unendliche Zeichenreihen, *Norske Vid. Selsk. Skr. I, Mat. Nat. Kl. Christiana* **7** (1906) 1–22.

[141] A. Thue, Über die gegenseitige Lage gleicher Teile gewisser Zeichenreihen, *Norske Vid. Selsk. Skr. I, Mat. Nat. Kl. Christiana* **12** (1912).

[142] J. Ziv and A. Lempel, A universal algorithm for sequential data compression, *IEEE Transactions on Information Theory* **23** (1977) 337–343.

Index

Milton Keynes UK
Ingram Content Group UK Ltd.
UKHW020319111024
449327UK00040B/1382

9 780367 388256